No. 1765
$21.95

HOW TO DESIGN OP AMP CIRCUITS WITH PROJECTS & EXPERIMENTS

BY DELTON T. HORN

TAB
TAB BOOKS Inc.
BLUE RIDGE SUMMIT, PA. 17214

Other TAB Books by the Author

1167 *Electronic Music Synthesizers*
1228 *34 MORE Tested, Ready-To-Run Game Programs in BASIC*
1338 *Basic Electronics Theory—with projects & experiments*
1438 *The Beginner's Book of Electronic Music*
1565 *Music Synthesizers: A Manual of Design & Construction*
1645 *Using Integrated Circuit Logic Devices*

FIRST EDITION

FIRST PRINTING

Printed in the United States of America

Library of Congress Cataloging in Publication Data

Horn, Delton T.
How to design op amp circuits, with projects and experiments.

Includes index.
1. Operational amplifiers. 2. Electronic circuit design. I. Title.
TK7871.58.06H67 1984 621.3815'35 84-6539
ISBN 0-8306-0765-X
ISBN 0-8306-1765-5 (pbk.)

Contents

Introduction

Integrated circuits hold an increasingly important place in all aspects of electronics. Circuits that would scarcely be practical in terms of size and cost using discrete components can now fit into the palm of your hand for just a few dollars.

One of the most important types of analog ICs is the operational amplifier, or op amp. No one who works with electronics either professionally or as a hobby can afford to be unfamiliar with this fascinating and versatile device.

This book presents a thorough look at the op amp and many of its more common applications. The emphasis is on designing circuits using op amps. A number of experiments are included to give the reader practical hands-on experience.

Electronics hobbyists and professional technicians will find this book useful, either for learning more about op amps, or as a refresher course and reference. While this book is aimed primarily towards people interested in circuit design, anyone involved with electronics can profit from learning the principles discussed here. Some basic knowledge of electronics is assumed in this book, but no prior familiarity with op amps or ICs is required.

In any type of circuit design, some math is essential. I have made every possible attempt to minimize the mathematics and make the formulas as painless as possible.

The experiments presented in Chapters 4, 8, and 15 are an important part of the learning experience offered by this book.

While you could just read the main text, it is highly recommended that you also perform the experiments. The best way to learn is to see the principles in action.

To make this book cover as broad a range as possible, many op amp applications are presented and discussed. After reading this book, you should be able to design your own custom op amp circuits with reasonable confidence, even if you've never done any circuit design before.

1

The Ideal Op Amp

The op amp is probably the most widely used and versatile of all modern integrated circuit types. It can be used in countless applications: amplification, filtering, mathematical functions, audio applications, signal generation, and many others. All of these topics will be discussed throughout this book.

The full name for an op amp is *operational amplifier.* Op amp circuitry was originally designed to perform mathematical operations for analog computers. The first op amp circuits were expensive and unwieldy, using dozens, or even hundreds of vacuum tubes. Transistorized versions were a little more reasonable, but still not very practical for experimentation.

In the mid-seventies, op amps began to appear in IC form. Integrated circuits created a major revolution within the electronics field. This revolution is especially apparent when dealing with the op amp. Now you can pick up surplus op amps at a dozen for a dollar. Many new op amp chips cost less than a dollar.

The vastly decreased size and price has encouraged a great deal of experimentation. This once esoteric circuit has been discovered by the hobbyist and put to work in thousands of ways that were never considered by the original designers.

Basically an IC op amp is a dc amplifier with very high gain. Its response can be controlled with external feedback networks. How useful is this? Extremely useful, as you will soon see.

CHARACTERISTICS

Before discussing how you can use practical op amp ICs, you need to understand the principles of the ideal op amp.

The schematic symbol for an op amp is shown in Fig. 1-1. Note that dual power supplies are required. One is positive with respect to ground and the other is negative with respect to ground. There is no direct ground connection.

The dual-polarity power supplies are necessary to allow the output to swing in both positive and negative directions, as determined by the signal(s) at the input(s).

In most circuits, the two power supplies will have equal voltages.

Figure 1-1 also shows that the op amp has two inputs. One is marked with a negative sign (−), and is called the *inverting input.* The other is marked with a positive sign (+) and is called the *noninverting input.* The output is controlled by the difference between the signals fed to these two inputs.

If the signal at the inverting input is greater than that at the noninverting input, the output signal will be negative (assuming that both input signals are of positive polarity).

Again assuming both inputs are being fed positive signals, if the signal at the inverting input is less than the signal at the noninverting input, the output will be positive. Notice that the signal fed into the inverting input has its polarity reversed, or inverted, by the time it reaches the output. The polarity of the signal fed to the noninverting input is not inverted.

The final possibility is that both input signals are equal. In this case they would cancel each other out, and the output would be 0.

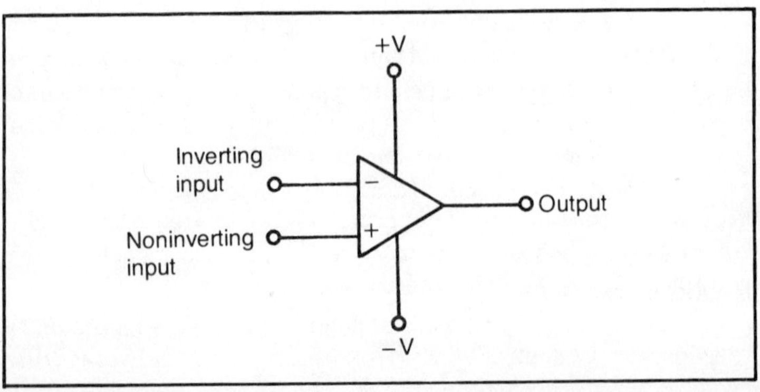

Fig. 1-1. This is the symbol most commonly used to represent an op amp in a schematic diagram.

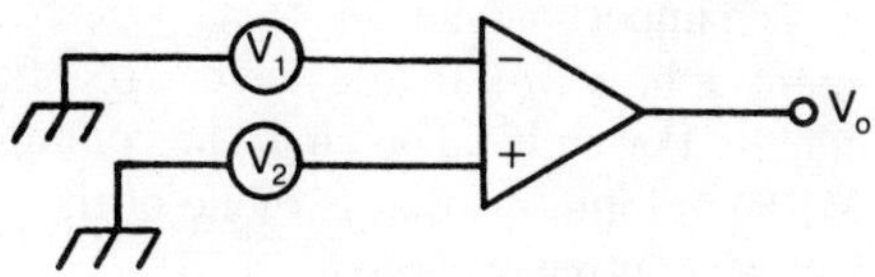

Fig. 1-2. For a differential amplifier, separate input voltages are fed to the inverting and noninverting inputs of an op amp.

What I have just described is a differential amplifier. The difference between the two inputs is amplified. The basic circuit is illustrated in Fig. 1-2. Notice that the power supply connections are omitted for simplicity. This omission is typical in schematic diagrams. If the power supply connections are not shown, the connections are assumed, as with any integrated circuit.

I will call the output voltage V_o. The signal fed to the inverting input is V_1, and V_2 is the signal fed to the noninverting input. For any combination of input values, the output may always be calculated with this simple formula:

$$V_o = (V_2 - V_1)G$$

G is the amount of gain. I will discuss gain in a few pages. For now I will assume unity gain—that is, G = 1. Since G is multiplied in this formula, a value of 1 means that I may ignore it altogether.

Let's try a few examples. Assume V_1 is 5 volts, and V_2 is 6 volts.

$$V_o = (V_2 - V_1) = (6 - 5) = 1$$

If I reverse these signals so that V_1 now equals 6 volts, and V_2 equals 5 volts, I get the following result:

$$V_o = (5 - 6) = -1$$

The polarity is inverted because the inverting input is dominant.

If both input voltages are 5 volts, they cancel out, just as I mentioned earlier.

$$V_o = (5 - 5) = 0$$

When the same signal is applied to both of the inputs of an op amp, they are said to be in the common mode. The common-mode rejection ratio is an important specification for practical op amps. It is a measurement of how well the device can reject, or cancel out equal input signals. For an ideal op amp, the common-mode rejection ratio (CMRR) is infinite—that is, none of the common signal leaks through to the output.

So far I have only considered positive input signals. An op amp can also accept negative input signals. The principles of operation remain exactly the same.

For instance, if V_1 is −3 volts, and V_2 is 5 volts, the output will be:

$$V_o = (5 - (-3)) = (5 + 3) = 8$$

Remember that double negative signs (−(−3)) cancel each other out to become positive.

If V_1 is 3 volts and V_2 is −5 volts, then the output will be:

$$V_o = (-5 - 3) = -8$$

Finally, let's look at an example when both inputs are negative. Let's say V_1 equals −7 volts, and V_2 equals −4 volts. In this case, the output works out to:

$$V_o = (-4 - (-7)) = (-4 + 7) = 3$$

As long as you keep track of the negative signs, you should have no problems with the differential amplifier. Some additional examples are given in Table 1-1.

The ideal op amp also has important impedance characteristics. The input impedance is infinite. This means there is no current flow into either input terminal. According to Ohm's Law:

$$I = E/R$$

Since R is infinity,

$$I = E/\infty$$

I (current) must be 0 (or extremely close) for any finite value of E. Since the input voltages to an op amp are likely to be low (usually

Table 1-1. Typical Output Signals for Differential Inputs.

Inverting Input Voltage	Noninverting Input Voltage	Output Voltage
1	1	0
−1	1	2
1	−1	−2
−1	−1	0
3	−2	−5
−3	2	5
−3	−2	−1
4	3	−1
6	5.75	−0.25
3.45	4.82	1.37
0.33	2.64	2.31
0	4	4
−6	0	6
−4.22	1.55	5.77

under 15 volts), current flow will be zero. The inputs are placed at virtual ground as far as current flow is concerned.

The output impedance of the ideal op amp, on the other hand, is zero.

Figure 1-3 shows an equivalent circuit for the ideal op amp. The output voltage, as stated earlier, is equal to the difference voltage, multiplied by the gain. For the ideal op amp the open loop gain (no feedback path) is infinite.

The ideal op amp also has an infinite bandwidth. That is, the gain is the same for all frequencies.

Zero offset is another characteristic of the ideal op amp. Zero offset means that when the input voltage is exactly zero, the output will be exactly zero. An ideal op amp also exhibits zero drift.

Practical op amps do not quite meet these ideal specifications, of course. The limitations of practical op amps will be discussed in the next chapter.

For now, I will consider the way the ideal op amp would behave in basic applications. I will concentrate on the ideal op amp because it makes the principles a little easier to understand.

THE BASIC INVERTING FOLLOWER

One of the most common op amp circuits is shown in Fig. 1-4. It is called an *inverting follower*. The input signal is fed into the

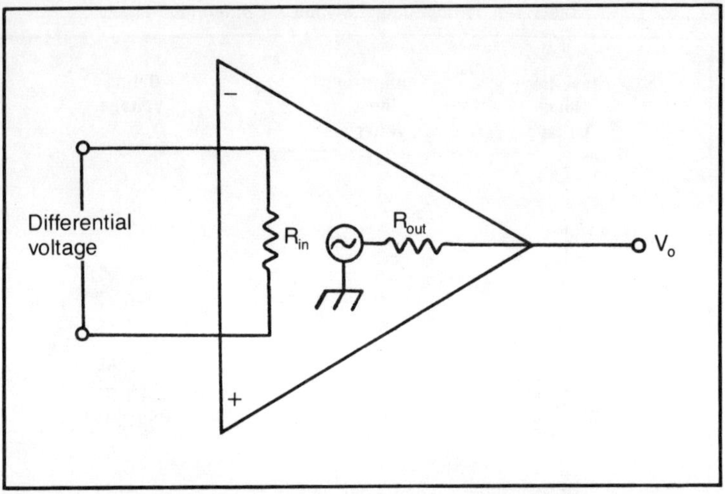

Fig. 1-3. This circuit is the functional equivalent of an operational amplifier.

inverting input only. The noninverting input is grounded. The feedback loop limits the infinite gain of the op amp's open loop configuration. In fact, this arrangement has a gain of 1, or unity. Since the inverting input is being used, the polarity will be reversed at the output. The output voltage will be:

$$V_o = -V_iG$$

In this circuit, the gain is 1.

$$V_o = -V_i \times 1 = -V_i$$

The output signal equals the input voltage with the polarity

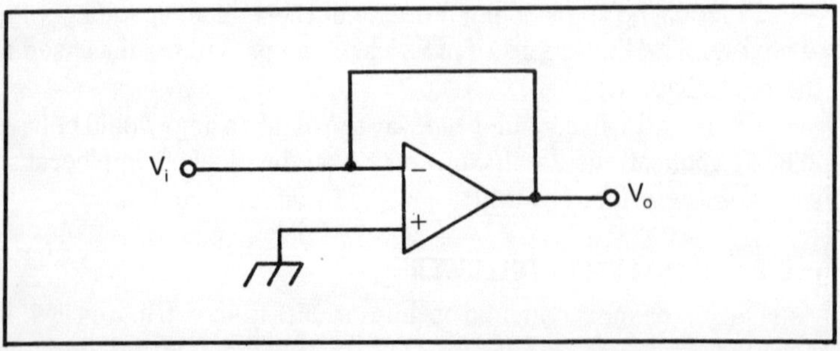

Fig. 1-4. Wiring an op amp in this manner creates an inverting voltage follower circuit.

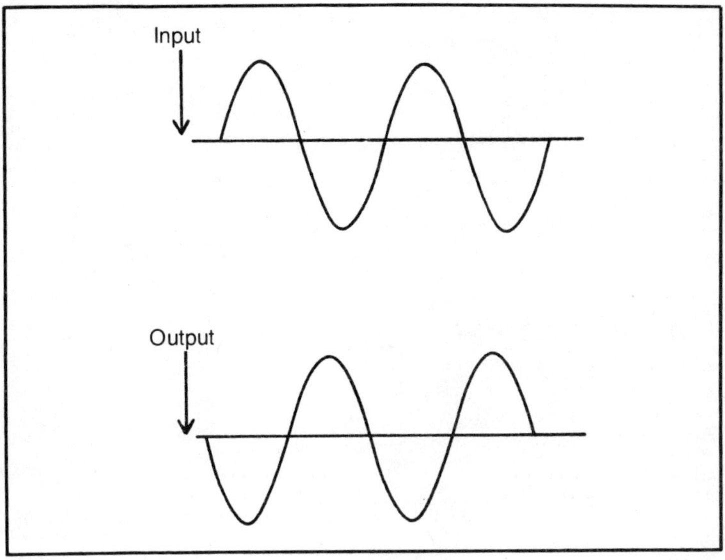

Fig. 1-5. An inverting voltage follower shifts the output signal 180° out of phase with the input signal.

inverted. If the input is 3.5 volts, the output will be −3.5 volts. If the input is −2.75 volts, the output will be +2.75 volts.

In other words, the output is 180° out of phase with the input. Typical input and output signals are illustrated in Fig. 1-5.

As useful as a unity gain inverting follower can be, I can create an even more useful circuit by adding the two resistors shown in Fig. 1-6.

R1 is the input resistor (R_i), and R2 is the feedback resistor (R_f). The ratio of these two resistances determines the gain of the circuit. The gain may be negative (less than 1 – R2 < R1), unity (equal to 1 – R2=R1), or positive (greater than 1 – R2 >R1). The gain can be determined with this formula:

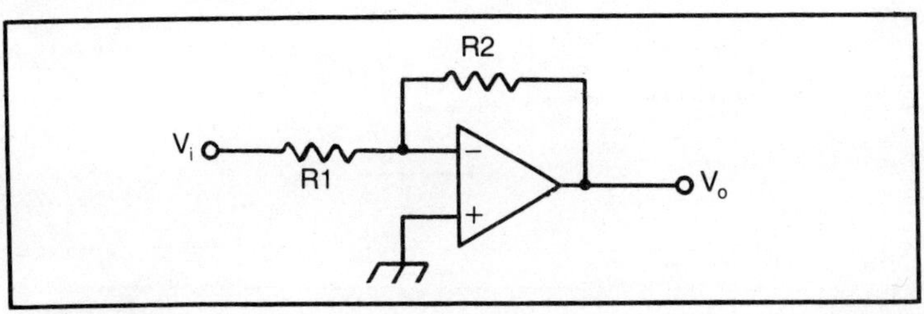

Fig. 1-6. Adding a pair of resistors in the feedback loop adds gain to an inverting amplifier circuit.

$$G = \frac{-R2}{R1}$$

The negative sign simply indicates the polarity inversion.

Earlier I stated that the output is the product of the input and the circuit gain:

$$V_o = V_i G$$

(Since the noninverting input is grounded, its input signal is 0 and can be ignored in the formula). I can combine these two equations and get this equation:

$$V_o = V_i \left(\frac{-R2}{R1} \right)$$

Earlier you learned that the input impedance of an ideal op amp is infinite, placing the input at virtual ground.

The significance of this fact is that R1 and R2 determine the current flow. The input voltage according to Ohm's Law is:

$$V_i = I_i \times R1$$

The output voltage works out in a similar manner:

$$V_o = -I_o \times R2$$

Again, the negative sign indicates the polarity inversion.

For the ideal op amp that doesn't draw any current, I_o will be equal to I_i. This value is simply called I:

$$V_i = I \times R1$$
$$V_o = I \times R2$$

Since gain can be defined by the ratio of input and output voltages, I can make the following substitution:

$$G = \frac{V_o}{V_i} = \frac{-I \times R2}{I \times R1} = \frac{-R2}{R1}$$

The equal I values cancel each other out. Notice that this gives us the same gain equation presented earlier.

Let's try a few examples. First, let R1 = 1 K and R2 = 100 K. The gain would be −100.

$$G = \frac{-R2}{R1} = \frac{-100\ K}{1\ K} = -100$$

If the input is 1 mV, the output will be 100 mV, or 0.1 volt. Changing R2 to 47 K will give us a gain of −47.

$$G = \frac{-47\ K}{1\ K} = -47$$

If another 47 K resistor is substituted for R1, I get unity gain (R1 = R2).

$$G = \frac{-47\ K}{47\ K} = -1$$

Negative gain (attenuation) can be achieved when R1 is greater than R2. For example, let's make R1 100 K and R2 10 K:

$$G = \frac{-\ 10\ K}{100\ K} = \frac{-1}{10} = -0.1$$

Very large gains (or very small gains) are usually not practical because of the large resistance ratios. Resistor values should generally be kept between about 1 K to 1 megohm. Table 1-2 lists some typical gain values.

Table 1-2. The Gain of an Inverting Amplifier.

R1 (in kΩ)	R2 (in kΩ)	Gain
1	1	−1
1	3.3	−3.3
1	10	−10
1	47	−47
1	100	−100
3.3	1	−0.30303
3.3	3.3	−1
3.3	4.7	−1.42424
3.3	10	−3.030303
3.3	47	−14.2424
3.3	100	−30.30303
4.7	1	−0.212766
4.7	3.3	−0.7021277
4.7	4.7	−1
4.7	10	−2.12766
4.7	47	−10
4.7	100	−21.2766
10	1	−0.1
10	3.3	−0.33
10	10	−1
10	100	−10

THE BASIC NONINVERTING FOLLOWER

You're probably ahead of me here. By grounding the noninverting input and feeding the input signal into the inverting input, I made an inverting follower.

You guessed it! By feeding the input signal into the noninverting input you can build a noninverting follower. The unity gain circuit is shown in Fig. 1-7. Note the similarity to the inverting follower shown in Fig. 1-4.

The inverting input is not grounded. The feedback signal is fed into the inverting input. The output signal's polarity is not inverted, so the output is in phase with the input.

A unity gain noninverting amplifier might seem to be an exercise in uselessness, but there are numerous applications. This circuit can be used for impedance conversion. An ideal op amp has an extremely high input impedance and an extremely low output impedance. A unity gain noninverting follower allows a high impedance output device to be used with a low impedance input device. This circuit can also prove useful as a buffer.

As with the inverting amplifiers discussed earlier, I can add a pair of resistors to a noninverting follower to control the gain. This is shown in Fig. 1-8. Notice that the feedback network is connected to the inverting input instead of the noninverting input. With no feedback, the gain is infinite (open loop gain). There is little advantage to feeding the output back to the noninverting input for additional amplification. Feeding the output back to the inverting input creates negative (out of phase) feedback and cancels out some of the device's gain. This approach reduces the gain to the desired level.

Once again I may define the circuit gain by comparing the output voltage with the input voltage.

$$G = \frac{V_o}{V_i}$$

The output voltage is determined by the resistance values of the feedback network and the input signal.

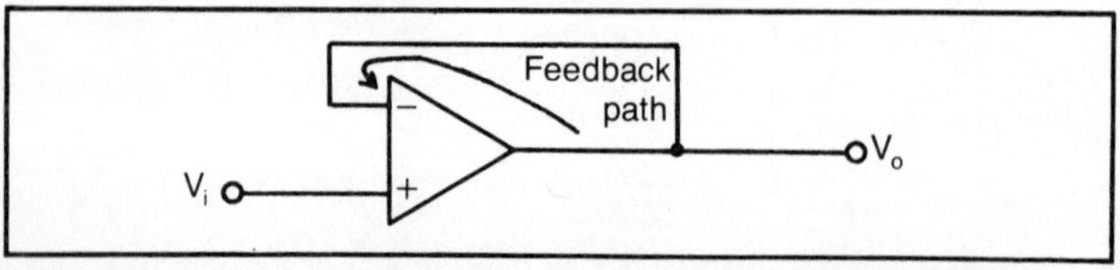

Fig. 1-7. The noninverting input of an op amp is used for a voltage follower circuit without phase shifting the output signal.

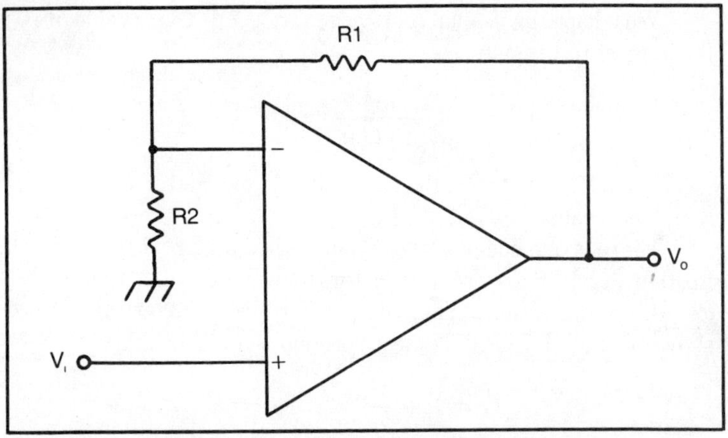

Fig. 1-8. For a noninverting amplifier with gain, the resistors are placed in the inverting feedback path.

$$V_o = \left[1 + \frac{R2}{R1}\right] V_i$$

I combine these two formulas to define the gain in terms of the resistance values.

$$G = \frac{V_o}{V_i} = \frac{\left[1 + \frac{R2}{R1}\right] V_i}{V_i} = 1 + \frac{R2}{R1}$$

Since 1 is added to the resistance ratio, the gain must always be greater than 1, no matter what values are assigned to R1 and R2. Let's try a few examples.

First make R1 much larger than R2. R1 will be equal to 100 K, and R2 will have a value of 1 K:

$$G = 1 + \frac{R2}{R1} = 1 + \frac{1,000}{100,000} = 1 + \frac{1}{100} = 1.01$$

When R1 is much greater than R2, the gain will be close to unity, but still slightly greater than 1.

Now make R2 larger than R1. Make R1 1 K and R2 10 K:

$$G = 1 + \frac{10,000}{1,000} = 1 + \frac{10}{1} = 11$$

When R1 is smaller than R2 the circuit gain will be significantly greater than unity.

What happens if equal resistances are used? Use a value of 10 K for both of the resistors:

$$G = 1 + \frac{10,000}{10,000} = 1 + 1 = 2$$

When R1 equals R2, the gain will be 2, no matter what specific resistance value is used.

There is one special case. That is when both resistors equal 0. In other words, there are no resistors in the feedback network at all, just direct connections. This, of course, is the noninverting follower shown in Fig. 1-7. In this case you find the following:

$$G = 1 + \frac{0}{0} = 1 + 0 = 1$$

For this case (R1 = R2 = 0), the gain will be exactly 1 (unity). For any other case, the gain will be greater than unity. Some additional examples are given in Table 1-3.

Since the input signal is fed directly into the op amp's noninverting input, I can conclude that the input impedance of the noninverting amplifier circuit is the same as the input impedance of the op

Table 1-3. The Gain of a Noninverting Amplifier.

R1 (in kΩ)	R2 (in kΩ)	Gain
1	1	2
1	3.3	4.3
1	10	11
1	47	48
1	100	101
3.3	1	1.30303
3.3	3.3	2
3.3	4.7	2.42424
3.3	10	4.030303
3.3	47	15.2424
3.3	100	31.30303
4.7	1	1.212766
4.7	3.3	1.70213
4.7	4.7	2
4.7	10	3.12766
4.7	47	11
4.7	100	22.2766
10	1	1.1
10	3.3	1.33
10	10	2
10	100	11

amp device itself. In other words, for an ideal op amp, the input impedance of a noninverting amplifier circuit is infinite.

Before solving the output impedance of this circuit, I need to define a factor called loop gain (A_l):

$$A_l = \frac{A_o}{G}$$

where A_o is the open loop gain for the op amp device, and G is the circuit gain defined in the earlier equations. The circuit's output impedance is defined by the op amp's intrinsic output impedance (Z_{oi}) and the loop gain (A_l):

$$Z_o = \frac{Z_{oi}}{A_l}$$

For the ideal op amp, you'll recall, the open-loop gain is infinity, and the output impedance is 0.

$$A_l = \frac{\infty}{G} = \infty$$

$$Z_o = \frac{0}{\infty} = 0$$

For an ideal op amp, the output impedance of a noninverting amplifier circuit is 0. For practical op amps, the open loop gain will have a finite value, and the output impedance will be greater than 0. The output impedance of a noninverting amplifier circuit will not necessarily be equal to the op amp's intrinsic output impedance.

2

Practical Op Amps

In the real world, op amp ICs do not quite measure up to the ideal specifications described in the last chapter. Their open loop gain and input impedance are not infinite, but merely very high. The output impedance is very low, but not quite 0. The common-mode rejection is not perfect.

In this chapter, I will explore the ways that practical op amp devices deviate from the ideal and discuss what these limitations may mean to the circuit designer.

SPEC SHEETS

Manufacturers of integrated circuits usually supply some kind of spec sheet or list of operating specifications for their devices. The precise information provided will vary according to the manufacturer and the device in question, but most spec sheets for op amp ICs include the following:

- ☐ A general description of the device.
- ☐ A pinout diagram.
- ☐ A schematic or block diagram of the equivalent internal circuitry.
- ☐ Absolute maximum ratings.
- ☐ Electrical characteristics.
- ☐ Typical performance curves.

Figure 2-1 shows a typical spec sheet for the 741, one of the most popular op amp ICs. I will discuss the 741 and other op amps specifically later. For now, let's consider what the specifications mean.

The absolute maximum ratings are pretty much self-explanatory. These are the highest acceptable values that may be applied to the chip without risking unreliable operation and/or permanent damage.

The *supply voltage* is the maximum negative and positive

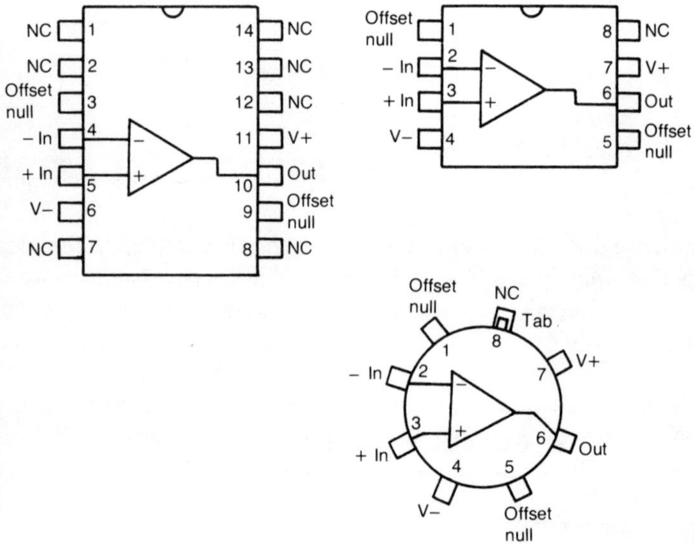

	Minimum	Typical	Maximum
Open-Loop Gain	50,000	200,000	--
Input Offset Voltage	--	1.0 mV	5.0 mV
Input Offset Current	--	20 nA	200 nA
Input Bias Current	--	80 nA	500 nA
CMRR	70 dB	90 dB	--
Slew Rate	--	0.5 V/μS	--
Input Resistance	1 megohm	6 megohms	--
Output Resistance	--	75 ohms	--
Differential Input Voltage	--	--	±30 volts

Fig. 2-1. A lot of information about an IC can be found in the manufacturer's spec sheets.

voltages that can be used to power the op amp. Remember that op amps require dual-polarity power supplies. Sometimes the positive and negative supply pins are labeled V_{SS} and V_{DD}.

Power dissipation is the amount of power (usually in milliwatts) that the device can safely dissipate. Exceeding this value could cause the chip to overheat and be permanently damaged. Generally the maximum power dissipation is a meaningful specification within a given temperature range. For example, 500 mW@< 75° C.

Differential input voltage is the maximum voltage applied across the inverting and noninverting inputs. This voltage should not exceed the voltage applied to the supply voltage pins.

The *input voltage* (sometimes called the common-mode voltage) is the maximum voltage that can be simultaneously applied between the two inputs and ground. Usually this specification is equal to the supply voltage.

The *output short-circuit duration* is the amount of time the device's output can be shorted directly to ground or either supply voltage. With many modern op amp ICs indefinite output short-circuit times may be withstood by the chip without damage.

Operating temperature is the range of ambient temperatures that will allow the IC to operate within the manufacturer's specifications. Military grade versions of most chips will have a wider acceptable temperature range than the somewhat less expensive commercial or hobbyist grade versions. For most applications the commercial/hobbyist chips will have a sufficient operating temperature range.

The *storage temperature range* is the range of temperatures that an unused IC may withstand without damage. This is usually a wider range than the operating temperature range. For hobbyist and experimenter applications, this specification will rarely matter. For almost all readily available chips, the storage temperature range will far exceed the temperature of wherever you store your components.

The electrical characteristics section of the spec sheet usually includes minimum, typical, and/or maximum values for each parameter. The electrical characteristics can be put into four broad categories:

- ☐ Input parameters.
- ☐ Output parameters.
- ☐ Dynamic parameters.
- ☐ Miscellaneous parameters.

Input Parameters

Remember from Chapter 1 that the ideal op amp features zero offset. That is, when the input is zero, the output will be zero. Most practical op amps have some degree of offset. *The input offset voltage* is the amount of voltage that must be applied to one of the inputs to get a true zero output. This will vary slightly from device to device, so the spec sheets usually include a typical and a maximum value. The typical value is the average for all similar devices, and in most applications, this typical figure will be close enough. No individual chip should exceed the maximum value specified. The lower the input offset voltage the better.

The *voltage drift* is how much the output voltage may drift, or deviate from its nominal value. This is usually dependent on temperature.

The input offset current is the difference between the two input bias currents (below) when the output voltage is at zero.

The *input bias current* is the average of the currents flowing into the inverting and noninverting inputs. Remember, the ideal op amp draws no current at all. This value should be as low as possible. The inverting bias current and the noninverting bias current should be equal, although this is not always the case with practical op amp ICs.

The *input resistance/impedance* is infinite in an ideal op amp. For a practical device the input resistance will be extremely high, usually on the order several megohms.

The *input voltage range* is the range of voltages common to both of the input and ground. This is also referred to as the common-mode input voltage.

There are a few other input parameter specifications but these are the most important and the most commonly used.

Output Parameters

The *output voltage swing* is the maximum peak output voltage that the op amp can generate without saturation or clipping. This value will depend on the load resistance and the supply voltages.

The *output short-circuit current* is the maximum output current the op amp can safely deliver to a load.

The *output resistance* of an ideal op amp is zero. For a practical op amp IC, the output resistance will be as low as possible.

Dynamic Parameters

The *open-loop voltage gain* is infinite for an ideal op amp. For a practical device this value should be very large.

The *large-scale voltage gain* is the ratio of the maximum voltage swing to the change in the input voltage that will drive the output from zero to a specified voltage.

Miscellaneous Parameters

A few of the specifications don't quite fit into any of the other categories, but they are still important enough to warrant inclusion here.

The common-mode rejection ratio (or CMRR) is a statement of the device's ability to cancel out identical signals simultaneously applied to both of the inputs. Only the difference between the two input signals should appear at the output. Later on you will see the importance of good common-mode rejection ratios. This specification is usually expressed in decibels (dB) and is the ratio of the common-mode input voltage to the op amp's output voltage.

The *supply current* is simply an expression of how much current the device will draw from the power supplies under normal operating conditions.

Another specification you may run across is only relevant for multiple op amp ICs, such as the 747 which contains two independent 741s on a single chip. A certain amount of crosstalk between op amps on the same slab of silicon is inevitable. If a input signal is fed to one of the inputs of op amp A, a small, but finite, output voltage will appear at op amp B, even if no input signals are being fed into B's inputs. The *channel separation* is a measurement of this effect. Obviously, this number should be as large as possible, indicating maximum separation and isolation between op amps.

COMMON-MODE REJECTION

An ideal op amp amplifies only the difference between the inverting and noninverting input signals. Any signal that is common to both inputs should be canceled out. In the circuit shown in Fig. 2-2, the output voltage should be zero. The common-mode gain of an op amp is the ratio of its output voltage to the common-mode input voltage.

$$G_{cm} = \frac{V_o}{V_{cm}}$$

Ideally V_o equals 0, so G_{cm} should also be 0. Practical op amp devices are not so perfect. V_o will have some finite nonzero value.

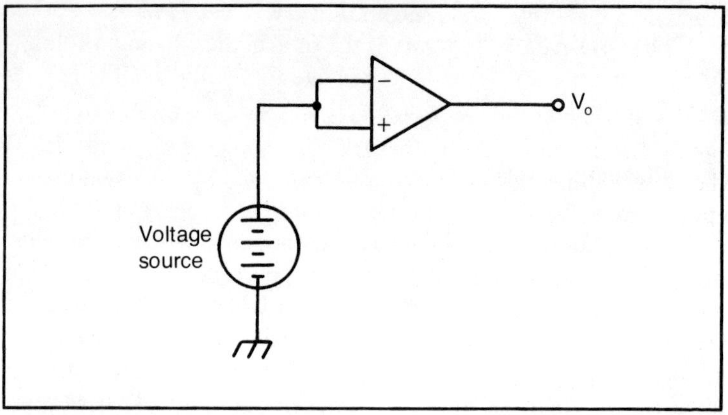

Fig. 2-2. The common-mode input signals cancel out and result in a 0 volt output.

One of the inputs will have slightly greater gain than its partner. If the inverting input is dominant, the output polarity will be reversed.

Of course, the common-mode gain should be as small as possible. Certainly it should be considerably less than 1.

Common-mode gain is not usually included in manufacturer's spec sheets. Almost all manufacturer's specify the common-mode rejection ratio, or CMRR. This is simply the ratio of the differential-mode gain and the common-mode gain. The differential-mode gain is simply the open-loop gain of the device. It is the measurement of the gain for the difference between the input signals:

$$CMRR = \frac{G_{o1}}{G_{cm}}$$

where G_{ol} is the open-loop gain (differential-mode gain), and G_{cm} is the common-mode gain.

For an ideal op amp this works out to infinity.

$$CMRR = \frac{\infty}{0} = \infty$$

The CMRR should be as large as possible. For a practical device the open-loop gain will be in the neighborhood of 200,000, and the common-mode gain will be about 6. The CMRR in this case will be:

$$CMRR = \frac{200,000}{6} \cong 33,000, \text{ or } 91 \text{ dB}$$

Even when the common-mode gain is very small, it can be of significance in some differential amplifier applications, especially where the difference signals are quite small and the common-mode input signal is large.

This problem might show up when a small signal voltage is carried along a fairly long input line through a strong electrical field. Standard 60 Hz ac house current can easily induce a volt or two into a length of wire.

Let's use the values given in the above example to illustrate this problem. Assume that the desired differential-mode input signal is 0.01 mV (0.00001 volt), and that the undesired common-mode input is 1 volt (the common-mode input will often be greater than the differential-mode input in many applications). The desired signal will depend on the differential-mode signal.

$$V_o = 200,000 \times 0.00001 = 2 \text{ volts}$$

The common-mode signal, will be 6 volts.

$$V_o = 6 \times 1 = 6 \text{ volts}$$

The actual output will be 8 volts, instead of the desired 2. This is a fairly extreme example, and in later chapters you will learn techniques to minimize the influence of the common-mode gain.

In closed-loop (gain reduced via a feedback network) circuits the common-mode gain will be reduced proportionately.

SUPPLY VOLTAGE REJECTION RATIO

Another potential source of output errors is leakage from the supply voltage. A good op amp device will have a high supply voltage rejection ratio. Ideally this will be infinite, but for practical ICs, it will have a finite value.

OFFSET VOLTAGES AND CURRENTS

When the input signal to an ideal op amp is exactly zero, the output will be exactly zero. This is referred to as zero offset. Practical devices usually exhibit some degree of offset. This comes from minor imbalances within the circuitry of the op amp itself.

In many circuits some sort of provision for correcting the input offset must be provided. This is only necessary for highly critical applications. Most practical op amp ICs have such small offset voltages that they can often be ignored.

If the offset is critical to the application, the output must be nulled by applying a small dc voltage between the inputs to pull the output to true zero when no input signal is applied. This correction voltage is called the *input offset voltage* for obvious reasons.

For general purpose op amp devices like the 741, the input offset will generally be in the 0.5 to 1.0 mV range. In some cases, however, it may be as high as 5 mV. The input error will get amplified right along with the desired input signal. This could force the op amp into saturation and clipping. If a dc offset is used, you must make sure that the maximum expected input signal added to the dc offset will not exceed the op amp's output capabilities, which is usually defined by the power supply voltage.

In critical applications where minimum offset voltages are essential, the best approach would be to use a high quality op amp IC with internal nulling.

Offset currents can also cause problems in some applications. The two input bias currents probably will not be exactly equal in a practical op amp IC. The difference between the inverting and noninverting input bias currents is referred to as the *input offset current*.

In a reasonably well-designed op amp device, the offset current should not be more than 10 percent of the input bias current. This offset current at the output will show up as an output voltage offset. The simplest cure for this problem is to add an offset-correcting resistor, as shown in Fig. 2-3. The value for this offset-correcting resistor is:

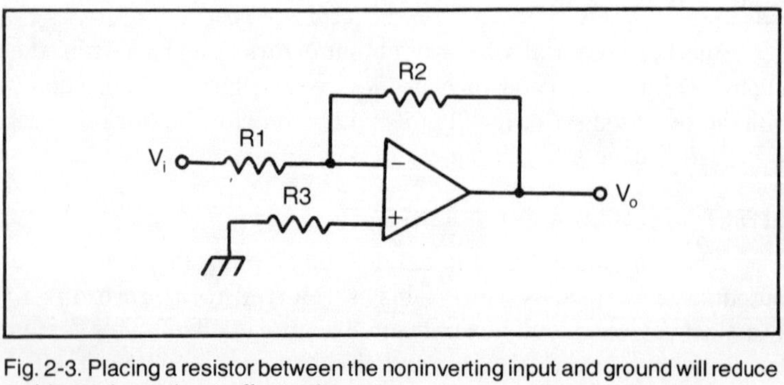

Fig. 2-3. Placing a resistor between the noninverting input and ground will reduce problems due to input offset voltages.

$$R3 = \frac{(R_s + R1) \times R2}{R_s + R1 + R2}$$

R_s is the output impedance of the signal source feeding the input of the op amp. In many cases, R_s will have such a small value that it can be ignored, simplifying the equation.

$$R3 = \frac{R1 \times R2}{R1 + R2}$$

Figure 2-4 illustrates the same technique for a noninverting amplifier circuit. If adding an offset-correcting resistor does not bring the output offset into an acceptable range, it may be necessary to add a more complex offset-current adjustment circuit or substitute a higher-grade op amp device.

FREQUENCY RESPONSE

Another characteristic of the ideal op amp is an infinite bandwidth. This means that all frequencies will be amplified equally, as shown in the frequency response graph of Fig. 2-5.

Practical op amp devices tend to rolloff at high frequencies. In fact, the open-loop gain may begin to rolloff at about 10 Hz, and steadily decrease to a maximum frequency of about 1 MHz. Some high-quality op amp devices may have a somewhat wider bandwidth. A practical frequency response graph is shown in Fig. 2-6.

The point where the gain is 3 dB from the full gain dc value is called the open-loop −3 dB point. It will typically be somewhere between 10 Hz and 1 kHz.

The slope of the high-frequency rolloff is 20 dB per decade (factor of 10), or 6 dB per octave (doubling of frequency). A 20 dB

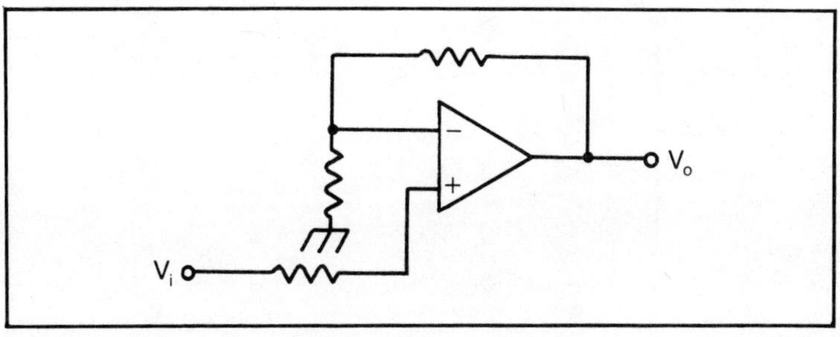

Fig. 2-4. A similar technique is used to correct for input offset voltages in a noninverting amplifier circuit.

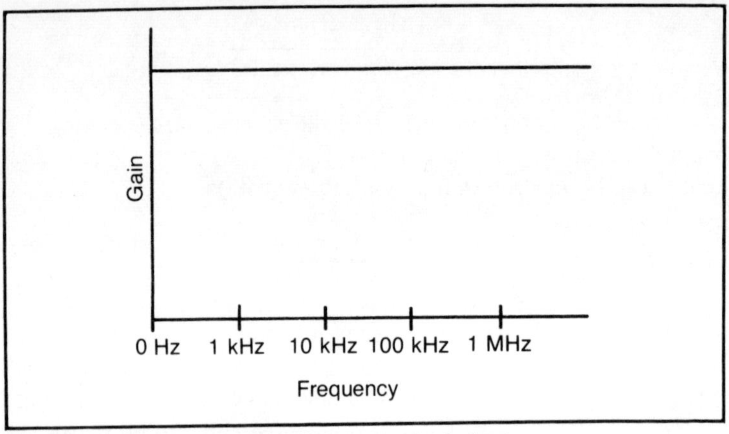

Fig. 2-5. An ideal op amp would have a perfectly flat frequency response.

rolloff translates to a voltage gain factor of 10. A 6 dB rolloff is a voltage gain factor of 2.

At some specific frequency, the open-loop gain will drop below unity. That is, this frequency, and higher frequencies, may be passed through the op amp, but they will be attenuated (reduced in level) rather than amplified.

The open-loop gain for any frequency is the maximum gain the device is capable of. A closed-loop feedback network can reduce, but not increase the gain. This means the useful bandwidth of an op amp will depend upon the desired application. The bandwidth on the spec sheet will usually be up to the unity gain point. For most general-purpose op amp devices, this will be about 1 MHz.

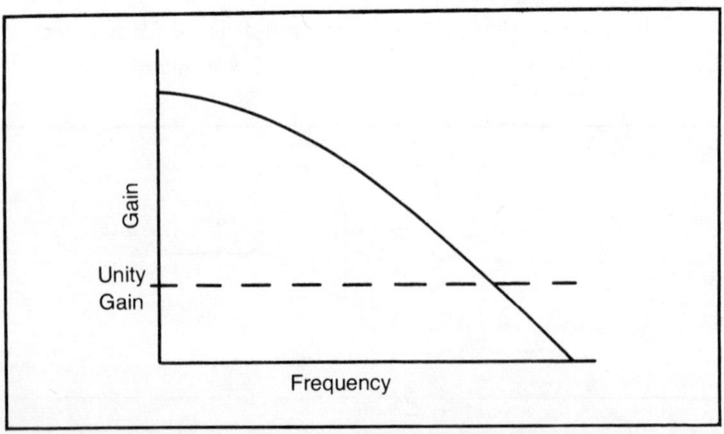

Fig. 2-6. A practical op amp doesn't amplify all frequencies equally.

There is no low frequency limitation, since the basic op amp is actually a dc amplifier. Therefore, the maximum open-loop gain will, by definition, be at dc, or 0 Hz.

Another important specification is the *gain bandwidth product*, or GBP. This is a constant for any point on the frequency response curve. Refer back to Fig. 2-6. This GBP value can be determined with the following formula:

$$GBP = G_{ol} \times BW$$

G_{ol} is the device's open-loop gain, and BW is the point where the closed-loop gain curve intersects with the open-loop gain curve.

Figure 2-7 shows a graph for a group of closed-loop gains. It is a simple matter to find the BW for any specific closed-loop gain. You simply need to extend a horizontal line from the desired gain level to intersect the rolloff of the open-loop gain curve. The intersection point marks the bandwidth at the specified gain level.

This value can also be found mathematically with the following formula:

$$BW = \frac{GBP}{G}$$

G is the desired closed-loop gain. In practical circuits, the closed-loop gain should be a factor of 1/20 to 1/10 of the open-loop gain at a specified frequency. For example, if the open-loop gain at 10 kHz is 100, then closed-loop gain should be set at 5 to 10. This is

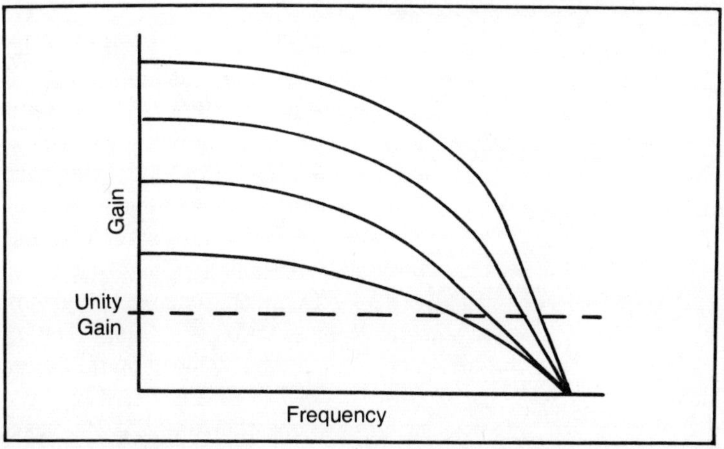

Fig. 2-7. This graph illustrates a family of closed-loop gains for a typical op amp.

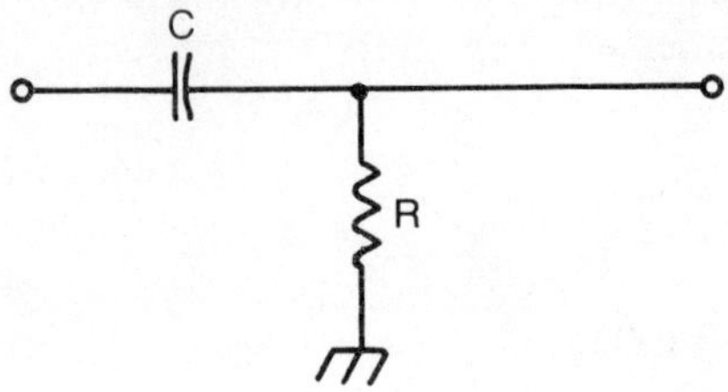

Fig. 2-8. A simple RC network is often used for frequency compensation in op amp circuits.

to ensure the proper functioning of the circuit and to minimize distortion.

At higher frequencies, the maximum allowable gain, according to this rule of thumb, drops off sharply. There is an eternal trade-off between bandwidth and gain. Standard 1 MHz GBP op amp devices are really suitable only for dc and low-frequency applications. A 1 MHz GBP unit has a maximum gain of about 50 (36 dB) at the upper audio frequencies. This may be insufficient for audio work. For high-frequency applications, it is usually necessary to use a more expensive high-quality op amp IC with a higher GBP rating.

FREQUENCY COMPENSATION

Op amp closed-loop circuits are dependent on negative feedback. Some of the output signal is fed back to the inverting input. At low frequencies this feedback signal is 180° out of phase with the input signal. This allows some of the input signal to be canceled out, limiting the gain of the device.

Unfortunately, as the input frequency increases, so does the phase shifting within the op amp. If the frequency is high enough to produce another 180 degrees of phase shift, the difference between the input and output signals will be 360°. Effectively, they will be in phase with each other. The feedback signal is now added to the input signal. If the closed-loop gain is greater than 1, this can lead to all sorts of problems. The op amp can be forced into saturation and clipping. In addition, instability and/or oscillation may result.

Manufacturers usually specify frequency compensation circuits for each device. Frequency compensation is accomplished by adding an RC network to special output pins included on the IC for just this purpose.

A basic RC network, such as the one shown in Fig. 2-8 has a maximum phase shift of 90°. A frequency compensated op amp therefore has a maximum phase shift of 90° + 180°, or 270°. This prevents the problems of instability and oscillation that can stem from a 360° phase shift of the feedback signal.

Generally frequency compensation is accomplished by adding a capacitor connected to the specified frequency compensation pins of the op amp IC. This capacitor combines with the ICs internal resistance to form the necessary RC network.

Many modern op amp ICs include the complete RC frequency compensation circuit on the chip itself. Op amps with built-in RC networks are said to be internally compensated. If an external capacitor is employed, the circuit is externally compensated.

SLEW RATE

Another characteristic of the ideal op amp is that it will be able to perfectly follow extremely rapid changes in the output signal. If the input signal changed instantly from 0 volt to 1 volt, the output should also change instantly from 0 to 1 volt times the circuit gain. This is illustrated in Fig. 2-9.

A practical op amp will exhibit some delay in switching between output levels. This is shown in exaggerated form in Fig.

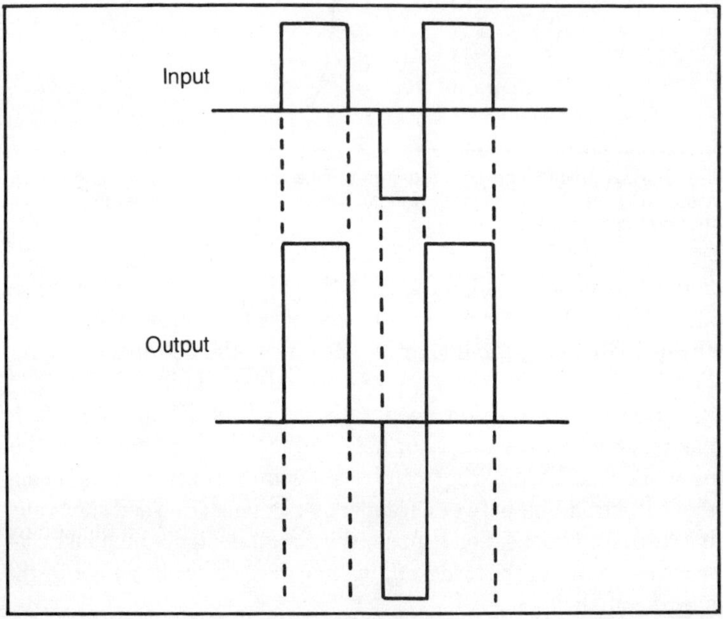

Fig. 2-9. An ideal op amp has an infinitely fast slew rate so the output changes in perfect step with changes in the input signal.

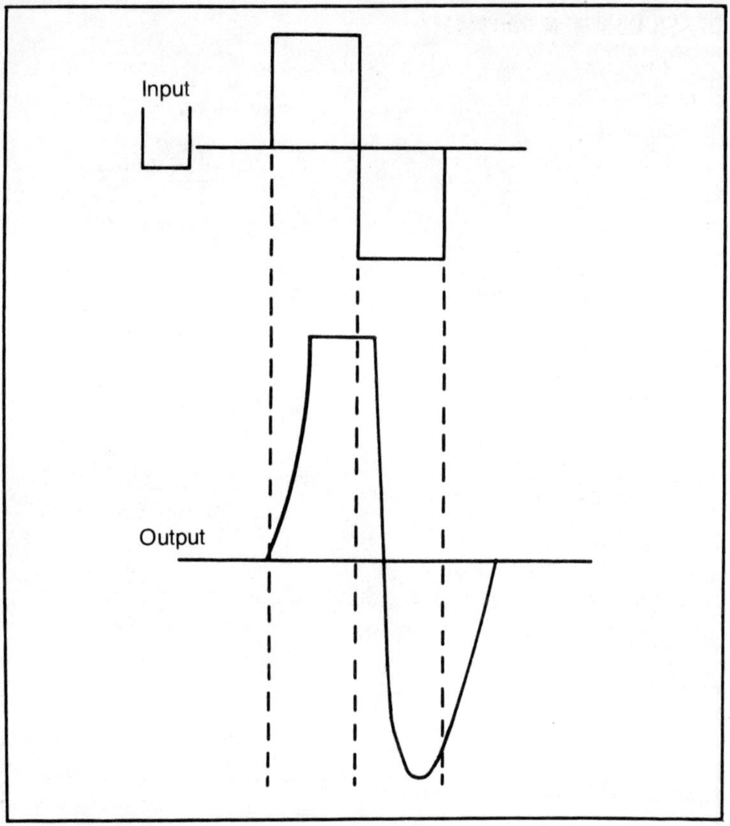

Fig. 2-10. A practical op amp requires a finite period of time for the output to respond to changes in the input voltage. The slew rate is greatly exaggerated in this sketch.

2-10. This delay is called the slew rate of the op amp. An op amp's slew rate is the maximum rate at which the output voltage can change. Obviously the higher the slew rate, the better the op amp. For an ideal op amp the slew rate would be infinite.

Lower slew rates are generally the result of capacitances. In many instances, the capacitor used in the frequency compensation network discussed earlier can be a significant factor. This means there is a trade-off between frequency compensation and slew rate. If external compensation is used, the circuit designer obviously has more control over the trade-off, making the best compromise for the specific application.

The load capacitance seen by the op amp's output can also be a contributing factor to a low slew rate.

OTHER COMMON PROBLEMS

Practical op amps have a number of additional limitations and potential problems that the circuit designer should give some consideration to.

Clipping can be a problem with any kind of amplifier, and op amps are no exception. There is a definite limit on the output voltage. For most op amp ICs, this limit is about one volt less than the power supply voltages. For example, if an op amp circuit is powered by a ±15 volt power supply, the output signal may range from about −14 volts up to about +14 volts. If the gain is 50 and the input signal is 0.5 volt, the theoretical output should be:

$$V_o = V_i \times G = 0.5 \times 50 = 25 \text{ volts}$$

This exceeds the +14 volt output limit by quite a bit. Anything above 14 volts will be clipped off, distorting the waveform as illustrated in Fig. 2-11.

A similar problem can show up with current. Newer op amp ICs include built-in current limiters in the output stage. This is to prevent excessive current draw and potential damage if the output

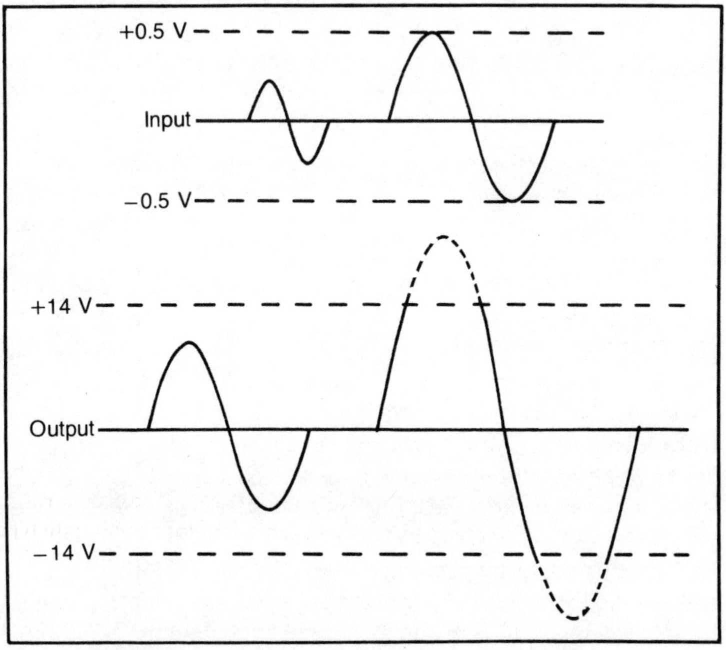

Fig. 2-11. Excessive gain can cause clipping of the output signal.

is short-circuited. The current limiter can cause current clamping if the limiting level is exceeded. This distorts waveforms in much the same way as voltage clipping.

If the maximum differential input voltage is exceeded, the output signal may get stuck at either the positive or negative maximum output voltage. This is called *latchup* or *lock up*. Latchup-problems generally show up only on older devices. More recent op amp ICs, such as the 741, have special circuitry to minimize latchup problems by increasing the maximum input voltage of the op amp device.

Other problems may be more serious, causing permanent damage to the IC and/or other circuit components. The damage is most commonly due to excessive current flow overheating the circuitry. When the op amp IC goes bad, the defect will probably be one of the following:

- ☐ Open input terminal(s).
- ☐ Open output terminal(s).
- ☐ Open power supply terminal(s).

An input terminal can be caused to open by an excessive input signal. A short-circuited output can cause excessive current flow through the op amp. This can overheat the device and cause the output terminal to open. Power supply terminals can be opened by high current transients, overvoltages, and/or polarity mismatch.

Because of the form of isolation to maintain separation between components within the IC chip, reversing the power supply polarities can cause damage. Ordinarily, components are separated by reverse biased junctions. Reversing the power supply polarity causes these junctions to become forward biased, allowing large currents to flow through the delicate isolation junctions, permanently burning them out.

Incorrect power supply polarity problems are particularly common in battery operated equipment. It is all too easy to install a battery backwards. Adding a couple of diodes as illustrated in Fig. 2-12 will protect the IC against incorrect polarity. If the batteries are installed backwards the circuit will not work, but no damage will be done.

The power supply voltage can also cause problems. Spec sheets for op amp ICs list minimum and maximum supply voltages. If less than the minimum supply voltage is applied the circuit will not function properly, but no permanent damage is likely.

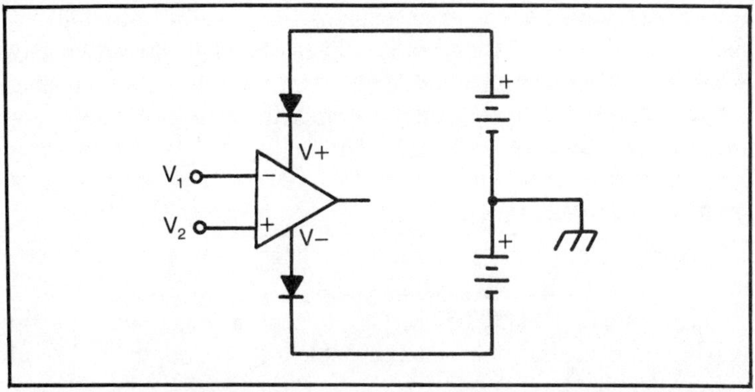

Fig. 2-12. Diodes can be used to protect against errors in power supply polarity.

Exceeding the specified maximum supply voltage can destroy an IC. The collector/base junctions of the transistors within the IC chip will break down under the excessive voltage. This is not harmful in itself, but it can cause large currents to flow through the device. This can cause damage due to overheating and/or due to the destruction of the small bonding and connecting lines on the chip because of the excessive current flowing through them.

Excessive differential input signals fed to the inputs of an op amp can break down the base/emitter junctions of the input stage circuitry. The large current flow because of this breakdown can destroy the op amp.

A practical circuit of any type exhibits some inefficiency. This causes heat to be generated to dissipate waste energy. The output stage of an op amp can generate quite a bit of heat when delivering power to an external load. A typical IC amp can safely dissipate about 500 mW at room temperature. Short circuiting the output can cause a very large current flow, exceeding the safe power dissipation value. This can damage the semiconductor chip.

Many modern op amp ICs have output current-clamping circuitry. This prevents excessive power dissipation levels, even if the output is accidentally short circuited for an extended period. This involves a trade-off, of course, against current-clamping distortion described earlier.

If you are using an op amp IC that does not have internal current clamping (such as the 709), a resistor can be added in series with the output to prevent accidental excessive power dissipation in the output stage. This is illustrated in Fig. 2-13.

Another technique would be to place a pair of back-to-back zener diodes across the output load. This will protect the IC against inadvertent application of an excessive voltage at the output terminal. The zener diode ratings should be greater than any reasonably expected output voltage, and lower than a potentially damaging voltage. This will cause the output voltage to be clipped if the zener breakdown voltage is exceeded.

LAYOUT CONSIDERATIONS

When planning the physical layout of an op amp circuit, it is a good idea to consider the circuit as an rf generator, even if only low frequencies will actually be used in the application. Even if the op amp IC being used is not normally capable of rf operation, this is still a very good idea.

The extra precautions of rf layout will certainly do no damage, and it can prevent unpleasant surprises. This is especially critical in the experimental/breadboarding stage. If the precautions are not taken, unreliable and possibly unrepeatable effects may occur.

Keep all leads as short as possible. Any terminals connected to the power supply should be capacitively bypassed to ground. Parasitic rf signals will be shorted to ground through the capacitor, while low frequencies will treat the bypass capacitor as an open circuit.

A common tie point for all grounds in the circuit should be used. Resonant circuits can be formed by poor ground connections or ground loop paths.

Shielded input and output leads should be used whenever possible, especially if long leads are unavoidable.

High-frequency transient spikes on the power supply lines can also cause unreliable operation. A decoupling capacitor across the power supply lines can short out such spikes. This is especially important in applications where high-speed switching (high slew

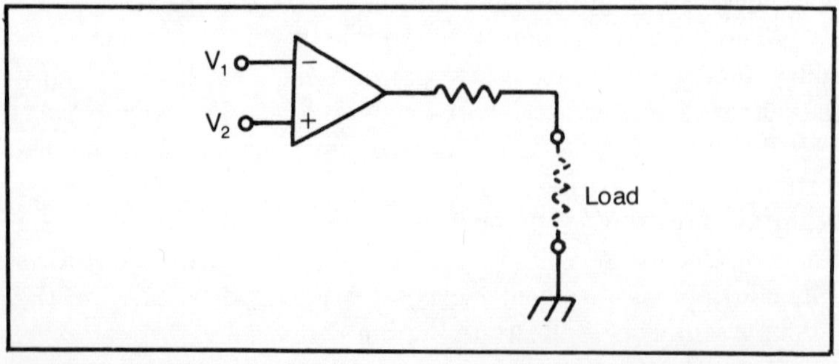

Fig. 2-13. A resistor in series with the output load can be used to limit power dissipation.

rate) is used. For maximum efficiency, the decoupling capacitors should be placed as close to the IC's power supply leads as physically possible.

Permanent circuits should be constructed on printed circuit boards, or on perfboard with wirewrap sockets.

Whenever an IC is soldered into a circuit, a socket is highly advisable. Of course for very expensive chips the advantage is obvious. A socket heads off the risk of overheating and damaging the IC device while soldering.

Even for less expensive ICs you may want to use sockets—even if the socket costs almost as much as the IC itself. If the IC ever has to be replaced, you'll be glad you used a socket. Desoldering and resoldering 8 to 16 (or even more) pins can be a major nuisance and a highly frustrating experience.

3

Amplifier Circuits

The very name *operational amplifier* indicates that an important class of applications for these devices is amplifier circuits.

In Chapter 1 basic inverting and noninverting amplifier circuits were introduced. Now that I have covered some of the limitations and special characteristics of practical op amp ICs I can describe these circuits in more depth.

INVERTING AMPLIFIERS

A basic inverting amplifier circuit is shown again in Fig. 3-1. Figure 3-2 shows an idealized equivalent circuit. Notice that the input and output circuits are isolated from each other.

According to Kirchhoff's Laws current I_i (input current) is equal in magnitude to current I_f (feedback current) but the polarity is reversed (inverted).

$$I_f = -I_i$$

Using Ohm's Law I can define the input voltage (V_i) as follows:

$$V_i = I_i R1$$

Similarly, the output voltage (V_o) works out to:

$$V_o = I_f R2$$

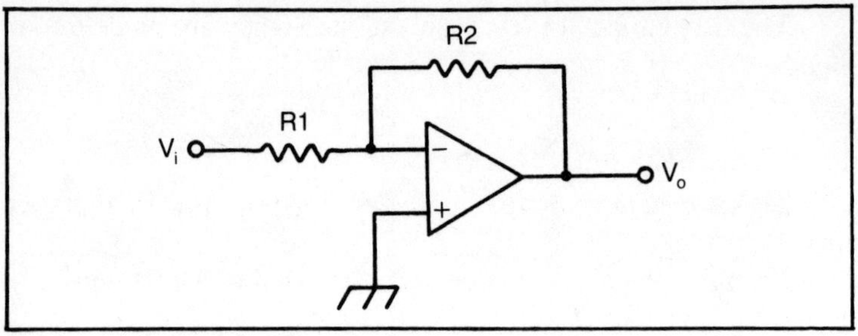

Fig. 3-1. A basic inverting amplifier is comprised of an op amp and a two resistor feedback network.

Since I_i and I_f are equal in magnitude but opposite in polarity, I can simply call I_i "I" and I_f "−I." The voltage equations become the following:

$$V_i = I\,R1$$
$$V_o = -I\,R2$$

The gain is the ratio between the input voltage (V_i) and the output voltage (V_o).

$$G = \frac{V_o}{V_i}$$

I can combine the gain and voltage equations to find the gain in terms of R1 and R2.

$$G = \frac{V_o}{V_i} = \frac{-I\,R2}{I\,R1}$$

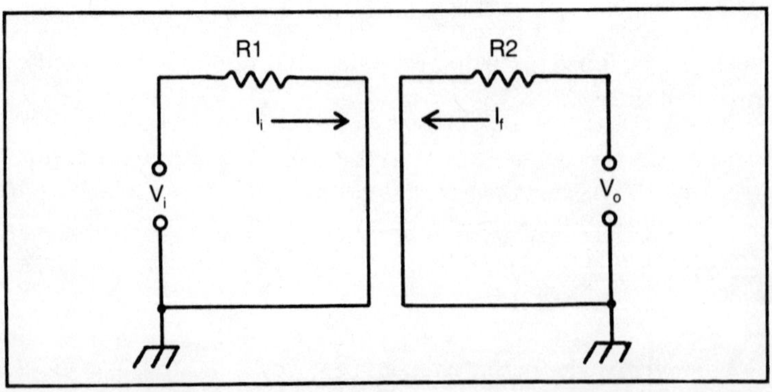

Fig. 3-2. This diagram shows the equivalent circuit for an ideal inverting amplifier circuit.

The current factors (I) cancel out since they are by definition equal.

$$G = \frac{-R2}{R1}$$

These equations may be juggled algebraically to find any missing factor.

For example, assume you have a circuit in which R1 equals 12 K and R2 equals 47 K. The gain of this inverting amplifier circuit would be −3.9.

$$G = \frac{-R2}{R1} = \frac{-47{,}000}{12{,}000} \cong -3.9$$

We can reasonably round this off to a gain of −4. Because of component tolerances, the equation won't be exactly precise anyway. The negative sign just indicates that the output is inverted (phase shifted 180°).

Now that I know the gain, I can predict the output voltage for any known input signal with the following equation:

$$V_o = V_i G$$

For instance, if the input is 1 volt, the output voltage will be −4 volts.

$$V_o = 1 \times -4 = -4 \text{ volts}$$

If the input voltage is −2.25 the output voltage works out to 9 volts.

$$V_o = -2.25 \times -4 = 9 \text{ volts}$$

Notice that the output voltage (V_o) always has the opposite polarity as the input voltage (V_i). A positive input produces a negative output, and a negative input generates a positive output.

I can also rearrange the equation to find the necessary input voltage to produce a specific desired output voltage:

$$V_i = \frac{V_o}{G}$$

If I need to generate a 5-volt output, what input signal will be needed, assuming an inverting gain of −4? I just plug the known values into the equation.

$$V_i = \frac{5}{-4} = -1.25 \text{ volt}$$

Let's run through another example. In this circuit, R1 is 680 K and R2 is 220 K. The gain would be equal to the following:

$$G = \frac{-R2}{R1} = \frac{-220{,}000}{680{,}000} \cong -0.3$$

Since R1 is larger than R2 the gain is less than unity. The signal will be attenuated as it passes through the amplifier (attenuator). If the input voltage is 2.7 volts, the output will be 0.87 volts.

$$V_o = V_i G = 2.7 \times -0.3 \cong -0.87 \text{ volt}$$

For an output voltage of 1.5 volt, the required input signal must be −4.6 volts.

$$V_i = \frac{V_o}{G} = \frac{1.5}{-0.3} \cong -4.6 \text{ volts}$$

The last possible combination would be if R1 is equal to R2. Let's say both resistors have a value of 33 K. The gain will now be −1.

$$G = \frac{-33{,}000}{33{,}000} = -1$$

For a 2-volt input signal, the output will be −2 volts.

$$V_o = 2 \times -1 = -2 \text{ volts}$$

Similarly, to produce a 3-volt output, an input signal of −3 volts is needed.

$$V_i = \frac{3}{-1} = -3 \text{ volts}$$

The input and output voltages are equal except for the polarity reversal due to the action of the op amp's inverting input. Such a circuit is called an inverting voltage follower.

In fact, the resistors can be eliminated altogether and replaced by a direct feedback connection, as shown in Fig. 3-3.

Notice that the current values (I_i and I_f) are not of importance in any of these equations. Since they are equal by definition they can be safely ignored in most design equations. The equal factors would only cancel each other out in any event.

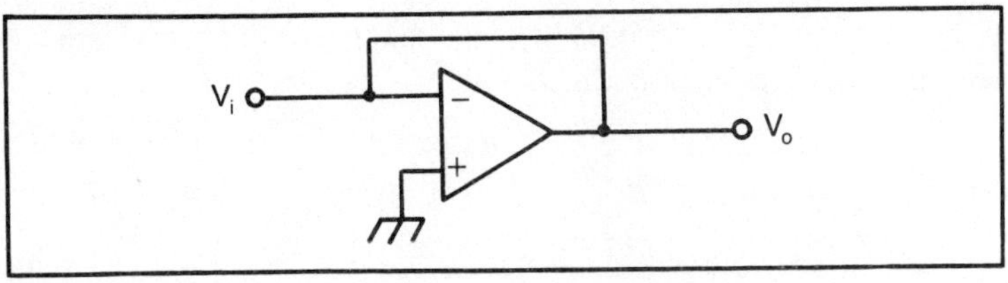

Fig. 3-3. For a unity gain inverting voltage follower, a direct feedback path is used.

In practical inverting amplifier circuits, a third resistor is usually added between the noninverting input and ground. This is illustrated in Fig. 3-4. Resistor R3 is used to correct for the device's offset current.

Remember—bias currents must be applied to the two inputs to ensure the proper biasing of the transistors in the differential input-stage. Bipolar transistors are current amplifiers so input bias currents must always be supplied even if all components are perfectly matched. This is done within the IC chip itself.

Some modern op amp ICs use unipolar or FET input stages. These are voltage amplifiers rather than current amplifiers, so bias currents are not always essential.

These mandatory bias currents can cause offset voltages in the output signal, as described in Chapter 2. If the input is exactly 0, the offset signal will cause the output to have some nonzero value.

Adding R3 as shown in Fig. 3-4 eliminates much of the problem. Because of the basic nature of the op amp, the voltage across R3 (caused by the bias current) will be equal to the voltage at the inverting input caused by the bias currents flowing through R1 and R2.

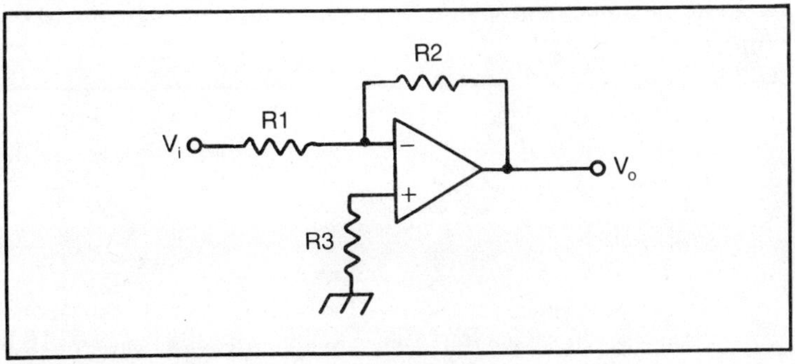

Fig. 3-4. A resistor between the noninverting input and ground can help eliminate problems due to input biases.

The value of R3 should be equal to the parallel combination of R1 and R2. The equation is as follows:

$$R3 = \frac{R1 \times R2}{R1 + R2}$$

If R1 has a value of 12 K and R2 is 47 K, R3's nominal value should be:

$$R3 = \frac{12{,}000 \times 47{,}000}{12{,}000 + 47{,}000} = \frac{564{,}000{,}000}{59{,}000} \cong 9560$$

The exact value is not absolutely critical. The nearest standard value (9.1 K or 10 K in this case) can be used. In a precision application, a ten-turn trimpot can be used to come up with the exact value.

For inverting voltage followers where R1 is equal to R2, R3 will have exactly half the value of either of the other resistors. For instance, let's make R1 = R2 = 22 K.

$$R3 = \frac{22{,}000 \times 22{,}000}{22{,}000 + 22{,}000} = \frac{484{,}000{,}000}{44{,}000} = 11{,}000 = 11\text{ K}$$

A correcting resistor is not used in a resistorless inverting voltage follower, like the one shown in Fig. 3-3. R1 equals R2 equals 0. One half of zero is zero, so R3's value is also 0, or a direct connection.

All of this assumes that the bias currents flowing into each of the input terminals are equal. Unfortunately, practical devices in the real world aren't so cooperative.

Even though the input bias currents may be very well matched, the odds are strongly against them being exactly equal. Thus there will still be some degree of offset in the output voltage due to the difference between the two bias currents. How significant this error will be depends on the application. Often it can be ignored.

This difference when the output voltage is zero is defined as the *input offset current.* It can be calculated by taking the ratio of the output offset voltage and R2.

$$I_{os} = \frac{V_{os}}{R2}$$

Since R2 will have a fairly large value (typically between 10 K and 1 M), and the output offset voltage will be fairly low (rarely more than a millivolt or two), the input offset current will be

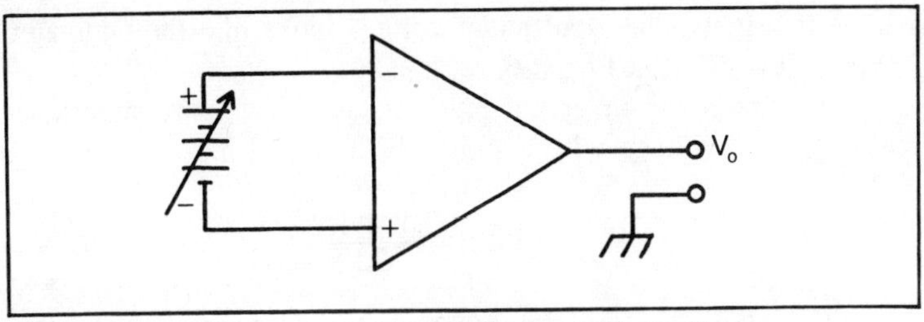

Fig. 3-5. This circuit can be used to test for input offset voltages.

extremely small. Still, the error can be of significance in precision applications.

Resistor R3 will eliminate most of the output offset voltage because the input offset current is considerably less than either of the bias currents.

The polarity of the output offset voltage will depend on the internal component mismatches within the IC.

In Chapter 2, input offset voltages were also discussed. This is another source of offset in the output voltage.

The input offset voltage is equal to the differential input voltage that must be applied across the two inputs of the op amps in order to force the output voltage to zero with no other input. A simple demonstration circuit is shown in Fig. 3-5.

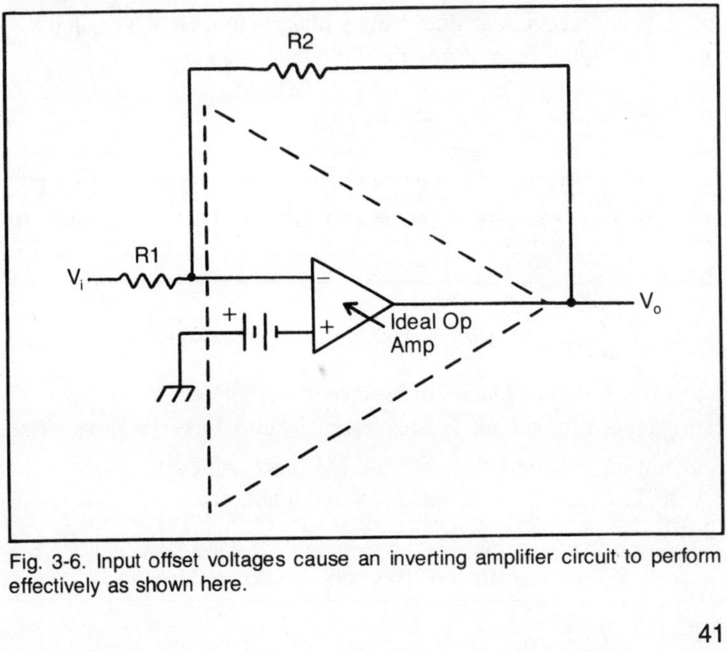

Fig. 3-6. Input offset voltages cause an inverting amplifier circuit to perform effectively as shown here.

Effectively, the input offset voltage looks like the equivalent circuit illustrated in Fig. 3-6.

Assuming the inverting input is grounded (zero input), the offset output voltage will be equal to the following equation:

$$V_{os} = V_{is} \times \frac{R1 + R2}{R1}$$

V_{is} is the input offset voltage. For a typical 741 op amp IC, the input offset voltage will be approximately 1 mV, or 10^{-3} volt.

In the following examples, I will assign R1 a value of 10 K, and R2 will be equal to 100 K. With these values, the gain will be −10.

$$G = \frac{-R2}{R1} = \frac{-100,000}{10,000} = -10$$

The output offset voltage due to the input offset voltage can now be calculated:

$$V_{os} = 10^{-3} \times \frac{10,000 + 100,000}{10,000} = 10^{-3} \times \frac{110,000}{10,000}$$

$$= 10^{-3} \times 11 = 11 \text{ mV} = 0.011 \text{ volt}$$

Since this offset voltage is due to input offset voltages, I will identify the result of this equation as V_{vos}.

The offset voltage due to input bias currents will be called V_{bos}. It can be found by multiplying the input bias current by R2.

$$V_{bos} = I_b \times R2$$

For the typical 741, the input bias current is about 80 nA, or 8 × 10^{-8} A. R2, from the previous example is 100 K. Therefore the offset output voltage due to the input bias current works out to 8 mV.

$$V_{bos} = 8 \times 10^{-8} \times 100,000 = 0.008 \text{ volt} = 8 \text{ mV}$$

When R3 is included in the circuit, the effects of the input bias current is eliminated. In this case, I substitute the input offset current into the equation. For the 741 this is typically 20 nA, or 2 × 10^{-8}. Now the offset voltage due to the input offset current becomes 2 mV.

$$V_{bos} = 2 \times 10^{-8} \times 100,000 = 0.002 \text{ volt} = 2 \text{ mV}$$

Note that this is a considerable improvement.

Since the two offset output voltages (V_{bos} and V_{vos}) are caused by different sources, they may compliment or oppose each other. In the worst case, the two offset output voltages will have identical polarities and will add.

$$V_{tos} = V_{bos} + V_{vos}$$

V_{tos} is the total offset output voltage.

For the example I have been using the total offset output voltage without R3 will be 19 mV.

$$V_{tos} = 8 + 11 = 19 \text{ mV}$$

If R3 is included in the circuit, this is reduced to 13 mV.

$$V_{tos} = 2 + 11 = 13 \text{ mV}$$

In the best case, these offset output voltages will have opposite polarities and oppose each other.

$$V_{tos} = | V_{bos} - V_{vos} | = | 8 - 11 | = | -3 | = 3 \text{ mV}$$

Or, with R3 in the circuit:

$$V_{tos} = | 2 - 11 | = | -9 | = 9 \text{ mV}$$

The polarity of the offset output voltage will depend on the specific mismatches within the IC. Polarity has been ignored in these examples.

Let's go back to the V_{bos} equations.

$$V_{bos} = I_b \times R2$$

Examine what happens when R2 is reduced to 10 K. Without R3 in the circuit, V_{bos} has a value of .8 mV.

$$V_{bos} = 8 \times 10^{-8} \times 10{,}000 = 0.0008 \text{ volt} = 0.8 \text{ mV}$$

When I add R3 to the circuit, this changes to .2 mV.

$$V_{bos} = 2 \times 10^{-8} \times 10{,}000 = 0.0002 \text{ volt} = 0.2 \text{ mV}$$

Instead of a difference of 6 mV, I only have a difference of 0.6 mV. The effects of R3 are lessened for lower values of R2.

Because there is no way to determine ahead of time whether the offsets will enforce or oppose each other, in critical applications it might be a good idea to measure the offset voltage of the individual IC directly, using the circuit of Fig. 3-7.

In an inverting amplifier, the closed-loop bandwidth will be equal to the unity-gain frequency (typically about 1 MHz, for popular op amp ICs like the 741) divided by the closed-loop gain plus 1. The formula is as follows:

$$BW_{cl} = \frac{F_{ug}}{G + 1}$$

where BW_{cl} is the closed-loop bandwidth, F_{ug} is the unity-gain frequency, and G is the closed-loop gain. Let's try a few examples using circuits built around the 741. The unity-gain frequency is 1 MHz in each of the following examples.

At unity gain:

$$BW_{cl} = \frac{1,000,000}{1 + 1} = \frac{1,000,000}{2} = 500,000 \text{ Hz} = 500 \text{ kHz}$$

For a gain of 5, I get;

$$BW_{cl} = \frac{1,000,000}{5 + 1} = \frac{1,000,000}{6} = 166,666 \text{ Hz} = 167 \text{ kHz}$$

Increasing the gain to 15 changes the closed-loop bandwidth to 62.5 kHz.

$$BW_{cl} = \frac{1,000,000}{15 + 1} = \frac{1,000,000}{16} = 62,500 \text{ Hz} = 62.5 \text{ kHz}$$

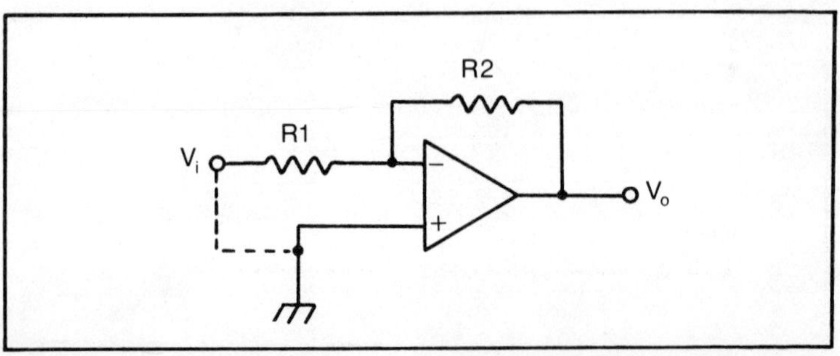

Fig. 3-7. Output offset problems can be found by shorting both inputs to ground.

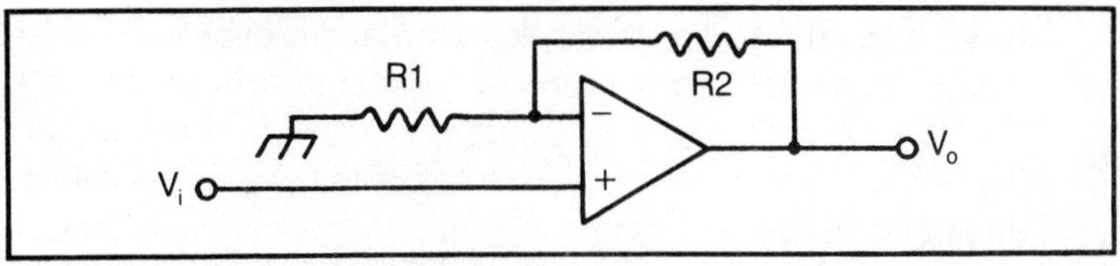

Fig. 3-8. The basic noninverting amplifier circuit includes two resistors in a feedback loop back to the inverting input to reduce circuit gain.

If the gain is below unity, say 0.75, the closed-loop bandwidth is 571 kHz.

$$BW_{cl} = \frac{1{,}000{,}000}{0.75 + 1} = \frac{1{,}000{,}000}{1.75} = 571{,}428 \text{ Hz} \cong 571 \text{ kHz}$$

The formula for the closed-loop gain in a noninverting amplifier is as follows:

$$G = \frac{R1 + R2}{R1}$$

Figure 3-9 shows an idealized equivalent circuit of a noninverting amplifier circuit. Since the output signal is referenced to the grounded end of R1, the voltage drop across R1 is in series with the feedback resistor (R2) and the output.

Obviously the same current (I) flows through the two sections of this circuit. By Ohm's Law I can define the input voltage as:

$$V_i = I \times R1$$

and the output voltage as:

$$V_o = I \times (R1 + R2)$$

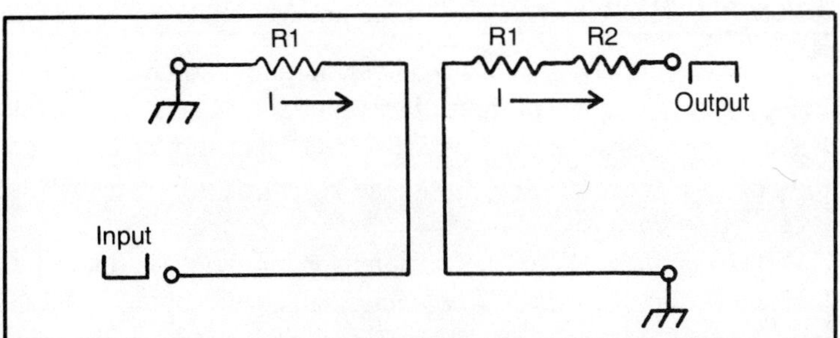

Fig. 3-9. This is the equivalent circuit for an ideal noninverting amplifier circuit.

Gain is written as a ratio of the input and output voltages.

$$G = \frac{V_o}{V_i}$$

If I combine these equations I come up with the following:

$$G = \frac{V_o}{V_i} = \frac{I \times (R1 + R2)}{I \times R1}$$

The identical I factors cancel each other out, leaving;

$$G = \frac{R1 + R2}{R1}$$

This, of course, is the formula given above for the closed loop gain of a noninverting amplifier.

As the gain is increased the closed-loop bandwidth is decreased.

The choice of values for R1 and R2 in an inverting amplifier is not entirely arbitrary. As a rough rule of thumb, R1 should be about 10 times the output impedance of the signal source. Of course this will not be true in cases where maximum power transfer between the source and the amplifier is of critical importance. The input impedance of an inverting amplifier circuit is for all intents purposes equal to the resistance of R1.

Since the output voltage is determined by the feedback current, which ideally is not affected by the load, the output impedance will be the inherent output impedance of the op amp IC itself.

NONINVERTING AMPLIFIERS

Noninverting amplifiers built around op amps are not quite as common as inverting amplifiers, but they are the preferred choice in some applications.

The closed-loop gain of a noninverting amplifier circuit (shown in Fig. 3-8) is always 1 or more. Unity gain ($G = 1$) is only possible when the resistors are replaced with direct connections ($R1 = R2 = 0$). For any nonzero value for these resistors, the gain will be greater than unity $G > 1$.

The input impedance of a noninverting amplifier circuit is extremely high. It is equal to the input impedance of the IC itself, multiplied by the circuit gain. The output impedance, on the other hand, is very low. The input impedance is not particularly important

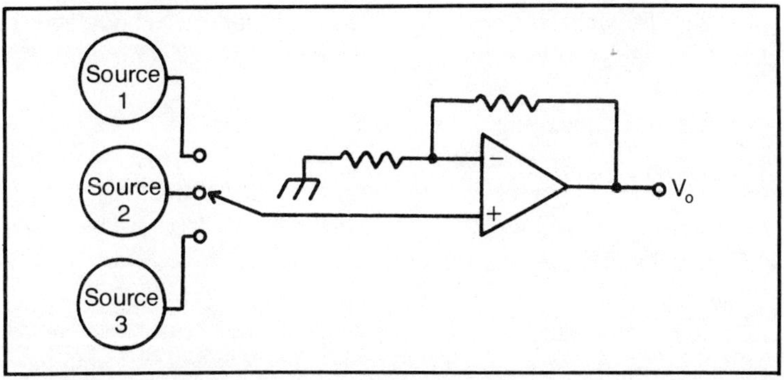

Fig. 3-10. Positive saturation latchup could be a problem when switching between various input signal sources.

in dc and low-frequency applications. The input current and voltage drop across the signal source's output impedance is more important.

A signal must be applied to the input of a noninverting amplifier at all times. If the input is allowed to float, the op amp will go into saturation and may latchup. This could be important in applications when the amplifier is switched between multiple sources, as illustrated in Fig. 3-10. Make-before-break type switches should be used to prevent momentary input floating.

If a direct feedback loop without resistors is used, as shown in Fig. 3-11, I have a noninverting voltage follower. The gain is exactly 1, so $V_o = V_i$. The output will be in phase with the input. This circuit may seem pointless—if the output is the same as the input, what's the point?

Actually a noninverting voltage follower can be an extremely useful device for buffering. Inserting such a buffer circuit between the signal source and the output load will provide isolation between the circuits, and reduce voltage errors that might be caused by loading the signal source.

The noninverting voltage follower can also be used for im-

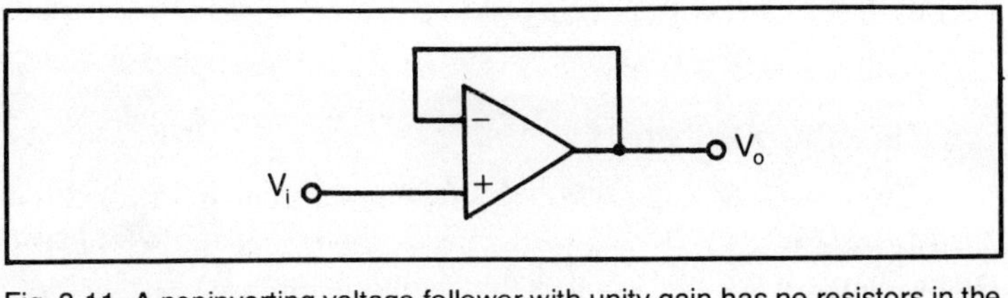

Fig. 3-11. A noninverting voltage follower with unity gain has no resistors in the negative feedback loop.

pedance transfer. The input impedance is very, very high, and the output is very, very low. High impedance signal sources can be isolated from later states.

Since the maximum possible amount of feedback is applied to the op amp, the noninverting voltage follower is a *worst case* circuit as far as stable operation is concerned. External components may be required to reduce the gain somewhat below unity at high frequencies. This is to prevent the circuit from breaking into oscillation.

Many modern op amp ICs are internally compensated. The 741 op amp device, for example, does not require external stabilizing components. They are fabricated directly onto the semiconductor chip.

Noninverting voltage followers are especially prone to output latchup. This can occur if the maximum common-mode voltage is exceeded.

If too large a signal is fed back to the inverting input, the input transistor will saturate, and the input signal will be fed directly through the collector circuit through the collector/base junction. This causes the inverting input to behave as a noninverting input. Instead of negative feedback holding the gain down, I now have positive (in phase) feedback running through the circuit. This causes the output level to rapidly rise, until the IC is in saturation. Since the output signal is fed directly back to the input the op amp will remain in this saturated state until power is removed from the circuit.

The input stage of a fully compensated op amp is usually designed to minimize the possibility of output latchup. A large common-mode input range also helps reduce the problem by allowing a larger range of output voltage excursions.

Highly accurate noninverting voltage following requires an op amp with a high open-loop gain and a high common-mode rejection ratio (CMRR). For a typical 741 device the open-loop gain is about 200,000, and the CMRR is 90 dB. This provides a dc accuracy of better than 0.003 percent. More deluxe IC op amp devices may do even better.

DIFFERENTIAL AMPLIFIERS

An inverting amplifier has the input signal fed only to the inverting input. By the same token, a noninverting amplifier has the input signal fed only to the noninverting input.

If separate input signals are simultaneously fed to both of the inputs you have a differential amplifier. The difference between the two input signals is amplified. A basic differential amplifier circuit is shown in Fig. 3-12. The output voltage (V_o) will be the difference between the two input voltages (V_1 and V_2) multiplied by the circuit gain.

$$V_o = G(V_2 - V_1)$$

If both input signals are positive, and V_2 is greater than V_1, the output will be positive. If both input signals are positive, but V_2 is less than V_1, the output voltage will be negative. Of course, this will be reversed if the two input signals are of negative polarity. If one input is positive and the other is negative, the output polarity will be determined by the absolute values of the input signals.

Op amps, or operational amplifiers, were originally designed to perform mathematical operations. The differential amplifier performs subtraction, one of the most basic mathematical functions. Later, we will learn how an op amp can be used to add signals.

Difference amplifiers are most valuable in applications where small signals must be detected and amplified without being influenced by large interfering signals. This type of circuit is based on the common-mode rejection ratio (CMRR) of the op amp.

The interfering common-mode signal can come from a number of sources. The most common and troublesome source is electrical pick-up in connecting leads. This can be especially troublesome when fairly long, unshielded wires must carry the signal from the source to the amplifier.

Under most conditions, if both input wires are of equal length, the interference signals picked up by each wire will have an equal level. The same interference signal is applied to both of the op amp's inputs, and will be rejected as a common-mode signal. Only the difference between the two lines (the desired signal) will be amplified. This can work even if the interfering signal is considerably larger than the desired signal.

In most applications, I won't need or want the full open-loop

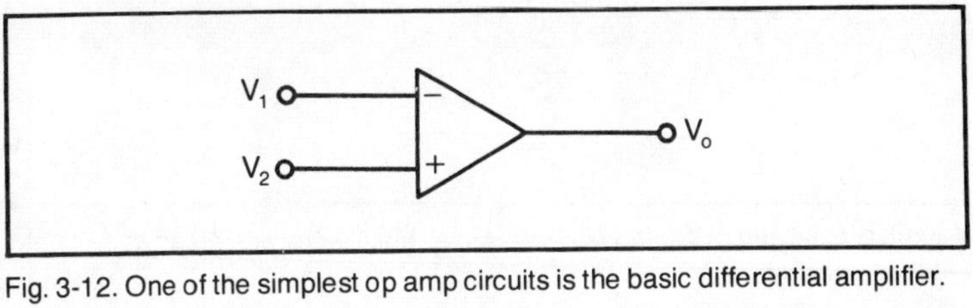

Fig. 3-12. One of the simplest op amp circuits is the basic differential amplifier.

gain provided by the circuit shown in Fig. 3-12. As with inverting and noninverting amplifier circuits, I can set up a controllable closed-loop gain by adding a feedback resistor network, as illustrated in Fig. 3-13.

In most cases, R3 will be equal to R4 and R1 will be equal to R2. This doesn't always have to be true, but it is a good rule of thumb for the majority of applications. In any case, the R3-R1 and R4-R2 ratios must be equal.

$$\frac{R3}{R1} = \frac{R4}{R2}$$

Incidentally, these ratios also define the closed-loop gain of the circuit.

$$G = \frac{R3}{R1} = \frac{R4}{R2}$$

Let's take a moment to try a few examples. Let's start out by assuming that R1=R2 and R3=R4. First, let's say all four resistors have a value of 22 K (22,000 ohms).

$$G = \frac{22{,}000}{22{,}000} = 1$$

If all of the resistors have the same value, the circuit will exhibit unity gain. This will also be true when R1=R3 and R2=R4. Let's say R1 and R3 equal 10 K, and R2 and R4 equal 33 K.

$$G = \frac{R3}{R1} = \frac{10{,}000}{10{,}000} = 1$$

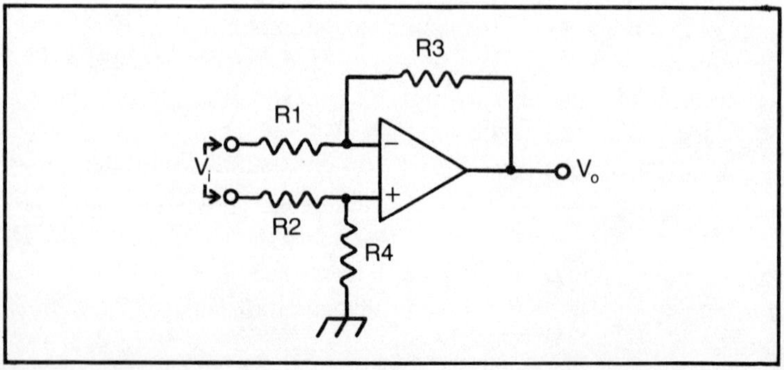

Fig. 3-13. Gain can be limited in a differential amplifier by adding an appropriate feedback network.

$$G = \frac{R4}{R2} = \frac{33,000}{33,000} = 1$$

The ratios match. Both are equal to 1, so the circuit again has unity gain.

Now assign a value of 33 K to R1 and R2 and 100 K to R3 and R4. The R3/R1 and R4/R2 ratios will be identical by definition, since R3=R4 and R1=R2.

$$G = \frac{100,000}{33,000} \cong 3$$

Next assume R1 = R2 = 68 K, and R3 = R4 = 22 K. This time I find the gain equal to 0.3.

$$G = \frac{22,000}{68,000} \cong 0.3$$

If the ratios work out to greater than 1, the circuit gain will be greater than unity. That is R3 > R1 and R4 > R2.

On the other hand, reversing this relationship by making R3 < R1 and R4 < R2 will result in a gain of less than unity (attenuation).

R1 and R2 and R3 and R4 do not necessarily have to be equal as long as their ratios match. For example, it is permissible to assign the following resistor values:

R1 = 10 K   R3 = 100 K
R2 = 47 K   R4 = 470 K

The relevant ratios match up and offer a predictable gain.

$$G = \frac{R3}{R1} = \frac{100,000}{10,000} = 10$$

$$G = \frac{R4}{R2} = \frac{470,000}{47,000} = 10$$

If the resistance ratios do not match within a reasonable tolerance, the circuit will be unstable, and probably will not work.

Since resistor values sometimes deviate from their nominal values, and because of imbalances between the op amp's inputs, R4 is often made variable so it can be adjusted for the correct ratios. A miniature trimpot is generally used.

Figure 3-14 shows how the circuit should be hooked up to adjust R4 for the maximum CMRR. The same signal is applied to

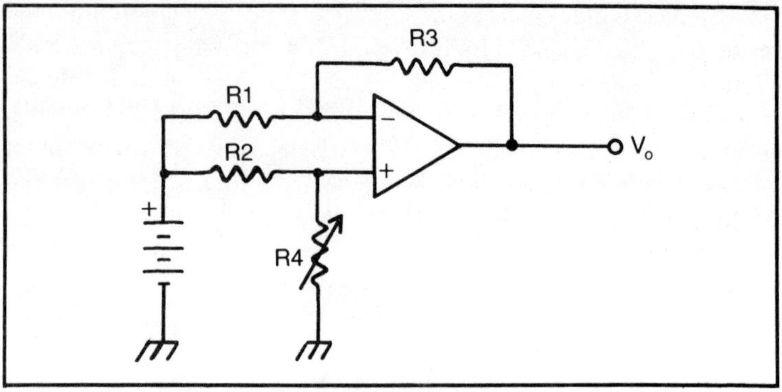

Fig. 3-14. This circuit can be used to adjust a differential amplifier for maximum CMRR.

both of the inputs. R4 is adjusted until the output is exactly zero. This compensates for imprecision in the resistor values and for internal offsets within the op amp IC itself. I want the circuit to reject as much of the common-mode signal (the only input in this case) as possible.

For most applications the fixed resistors should be 1 percent types. You may be able to use 5 percent tolerance resistors in some noncritical applications, but R4 may be difficult to adjust.

For high precision applications, especially when dealing with very small signals (a fraction of a millivolt), it may be necessary to use resistors with tolerances of 0.1 percent or even 0.05 percent. In this case, the adjustable R4 should be a high-grade 10-turn trimpot. Resistors with wide tolerances, such as 10 percent or 20 percent are generally not very suitable for use in differential amplifier circuits.

Some common-mode signals may still manage to sneak through to the output, especially if the interference is strong. There are some tricks to minimize this.

One of the simplest approaches is to simply twist the two input wires around each other, as illustrated in Fig. 3-15. Such twisted pair wiring can significantly reduce interference pickup if the input lines are not too long. For very long lengths, this technique will be insufficient.

Fig. 3-15. A twisted cable pair can reduce common-mode input signals.

Shielded cables can also be used. However, the shielding could make transmission problems even worse in some applications. Figure 3-16 shows an equivalent circuit for a shielded cable. Resistances R1 and R2 represent the resistances in the input circuit, including the impedance and the resistance of the wires themselves. The input impedances of the op amp circuit are represented by R3 and R4.

Stray capacitances will inevitably appear between the shield (which is grounded) and the two signal lines. A capacitor is nothing more than two conductors separated by an insulator, and that's just what a shielded cable is. These stray capacitances are represented in the diagram as C1 and C2.

If the product of R1 and C1 equals the product of R2 and C2 (R1C1 = R2C2), there should be no problem. However, even a slight imbalance between these values can deteriorate the common-mode rejection, allowing undesired interference signals to reach the inputs of the op amp.

A very small error can cause major output errors to occur. These problems are particularly common at high frequencies and very long input cables.

Perhaps the best, but most expensive, solution in critical applications is to use baluns. This is commonly done in professional data transmission systems.

A pair of three winding baluns (one at each end of the input line) can cancel out much of the interference noise. The balun grounds are connected to oppose the signal coils. This technique is illustrated in Fig. 3-17.

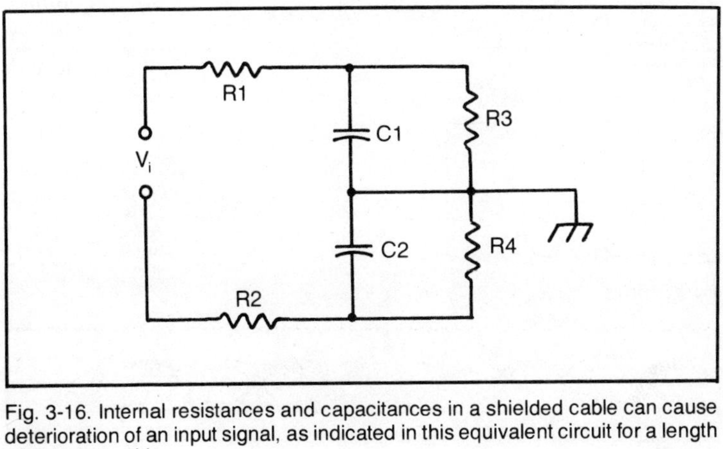

Fig. 3-16. Internal resistances and capacitances in a shielded cable can cause deterioration of an input signal, as indicated in this equivalent circuit for a length of shielded cable.

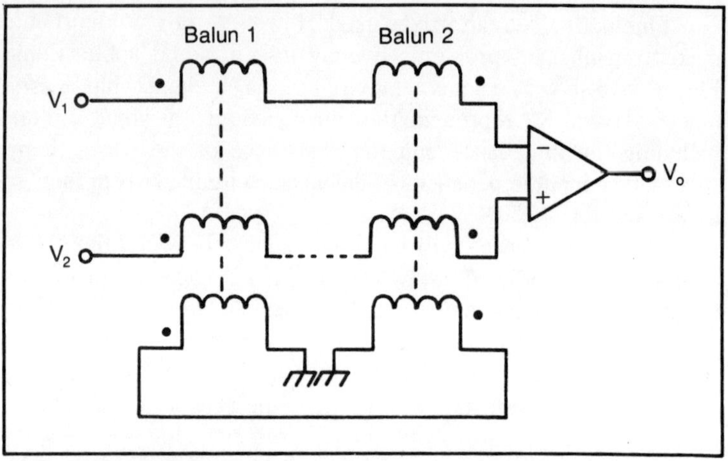

Fig. 3-17. Baluns can be used in input lines for noise suppression.

SUMMING AMPLIFIERS

Suppose you need an amplifier that will accept more than one input, but with the input signals added to each other, rather than subtracted. An op amp can also be wired as a summing amplifier. A typical circuit is shown in Fig. 3-18.

This circuit has all of the features and characteristics of the basic inverting amplifier discussed earlier in this chapter. In addition, the output signal is equal to the inverted sum of all the inputs times their individual gains.

$$V_o = -\left(\frac{R1}{R4} \times V_1 + \frac{R4}{R2} \times V_2 + \frac{R4}{R3} \times V_3\right)$$

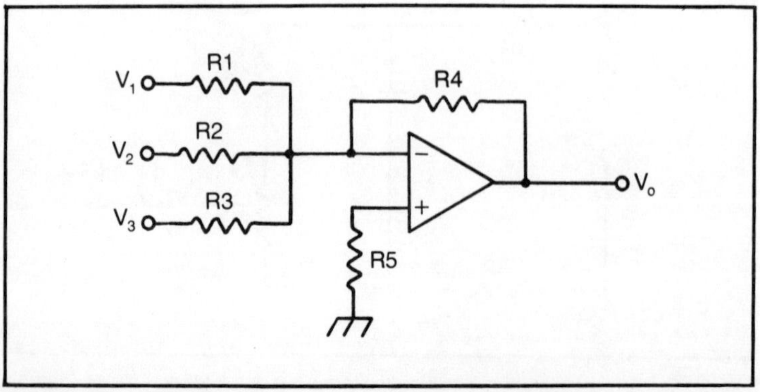

Fig. 3-18. An inverting summing amplifier accepts multiple-input signals, each through its own input resistor.

If R1, R2, and R3 have equal values, the equation can be simplified.

$$V_o = \frac{-R4}{R} \times (V_1 + V_2 + V_3)$$

$$R = R1 = R2 = R3$$

Let's try a few examples. First, assume all of the gain determining resistors (R1 through R4) have a value of 100 K. Since R1=R2=R3 I can use the short form of the equation;

$$V_o = -\frac{R4}{R}(V_1 + V_2 + V_3) = -\frac{1}{1}(V_1 + V_2 + V_3)$$

$$V_o = -1(V_1 + V_2 + V_3) = -(V_1 + V_2 + V_3)$$

The input voltages are simply added together and inverted. All inputs are treated to unity gain. For instance, if all of the input voltages are 1 volt:

$$V_o = -(1 + 1 + 1) = -3$$

The input voltages are more likely to take on different values. Set V_1 at 1.25 volt, V_2 at 3.4 volt, and V_3 at 2.35 volt.

$$V_o = -(1.25 + 3.4 + 2.35) = -7.0 \text{ volts}$$

The input voltages do not necessarily have to have the same polarity. For example, if $V_1 = 2$ volts, $V_2 = -3$ volts, and $V_3 = 4$ volts, V_o equals -3 volts.

$$V_o = -(2 + (-3) + 4) = -(2 - 3 + 4) = -3 \text{ volts}$$

Any combination of input voltages can be combined in this way.

A summing amplifier may also have nonunity gain, of course. Leave R1=R2=R3=100 K, and change R4 to 1 MΩ. Now the general equation looks like this:

$$V_o = -\frac{R4}{R}(V_1 + V_2 + V_3)$$

$$V_o = \frac{-1{,}000{,}000}{100{,}000}(V_1 + V_2 + V_3) = -10(V_1 + V_2 + V_3)$$

The sum of the inputs is simply multiplied by the gain.

Now let's continue with the same example, but reduce R4 to 10 K. Now the output signal works out to:

$$V_o = -\frac{10,000}{100,000}(V_1 + V_2 + V_3) = -0.1(V_1 + V_2 + V_3)$$

Less than unity gain is also possible. Here the output is one tenth of the inverted sum of the input voltages.

Weighted averaging is easily achieved with a summing amplifier circuit by selecting the right values for resistors R1 through R2. As a typical example, let's use the following component values:

R1 = 25 K	R3 = 100 K
R2 = 50 K	R4 = 100 K

Since the values of R1, R2, and R3 are not equal, I must use the long form of the output voltage equation.

$$V_o = -\left(\frac{R4}{R1} \times V_1 + \frac{R4}{R2} \times V_2 + \frac{R4}{R3} \times V_3\right)$$

$$V_o = -\left(\frac{100,000}{25,000} \times V_1 + \frac{100,000}{50,000} \times V_2 + \frac{100,000}{100,000} \times V_3\right)$$

$$V_o = -(4 \times V_1 + 2 \times V_2 + 1 \times V_3)$$

Input signal V_1 will have much more influence on the output than input signal V_3. Each input has its own individual gain. This is called weighting.

To see how this works, let's run through a few quick examples.

First, let's say that all three input voltages are 2 volts. Solving for the output voltage I find;

$$V_o = -(4\ V_1 + 2\ V_2 + V_3) = -(4 \times 2 + 2 \times 2 + 2)$$

$$V_o = -(8 + 4 + 2) = -14 \text{ volts}$$

Naturally the input signals won't always be of equal value. This time, let's make $V_1 = 1.25$, $V_2 = 1.5$, and $V_3 = 1.75$. This time the output voltage comes out to -9.75 volts.

$$V_o = -(4 \times 1.25 + 2 \times 1.50 + 1.75)$$
$$V_o = -(5 + 3 + 1.75) = -9.75 \text{ volts}$$

The input polarities may also be mixed. For example, if $V_1 = -2$ volts, $V_2 = 3$ volts, and $V_3 = 2$ volts, I get an output signal of zero.

$$V_o = -(4 \times -2 + 2 \times 3 + 2) = -(-8 + 6 + 2) = 0$$

Because of its stronger weighting, input V_1 cancels out V_2 and V_3.

For our next example, let's use the following input signals:

$$V_1 = 1.5 \text{ volt}$$
$$V_2 = -1.5 \text{ volt}$$
$$V_3 = 2 \text{ volts}$$

Plugging these values into the equation, I find the following.

$$V_o = -(4 \times 1.5 + 2 \times -1.5 + 2)$$
$$V_o = -(6 - 3 + 2) = -5$$

Let's try one last example. I will use the following input values:

$$V_1 = -2 \text{ volts}$$
$$V_2 = 3 \text{ volts}$$
$$V_3 = -2.5 \text{ volts}$$

The output voltage this time will be positive.

$$V_o = -(4 \times -2 + 2 \times 3 + -2.5) = -(-8 + 6 - 2.5)$$
$$V_o = -(-4.5) = 4.5 \text{ volt}$$

The negative sum is inverted into a positive output voltage.

Resistor R5 in Fig. 3-18 is used for offset correction. Its value is equivalent to the parallel combination of the other four resistors.

$$\frac{1}{R5} = \frac{1}{R1} + \frac{1}{R2} + \frac{1}{R3} + \frac{1}{R4}$$

R5 may be made variable for fine adjustment of the output offset voltage.

In a summing amplifier, the inputs are independent. In other words, no interaction takes place between the various input signals.

Source V_2, for example, will not load down, or otherwise affect the operation of source V_1 or source V_3.

It is very easy to exceed the maximum output voltage in a summing amplifier. For instance, consider a circuit with the following values:

R1 = 10 K	V_1 = 1 volt
R2 = 33 K	V_2 = 1.25 volt
R3 = 47 K	V_3 = 0.5 volt
R4 = 100 K	

Theoretically, the output should work out to −14.75 volts.

$$V_o = -\left(\frac{100{,}000}{10{,}000} \times 1 + \frac{100{,}000}{33{,}000} \times 1.25 + \frac{100{,}000}{47{,}000} \times 0.5\right)$$

$$= -(10 \times 1 + 3 \times 1.25 + 2 + 0.5)$$

$$= -(10 + 3.75 + 1) = -14.75$$

If a ±15-volt power supply is used to operate the circuit, the maximum output voltage swing will be ±14 volts. This output signal would be clipped. If ac signals were being fed through the summing amplifier, their waveforms would be distorted.

It is a good rule of thumb to keep the gains in a summing amplifier relatively low to prevent clipping of the output signal. In many applications, it would probably be a reasonable precaution to keep all of the circuit gains to less than unity. The output signal may be reamplified by a later stage if necessary.

A summing amplifier may have virtually as many inputs as the application requires, although I wouldn't advise over doing it. A hundred input summing amplifier is just begging for trouble.

4

Experiments I

Certainly the best way of learning is by doing. Performing the experiments in this and later chapters is highly recommended. It will help you understand and absorb the material you've read about in the text.

In this section you will explore some of the characteristics of an op amp IC and try out a few basic amplifier circuits.

BREADBOARDING

Some sort of breadboarding system will be needed to perform the experiments. This is simply a means of temporarily hooking up electronic circuits for purposes of testing and experimenting.

If you go in for doing things the hard way, you could hard-wire and solder each of the test circuits as if it was intended for permanent use. Many of the experiments will require substituting several components a number of times. This will require extensive desoldering and resoldering. Besides being tedious, this approach can also be rather expensive, since many of the components you will be working with will stand up to just so many solderings and desolderings before they must be replaced with new parts. It is much easier and more practical to breadboard the circuits temporarily.

In its simplest form, a breadboard is merely a solderless socket that will allow the various component leads to be quickly and easily plugged in and pulled out as often as desired. A typical solderless socket suitable for IC work is shown in Fig. 4-1. The various holes

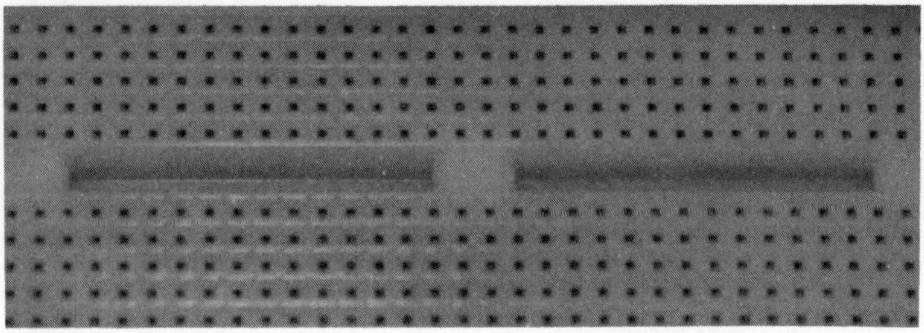

Fig. 4-1. A solderless socket makes experimentation and circuit testing very easy.

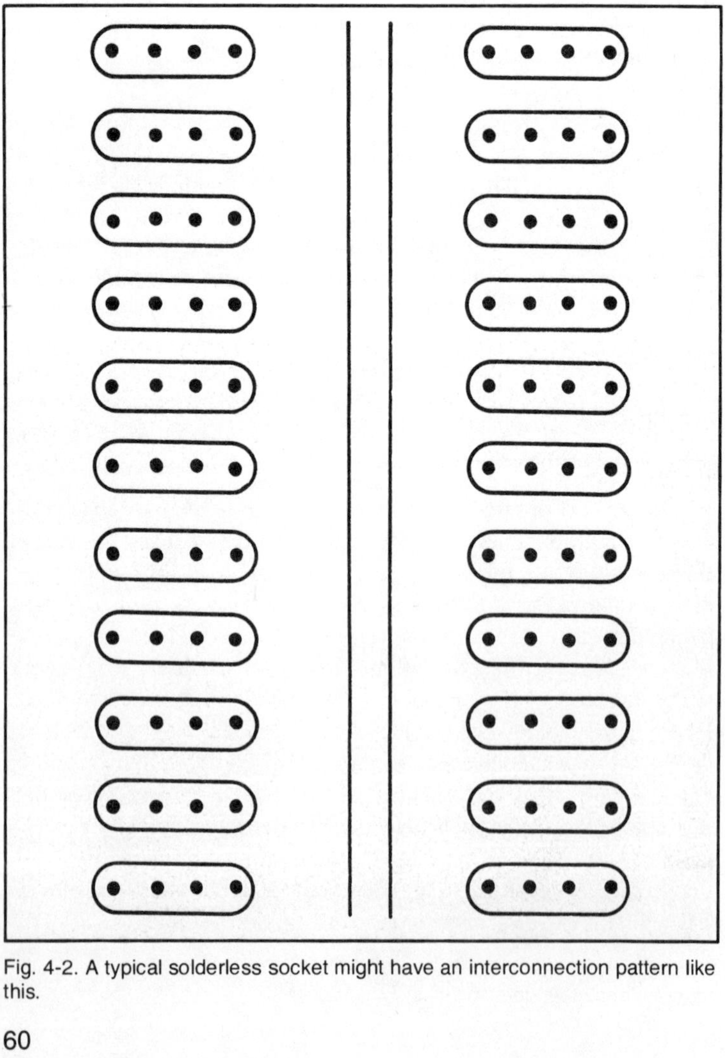

Fig. 4-2. A typical solderless socket might have an interconnection pattern like this.

in this socket are electrically interconnected in a pattern. Probably the most useful pattern for our purposes (and the most commonly available) is shown in Fig. 4-2.

Solderless sockets, which can be purchased from many electronics dealers for less than $10 make experimentation and circuit design much easier than it was before they became available. Before the seventies, experimenters and designers had to fiddle about with tiny spring clips or some kind of jury-rigged breadboarding system which was generally as awkward as it was useful. Many impatient experimenters simply bypassed the breadboarding stage and soldered the first version of their circuits together, making any necessary changes with the soldering iron.

These days there is really no excuse for making more work for yourself. Get a solderless socket and do the job right.

Solderless sockets can be made even more useful when they are part of a complete breadboarding system such as the one shown in Fig. 4-3. A complete breadboarding system includes a solderless socket and various commonly used subcircuits. (Power supplies to replace batteries and signal generators to produce ac signals.) These subcircuits could be separate, self-standing modules, but it is

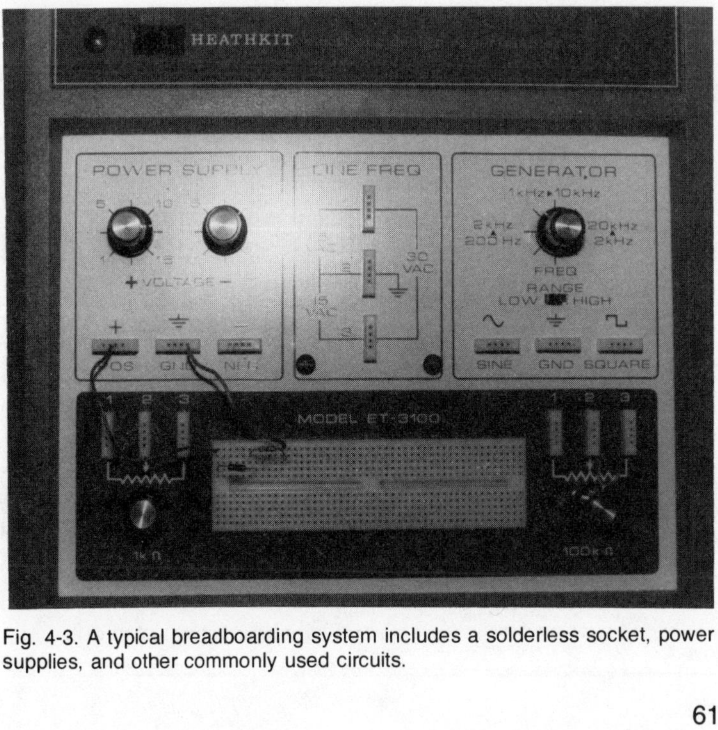

Fig. 4-3. A typical breadboarding system includes a solderless socket, power supplies, and other commonly used circuits.

usually more convenient to have them grouped together within a single, compact unit. These subcircuits will be needed far too often to breadboard them each time they are needed.

If you don't want to buy a commercial breadboarding system, you can build your own, using circuits described throughout this book.

The materials needed to perform all of the experiments in this chapter are summarized in Table 4-1. Each experiment will have its own individual parts list.

EXPERIMENT 1—NONINVERTING VOLTAGE FOLLOWER

The experiments in this chapter will be built around the 741 op amp IC. This device is available in a number of different packages, but the most common is the 8-pin DIP package. The pinout diagram for an 8-bit DIP 741 op amp IC is shown in Fig. 4-4. If you use another type of IC, or a 741 in another type of package, be sure to check the pinout diagram and change the pin numbers in the schematics accordingly.

Collect the parts listed in Table 4-2, and connect them as shown in Fig. 4-5. You may use a single voltmeter by moving it back and forth between points A and B at each stage of the experiment.

Table 4-1. Components and Equipment List.

Item	Quantity
Breadboard system	
solderless socket	
+V power supply (12 to 18 volts)	
−V power supply (−12 to −18 volts)	
Components	
741 op amp IC (8-pin DIP—see text)	
100 ohm ½ watt resistor	
1 K ½ watt resistor	(need 2)
4.7 K ½ watt resistor	(need 2)
6.8 K ½ watt resistor	
10 K ½ watt resistor	(need 4)
22 K ½ watt resistor	
47 K ½ watt resistor	(need 2)
100 K ½ watt resistor	(need 4)
220 K ½ watt resistor	
1 K linear taper potentiometer	(need 2)
Other	
hook-up wire	
dc voltmeter (1 or 2 — see text)	
ac voltmeter or oscilloscope	
1.5 volt AA batteries with holders	(need 4)

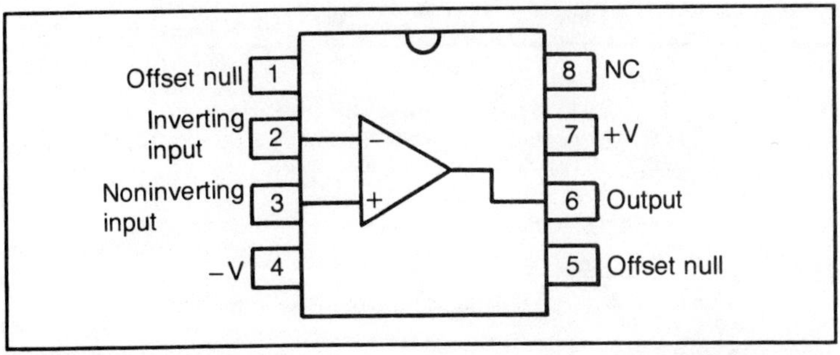

Fig. 4-4. This is the pinout diagram for a 741 op amp IC in an 8-pin DIP package.

Point A is the input and point B is the output of the circuit.

You should recognize this circuit as a basic noninverting voltage follower. Resistor R1 and potentiometer R2 form a simple voltage divider to feed different voltages to the input of the op amp. By adjusting the setting of R2, you can set the voltage being fed to the circuit's input.

A voltage follower, you should recall has an output that is the same as the input. In other words, $V_o = V_i$. The gain (G) equals 1.

Double-check the circuit before applying power. An incorrect connection could damage the op amp, or even the power supply circuitry of your breadboarding system.

Once you are sure the circuit is correctly wired, apply power and read the voltmeter at point A (input). Use the volt section of a vom (voltohm-milliammeter).

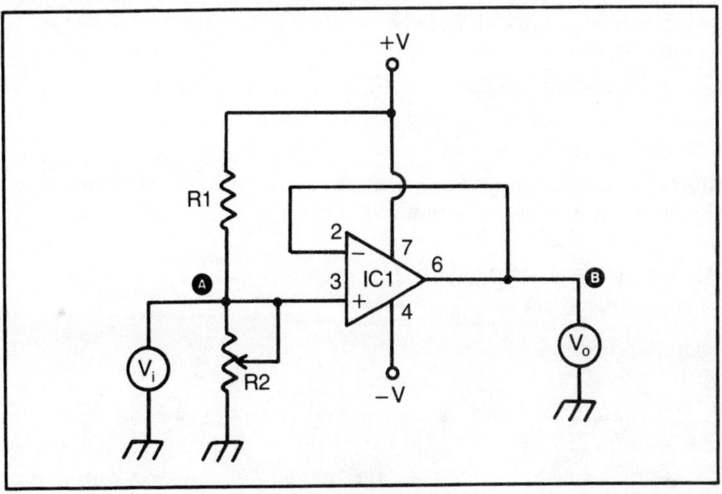

Fig. 4-5. This is the schematic for Experiment 1.

Table 4-2. Parts List for Experiment 1.

Breadboard system	
IC1	741 op amp IC (8-pin DIP—see text)
R1	1 K resistor ½ watt
R2	1 K potentiometer, linear taper
Hook-up wire	
Voltmeters (2)	
	A single voltmeter may be used—see text.

Adjust R2 until the reading on this voltmeter is one volt. Read the voltage at point B (output). It should read one volt also, or very close to it. There may be a small error due to offsets within the chip.

Now go back to the voltmeter at point A and readjust R2 for a reading of three volts. The voltmeter at point B should also be reading (approximately) three volts.

Repeat this procedure for a number of different voltages at point A. Point B and point A should always give the same readings. The output voltage *follows* the input voltage.

If you have a signal generator or variable frequency oscillator and an oscilloscope, you can repeat the same experiment with ac signals. The output signal should match the input signal. Experiment with various ac voltages and frequencies.

Disconnect R1 and R2 and feed the ac signal directly into pin 3 of the IC. If you apply very high frequencies (above a few hundred kilohertz), the output level may start to drop off a little. The gain will dip below unity. This is because of the limitations of the 741 IC itself. Don't worry about it.

The circuit may also break into oscillation at high input frequencies, because I haven't included any correction components. If the output starts looking weird at high frequencies, just relax for now. I will deal with these problems later. For now concentrate on lower frequencies. You shouldn't encounter any difficulties.

EXPERIMENT 2—NONINVERTING AMPLIFIER

Now let's take a hands-on look at a noninverting amplifier with gain. The complete parts list for this experiment is given in Table 4-3.

Carefully wire up the circuit shown in Fig. 4-6, using a 10 K resistor for R3 and a 22 K resistor for R4. Resistor R1 and potentiometer R2 again form a voltage divider to provide a variable dc input voltage.

Breadboard system	
IC1	741 op amp IC (8-pin DIP)
R1	1 K ½ watt resistor
R2	1 K potentiometer, linear taper
R3	10 K ½ watt resistor (2)
R4	22 K ½ watt resistor
	4.7 K ½ watt resistor
Hook-up wire	
Voltmeter(s)	

Table 4-3. Parts List for Experiment 2.

Once again, carefully go back over your wiring before applying power to the circuit. Why go out of your way to invite trouble?

The output of this circuit should equal the input voltage multiplied by the gain.

$$V_o = V_i \times G$$

The gain (G) is determined by the ratio of R3 and R4 according to the following formula:

$$G = 1 + \frac{R4}{R2}$$

Right now our circuit has a value of 10 K for R3 and 22 K for R4. Therefore, the gain works out to 3.2.

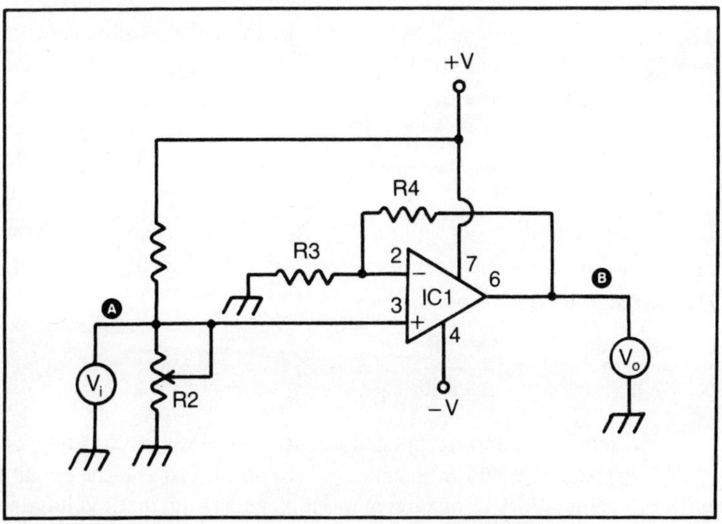

Fig. 4-6. Experiment 2 uses this noninverting amplifier circuit.

$$G = 1 + \frac{22,000}{10,000} = 1 + 2.2 = 3.2$$

Therefore, the output should be predicted by the following equation:

$$V_o = V_i \times 3.2$$

Adjust R2 so that there is an input of one volt at point A. The output should be very close to 3.2 volts.

$$V_o = 1 \times 3.2 = 3.2 \text{ volts}$$

Check the voltage at point B to confirm this.

Increase the input voltage (A) to 2 volts. The output (B) should now be 6.4 volts.

$$V_o = 2 \times 3.2 = 6.4 \text{ volts}$$

If the input (A) is three volts, the output (B) should be three times the gain.

$$V_o = 3 \times 3.2 = 9.6 \text{ volts}$$

There may be some slight differences in the output readings because of minor inaccuracies within the IC and resistor tolerances. The measured values should be close to the calculated values.

Replace R4 with a 10 K resistor. R3 and R4 should now be equal. The gain is now 2.

$$G = 1 + \frac{10,000}{10,000} = 1 + 1 = 2$$

Some typical inputs and outputs are listed in Table 4-4. Now replace R4 with a 4.7 K resistor. R4 is less than R3, but the gain is still greater than unity.

$$G = 1 + \frac{4700}{10,000} = 1 + 0.47 = 1.47 = 1.5$$

Typical input and output values are listed in Table 4-5.

Experiment with other values of R3 and R4 to see the results with different gains. If the gain is too high and a high input voltage is applied, you may see some clipping of the output signal.

Input	Output
0.5 volt	1 volt
1 volt	2 volts
2 volts	4 volts
2.25 volts	4.5 volts
2.5 volts	5 volts
3 volts	6 volts
4 volts	8 volts
5 volts	10 volts

Clipping may occur at higher input voltages.

Table 4-4. Typical Input Voltages.

Notice that no combination of R3 and R4 can ever give the circuit a gain of less than unity (1).

EXPERIMENT 3—INVERTING AMPLIFIER

The next experiment looks at the inverting amplifier. The circuit is shown in Fig. 4-7. The parts list is given in Table 4-6.

Wire the circuit carefully on your breadboard, and double-check all connections.

R1 and R2 are a voltage divider network to provide a variable dc input voltage to the circuit.

Remember the polarity of the output of an inverting amplifier is the opposite of the input, so be careful with the voltmeter polarities at points A and B.

The gain for this circuit can be found with the following equation:

Table 4-5. Output Voltages.

Input	Output
0.5 volt	0.75 volt
1 volt	1.5 volt
1.5 volt	2.25 volts
2 volts	3 volts
3 volts	4.5 volts
3.5 volts	5.25 volts
4 volts	6 volts
5 volts	7.5 volts
6 volts	9 volts
7 volts	10.5 volts

Clipping may occur at higher input voltages.

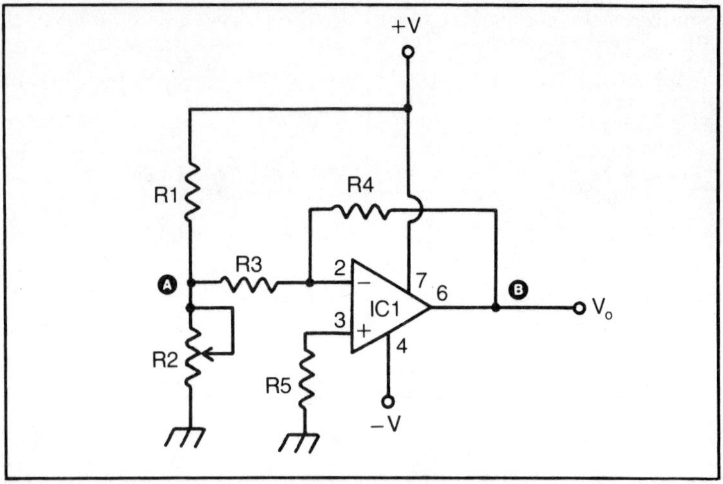

Fig. 4-7. In Experiment 3, you will work with this inverting amplifier circuit.

$$G = \frac{-R4}{R3}$$

Let's begin with R3 equal to 10 K and R4 equal to 47 K. The gain should be about 5.

$$G = \frac{-47,000}{10,000} = 4.7 \cong 5$$

Set R2 for an input of one volt (point A) and measure the output voltage (point B). Remember that the polarity should be reversed at the output.

The measured output should be between about −4.5 and −5 volts. Ideally it should be −4.7 volts, but resistor tolerances and

Breadboard system	
IC1	741 op amp (8-pin DIP)
R1	1 K ½ watt resistor
R2	1 K potentiometer, linear taper
	4.7 K ½ watt resistor
	10 K ½ watt resistor (2)
	47 K ½ watt resistor
	100 K potentiometer
R5	4.7 K ½ watt resistor
R6	100 ohm ½ watt resistor
Hook-up wire	
Voltmeter(s)	

Table 4-6. Parts List for Experiment 3.

minor inaccuracies within the IC itself may cause a slight error in the output.

Increase the input voltage to 2 volts. The output voltage should now equal −9.4 volts.

$$V_o = V_i \times G = 2 \times -4.7 = -9.4 \text{ volts.}$$

A reading between −9 and −9.75 volts is acceptable.

Try calculating and measuring the output voltage for various other input voltages.

Rewire the circuit as shown in Fig. 4-8. This version of the circuit will allow you to feed a negative voltage into the inverting input of the op amp. The output voltage should be positive. Watch the polarity of the voltmeter.

Set the input (point A) to −1.5 volts, and measure the output. The nominal output should be 7.05 volts.

$$V_o = -1.5 \times -4.7 = 7.05 \text{ volts}$$

Notice how the negative input voltage (V_i) and the negative gain (G) cancel each other out, resulting in a positive output voltage (V_o). Experiment with other input values. Then return the circuit to the original configuration (Fig. 4-7), and replace R4 with a 10 K resistor. Now, since R3 = R4:

$$G = \frac{-R4}{R3} = \frac{-10,000}{10,000} = -1$$

This gives us a unity gain inverting amplifier. The output voltage should equal the input voltage with the polarity reversed. If the input is 2 volts, the output should be −2 volts. If the input is −3 volts, the output should be +3 volts, and so forth. Try measuring the output voltage for a number of input values. Be sure to watch the voltmeter polarity.

You might want to try the circuit of Fig. 4-8 also to prove that a negative input voltage produces a positive output voltage.

If you have a signal generator and an oscilloscope, you might want to check the operation of an inverting voltage follower on ac signals. You will need a dual trace oscilloscope to simultaneously display the input and output signals. The output should be 180° out of phase with the input, as illustrated in Fig. 4-9.

Next, if you have made any changes in the circuit, return to the circuit of Fig. 4-7 and replace R4 with a 4.7 K resistor.

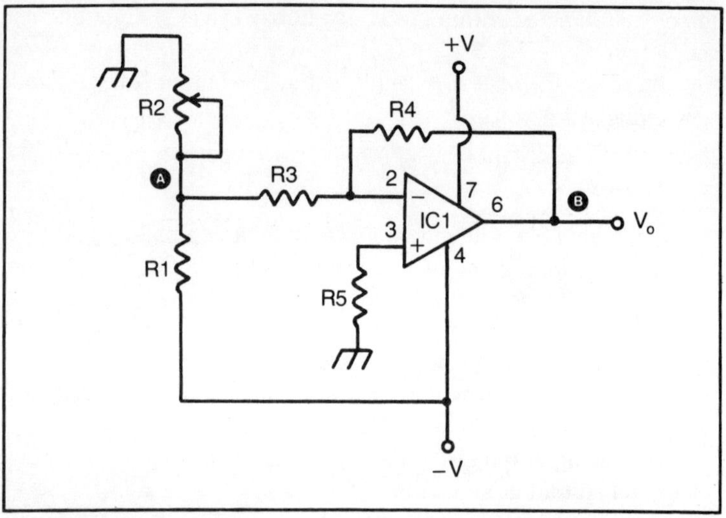

Fig. 4-8. The basic inverting amplifier circuit can be modified as shown here to set up a negative input signal.

Set the input voltage (point A) by adjusting potentiometer R2 to a value of 3 volts. What reading do you get at the output? It should be in the neighborhood of −1.5 volts. The gain for the circuit at this point is −0.5.

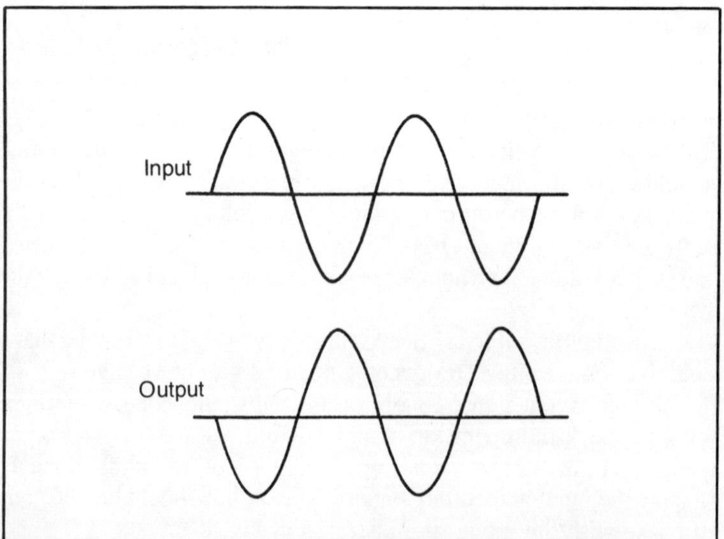

Fig. 4-9. The output of an inverting amplifier is 180° out of phase with the input signal.

$$G = \frac{-R4}{R3} = \frac{-4700}{10{,}000} = -0.47 \cong -0.5$$

When the feedback resistor (R4) is smaller than the input resistor (R3), the gain is less than unity. In other words, the signal is attenuated.

Try different input voltages and see if the output is about half the input voltage (with the polarity inverted, of course) each time.

Figure 4-10 shows yet another variation on the basic inverting amplifier circuit of Fig. 4-7. This time the feedback resistor (R4) is a variable potentiometer, allowing you to vary the gain of the amplifier without rewiring the circuit.

Adjust R2 for a medium input voltage (point A), say about 3 volts. Slowly move the shaft of R4 through its range several times while monitoring the output voltage. The output voltage (point B) should range from near zero to the clipping point of the amplifier (typically one volt less than the applied power supply voltage).

If you graphed the output at several points along R4's range, you should get a graph that looks something like the one shown in Fig. 4-11.

Before moving on, you may want to experiment further with the inverting amplifier circuits of Figs. 4-7 and 4-8. Substitute various values for R3 and R4. Do the measured outputs come close to the calculated values of $V_i \times G$?

Watch out for voltmeter polarity and large gains that could cause clipping of the output signal.

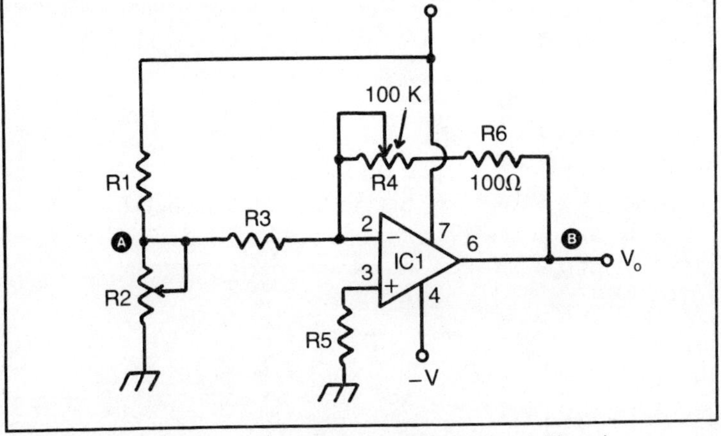

Fig. 4-10. This circuit is for an inverting amplifier with variable gain.

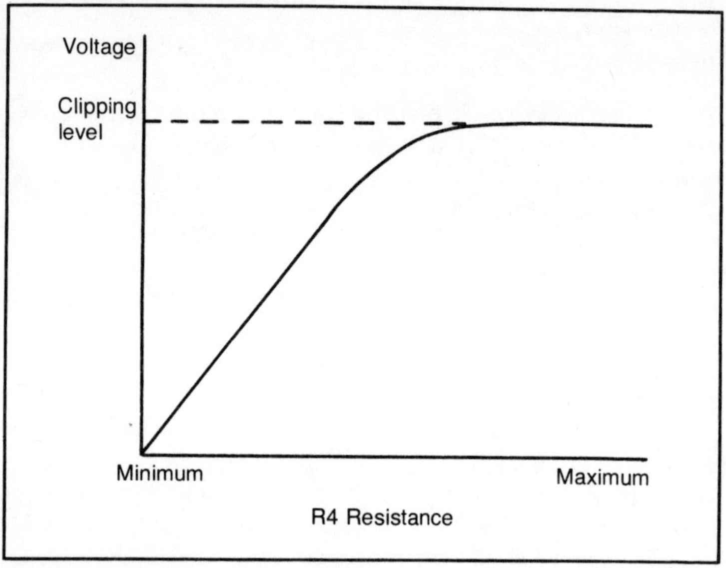

Fig. 4-11. Graphing the output voltage for the circuit of Fig. 4-10 produces this result.

EXPERIMENT 4—OFFSET

In the previous experiments, you may have come up with measured values that differed somewhat from the calculated values. There are various causes for this.

Resistor tolerances can cause deviations from the nominal calculated gain. Let's assume you are working with the inverting amplifier circuit of Fig. 4-7. R3 has a nominal value of 10 K (10,000 ohms), and R4 has a nominal value of 47 K (47,000 ohms). The gain ideally should be −4.7.

$$G = \frac{-R4}{R3} = \frac{-47{,}000}{10{,}000} = -4.7$$

If the resistors have a 5 percent tolerance, R3's actual value may range from 9.5 K to 10.5 K. Similarly, R4 may actually have a value between 44.65 K and 49.35 K. The result of this is that the actual gain could range between −4.25 and −5.2.

$$G = \frac{-44{,}650}{10{,}500} = -4.25$$

$$G = \frac{-49{,}350}{9500} = -5.2$$

If the input voltage is 2 volts, the calculated output voltage should be −9.4 volts.

$$V_o = V_i \times G = 2 \times -4.7 = -9.4 \text{ volts}$$

The resistor tolerances may cause the actual measured output voltage to be as low as −8.5 volts or as high as −10.4 volts.

$$V_o = 2 \times -4.25 = -8.5 \text{ volts}$$
$$V_o = 2 \times -5.2 = -10.4 \text{ volts}$$

The problem can be even worse if 10 percent tolerance resistors are used. R3's nominal value of 10 K could actually be anything from 9 K to 11 K. The actual value of R4 (nominally 47 K) will fall somewhere in the range of 42.3 K (42,300 ohms) to 51.7 K (51,700 ohms). The worst case output error could be as much as ±20 percent of the calculated value.

Fortunately, most resistors will be fairly close to their nominal value. Few will be at either extreme of their tolerance range. Still, you can never know unless you actually measure the specific resistor's actual value with a good ohmmeter.

In applications where tolerances are critical, it would obviously be a good idea to use low tolerance (1 percent or 0.5 percent) resistors. These precision components can be rather expensive, however, so use 5 percent resistors if you can get away with the error. The error will be constant for any input, so in many applications it won't be too critical, since the output will be proportional to the input signal.

Resistors with a 10 percent tolerance should probably be avoided in most op amp circuits, unless the requirements are very loose. Even then I'd measure each resistor used with an ohmmeter to make sure it is reasonably close to the nominal value. Do not bother with 20 percent resistors at all.

Other output errors stem from internal inaccuracies within the op amp IC itself. The more important types of internal circuitry errors were discussed in Chapter 2.

Offsets are one of the biggest types of contributers to output errors. In this experiment, you will measure the output offset voltage of a 741 op amp IC. The parts you will need are listed in Table 4-7.

Carefully construct the circuit illustrated in Fig. 4-12. Double-check everything before applying power to the circuit.

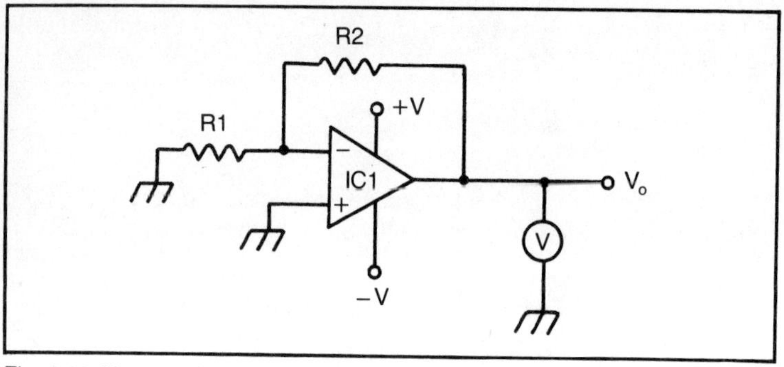

Fig. 4-12. Use this circuit to measure the output offset voltage in Experiment 4.

You should recognize this circuit as another basic inverting amplifier, except that the input is grounded. This means, by definition, that the input signal is zero volts. This implies that in an ideal circuit the output voltage should also be zero.

$$V_o = V_i \times G = 0 \times G = 0$$

Multiplying by zero results in zero, no matter what value G may have.

When you measure the output voltage of this circuit (you may have to reverse the voltmeter polarity), you will almost surely get some nonzero output voltage. It may have a positive or a negative polarity.

This output voltage is the input offset voltage multiplied by the gain. For the component values specified in the parts list, the gain of the circuit is −22.

$$G = \frac{-R2}{R1} = \frac{-220,000}{10,000} = -22$$

You should measure an output voltage of less than 130 millivolts (0.13 volt). If you check the spec sheet for a 741 op amp IC, you will see the maximum input offset voltage for this device is 6

Breadboard system	
IC1	741 op amp
R1	10 K ½ watt resistor
R2	220 K ½ watt resistor
Hook-up wire	
Voltmeter	

Table 4-7. Parts List for Experiment 4.

mV (0.006 volt). Multiplying this worst case figure by the gain (ignoring polarity which may go either way).

$$V_{os} = V_{is} \times G = 0.006 \times 22 = 0.132 \text{ volt} = 132 \text{ mV}$$

According to the spec sheet the typical input offset voltage figure is 2 mV (0.002 volt). This would make the typical output offset voltage take on a value of 44 mV.

$$V_{os} = 0.002 \times 22 = 0.044 \text{ volt} = 44 \text{ mV}$$

You may be lucky enough to have a chip with even less input offset voltage than this, although it's not particularly likely.

Try this experiment with a number of different 741 op amp ICs, preferably from different manufacturing batches, to get a better idea of variations from chip to chip.

EXPERIMENT 5—DIFFERENTIAL AMP

Figure 4-13 illustrates a circuit for a differential amplifier. The output is equal to the difference between the two inputs multiplied by the gain. Since all four resistors in the feedback network (R5 through R8) have equal values (100 K), the gain is 1. Therefore, the output will be simply equal to input A (noninverting) minus input B (inverting input). In mathematical terms, C = A − B.

The parts list for this experiment is given in Table 4-8.

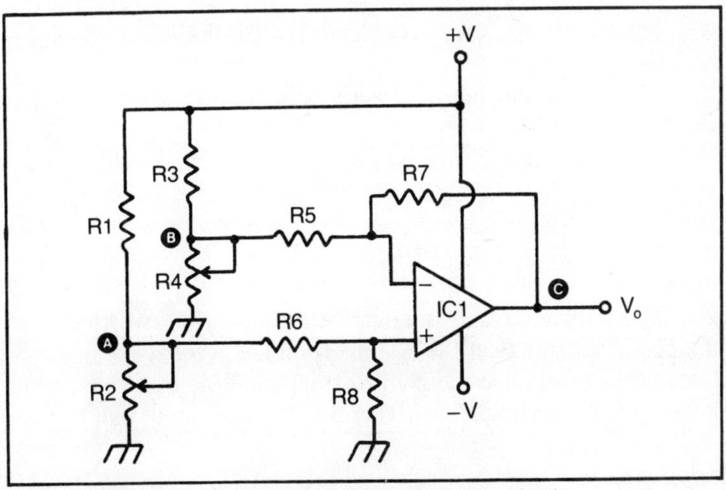

Fig. 4-13. Experiment 5 uses this differential amplifier circuit.

Table 4-8. Parts List for Experiment 5.

Breadboard system	
IC1	741 op amp
R1, R3	1 K ½ watt resistor
R2, R4	1 K potentiometer, linear taper
R5, R6, R7, R8	10 K resistor, ½ watt
Hook-up wire	
Voltmeters	

Resistor R1 and potentiometer R2 control the input voltage (A) for the noninverting input. Similarly, R3 and R4 are the voltage divided for the noninverting (B) input. In this simple arrangement, there may be some interaction between the two voltage dividers, but it will do for our experimental purposes.

Adjust potentiometers R2 and R4 so that both inputs are 3 volts. Because the dividers interact, you will probably have to move the voltmeter back and forth between points A and B several times to get both inputs set. If you have two voltmeters handy you can save time by simultaneously measuring both inputs while you set the controls.

The exact values at the inputs are not too critical here, as long as both input voltages are equal.

Once the input voltages are set, measure the output voltage at point C. Since input A = 3 volts and input B = 3 volts, output C should be zero.

$$C = A - B = 3 - 3 = 0 \text{ volts}$$

The output may not be exactly zero because of offsets within the IC.

If input A is now changed to 2 volts, the output should equal −1 volt.

$$C = 2 - 3 = -1 \text{ volt}$$

The output voltage is negative because the inverting input voltage is greater than the noninverting input voltage.

Now, if you raise the noninverting input (A) to 4 volts, while keeping the inverting input (B) at 3 volts, the output will become positive 1 volt.

$$C = 4 - 3 = 1 \text{ volt}$$

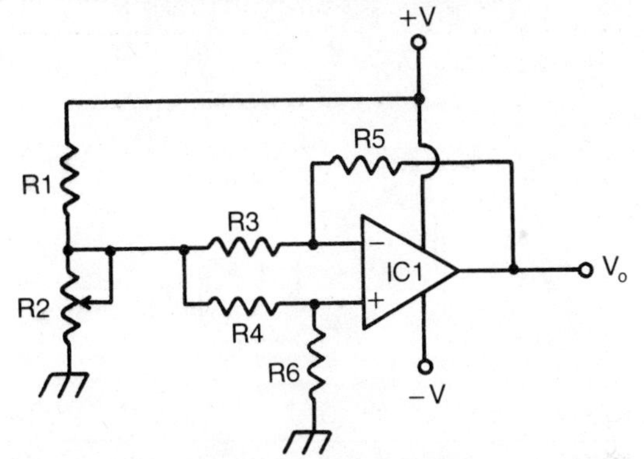

Fig. 4-14. Use this circuit to measure CMRR for Experiment 6.

Since the noninverting input voltage is greater than the inverting input voltage, the output voltage will be positive.

Experiment with other combinations of input voltages. The output voltage should always be approximately equal to the difference between the two inputs.

EXPERIMENT 6—MEASURING CMRR

For this experiment you will use the circuit shown in Fig. 4-14. Notice that it is essentially the differential amplifier from the last experiment (Fig. 4-13).

The parts list for this experiment is given in Table 4-9.

Only a single voltage divider is used this time, so the same signal is fed into both inputs. Only the common-mode signal is applied to the op amp. Ideally, the output should be zero. Is it? Try various settings of R2. The output voltage should not change significantly.

Table 4-9. Parts List for Experiment 6.

Breadboard system	
Oscillator or signal generator	
IC1	741 op amp
R1	1 K ½ watt resistor
R2	1 K potentiometer, linear taper
R3-R6	100 K ½ watt resistor
Hook-up wire	
dc Voltmeter	
ac Voltmeter or Oscilloscope	

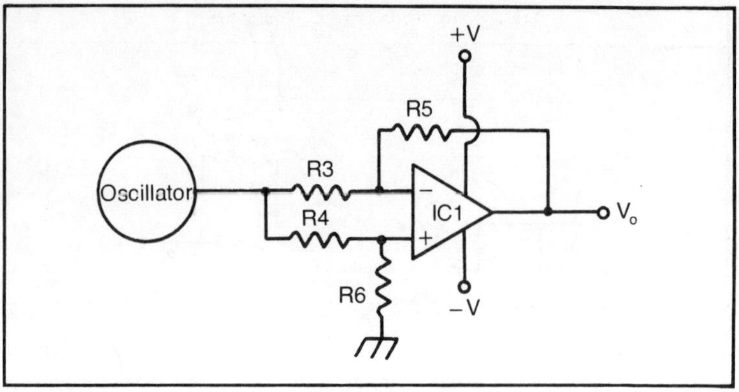

Fig. 4-15. In the second part of Experiment 6, ac CMRR is measured.

This experiment gives an indication of the op amp's CMRR (common-mode rejection ratio). It can be seen better with an ac signal. Use the circuit shown in Fig. 4-15. It is similar to the dc version, but the input signal comes from an oscillator or signal generator.

Measure the output with an ac voltmeter or an oscilloscope. A regular dc voltmeter will not give a reliable reading, and it might be damaged because of the polarity swinging back and forth.

EXPERIMENT 7— SUMMING AMPLIFIER

The fourth basic type of op amp circuit is the summing amplifier. Build the circuit shown in Fig. 4-16. The parts list for this experiment is given in Table 4-10.

The input signals may be derived from simple voltage dividers

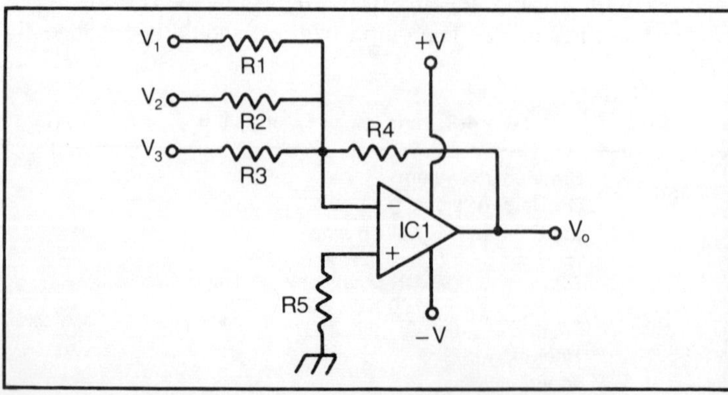

Fig. 4-16. This summing amplifier circuit is used for Experiment 7.

Table 4-10. Parts List for Experiment 7.

Breadboard system	
IC1	741 op amp
R1-R4	10 K ½ watt resistor (4)
	22 K ½ watt resistor
	47 K ½ watt resistor (2)
R5	6.8 K ½ watt resistor
Hook-up wire	
Voltmeter	
1.5 volt AA batteries (4) and holders	

as in the previous experiments, or from external batteries. In the text, I will assume standard 1.5-volt AA cells are used to provide the input voltages.

Start with a 1.5-volt signal to each of the circuit's inputs. Begin with resistors R1 through R4 all equal to 10 K.

The output voltage (V_o) should be equal to the inverted sum of the inputs multiplied by the gain. Since R1 = R2 = R3, the gain is identical for all of the inputs. You can calculate the output with this formula:

$$V_o = \frac{-R4}{R1}(V_1 + V_2 + V_3)$$

Since R4 has the same value as the input resistors, the gain is unity. The equation simplifies to the following:

$$V_o = -(V_1 + V_2 + V_3)$$

In our first example, each of the input voltages is 1.5-volt.

$$V_o = -(1.5 + 1.5 + 1.5) = -4.5 \text{ volts}$$

Reverse the polarity of the battery supplying the input voltage to input V_2. How does this affect the output? You should now measure an output of −1.5 volts.

$$V_o = -(1.5 + -1.5 + 1.5) = -1.5 \text{ volts}$$

Increase input voltage V_3 to 3 volts. Now you should get a measured output of:

$$V_o = -(1.5 + -1.5 + 3) = -3.0 \text{ volts}$$

Table 4-11. Typical Input and Output Values.

V_1	V_2	V_3	Output
1.5	1.5	0	−6.6
1.5	1.5	1.5	−9.9
3.0	1.5	1.5	−13.2 (may be clipped)
1.5	1.5	−1.5	−3.3
−1.5	1.5	−1.5	+3.3
1.5	1.5	−3.0	0

Gain = −2.2

You can have a nonunity gain by changing the value of the feedback resistor (R4). Let's increase this resistor to 22 K. The output should be equal to this equation:

$$V_o = \frac{-R4}{R1}(V_1 + V_2 + V_3) = \frac{-22,000}{10,000}(V_1 + V_2 + V_3)$$

$$= -2.2(V_1 + V_2 + V_3)$$

Experiment with various input voltages. Some typical results are listed in Table 4-11.

A summing amplifier can also be used with weighted inputs. Different inputs can have different amounts of influence on the output voltage. This is done simply by using different resistor values for the input resistors (R1, R2, and R3).

The equation for weighted input summing is as follows:

$$V_o = \left(\frac{-R4}{R1} \times V_1 + \frac{R4}{R2} \times V_2 + \frac{R4}{R3} \times V_3\right)$$

Use the following resistor values for this part of the experiment:

Table 4-12. Typical Values for Weighted Inputs.

V_1	$(\times G_1)$	V_2	$(\times G_2)$	V_3	$(\times G_3)$	V_o
1.5	−7.05	0	0	0	0	−7.05
0	0	1.5	−3.15	0	0	−3.15
0	0	0	0	1.5	−1.5	−1.5
1.5	−7.05	1.5	−3.15	1.5	−1.5	−11.7
1.5	−7.05	−1.5	3.15	3.0	−3.0	−6.9
1.5	−7.05	−1.5	3.15	−1.5	1.5	−2.4
−3.0	14.1	3.0	−9.45	1.5	−1.5	3.15
−3.0	14.1	1.5	−3.15	3.0	−3.0	7.95

$$R1 = 10\ K \quad R3 = 47\ K$$
$$R2 = 22\ K \quad R4 = 47\ K$$

Plugging these resistance values into the equation gives the following relationship for the output voltage:

$$V_o = \left(-\frac{R4}{R1} \times V_1 + \frac{R4}{R2} \times V_2 + \frac{R4}{R3} \times V_3\right) =$$

$$\left(-\frac{47{,}000}{10{,}000} \times V_1 + \frac{47{,}000}{22{,}000} \times V_2 + \frac{47{,}000}{47{,}000} \times V_3\right)$$

$$= -(4.7 \times V_1 + 2.1 \times V_2 + V_3)$$

Experiment with a number of input voltages. Some typical values and their predicted results are listed in Table 4-12.

5

Active Filters

So far I have only been using op amp ICs in various amplifier circuits. Within the limitations of the IC itself, all input frequencies have the same gain through the circuit. Sometimes a circuit is needed that is frequency dependent. That is, some frequencies are passed through at the circuit's full gain, while other frequencies are given less gain or are attenuated. Such a circuit is called a *filter*.

Op amp filters are created by replacing one or more of the resistors in the feedback network with a frequency dependent network.

In their simplest form, an op amp filter's feedback network contains a resistor and a capacitor. A capacitor blocks dc and low-frequency signals but passes high-frequency signals. It effectively acts like a resistor whose value changes with frequency. The ac resistance (reactance) is defined by the following formula:

$$X_c = \frac{1}{2\pi FC}$$

where X_c is the reactance of the capacitor at the specified frequency, F is the specified frequency, and C is the capacitance. The symbol π represents pi, a constant with a value of approximately 3.14. The 2π portion is sometimes written directly as 6.28, so you may occasionally run into the formula written like this:

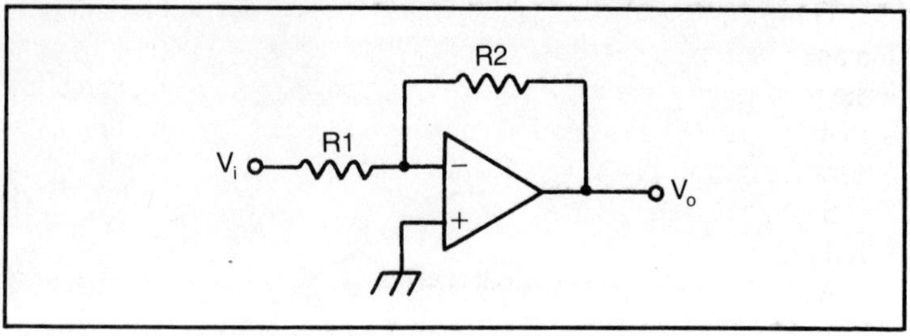

Fig. 5-1. The basic inverting amplifier circuit.

$$X_c = \frac{1}{6.28FC}$$

These formulas are identical.

Capacitors cause a 90° phase shift, which may introduce problems in some applications at certain frequencies.

INTEGRATOR

A basic inverting amplifier (see Fig. 5-1) has a feedback network made up of the input resistor (R1) and the feedback resistor (R2). Some of the output signal is fed back to the inverting input through R2, causing some canceling out of the signal and limiting the gain to a value determined by the ratio of the two resistor values.

In Fig. 5-2, I have replaced the feedback resistor (R2) with a capacitor. Since a capacitor passes high frequencies but opposes low frequencies, only the higher frequencies will be fed back to the inverting input for cancellation. The higher the frequency, the more it will be fed back and reduced in gain.

Low frequencies receive full gain. At higher frequencies the gain is reduced as the frequency is increased.

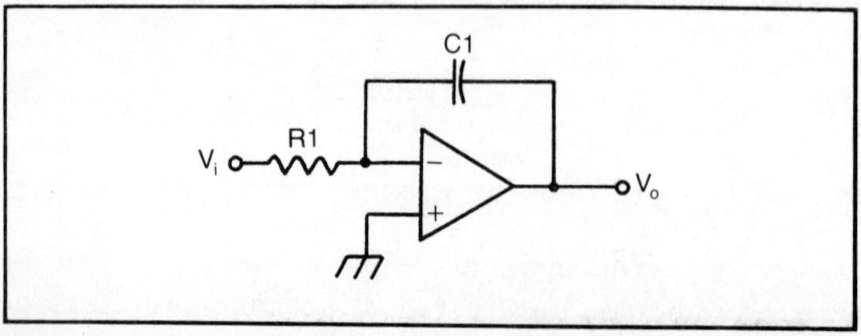

Fig. 5-2. Replacing the feedback resistor in an inverting amplifier with a capacitor creates a basic integrator circuit.

This circuit performs a mathematical operation called integration on the input signal. Basically, this is a process of taking the sum of the input signal over a period of time. Another way of putting this is that the circuit mathematically finds the area under the curve generated by the input waveform.

Since this circuit performs the integration operation, it is called an *integrator.*

A practical integrator circuit requires some external method of setting an initial condition. For most applications this initial condition is an output of zero.

Figure 5-3 shows how this can be accomplished with a DPDT switch. When the switch is set to the position shown in the diagram, the circuit performs the integration function on the signal being fed into the circuit (V_i).

Reversing the switch position (closing section A) grounds the input ($V_i = 0$) and the capacitor is shorted, permitting it to discharge. This forces the output to the initial condition of $V_o = 0$.

Returning the switch to the original position (section A closed) causes the integration operation to begin again.

If V_i is a constant dc signal, the output will be a ramp wave, as illustrated in Fig. 5-4. The dotted lines indicate the points where the switch is momentarily closed, resetting the circuit.

Of course, for a practical device, a maximum output voltage will soon be reached, clipping the output signal, as shown in Fig. 5-5.

Ordinarily the input signal to an integrator will be an ac waveform. The shape of the waveform will be altered by the integration process. For example, if V_i is a square wave, V_o will be a triangle wave, as shown in Fig. 5-6.

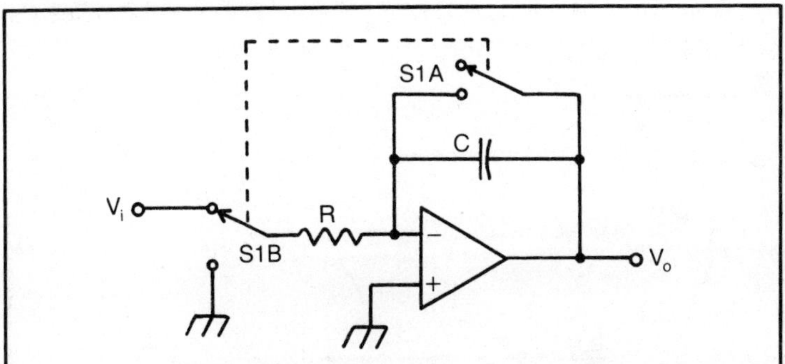

Fig. 5-3. A practical integrator requires some method of discharging the feedback capacitor.

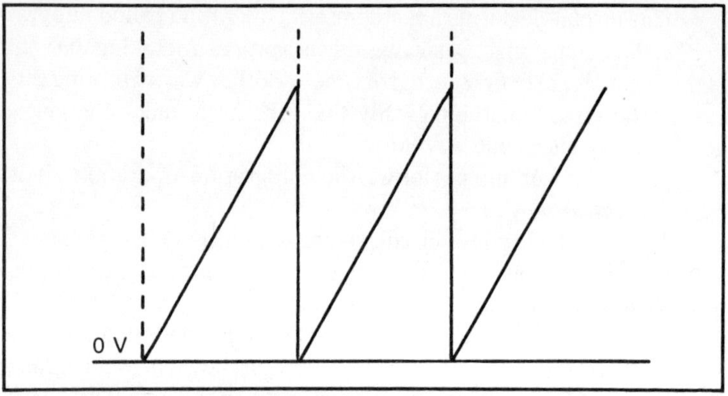

Fig. 5-4. If the input to an integrator is a continuous dc voltage, the output will be an ascending ramp wave, or sawtooth wave.

Using a sine wave as the input signal, the output will be a cosine wave. This is shown in Fig. 5-7. Conversely, Fig. 5-8 shows that a cosine wave input will produce a sine wave at the output. A cosine wave, incidentally, is simply a sine wave that has been phase shifted 90°.

The time constant of the resistor and capacitor used in the feedback network (RC) is critical to operation. It must not be less than the cycle period of the input waveform. For high frequencies R and C must be relatively small.

The RC time constant can be calculated with this formula:

$$RC = \frac{1}{2\pi F_m}$$

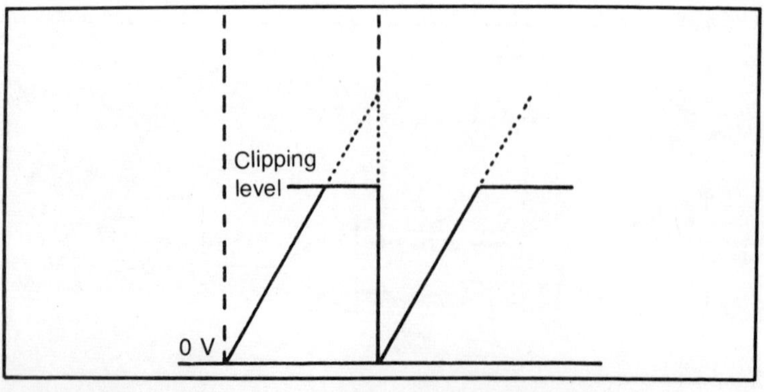

Fig. 5-5. If the maximum gain of the op amp is exceeded, the output ramp will level off at the postive saturation level.

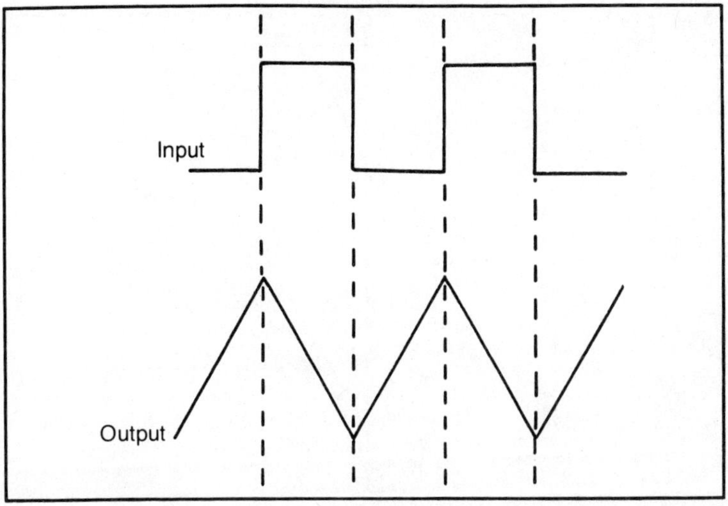

Fig. 5-6. Using a square wave at the input of an integrator produces a triangle wave at the output.

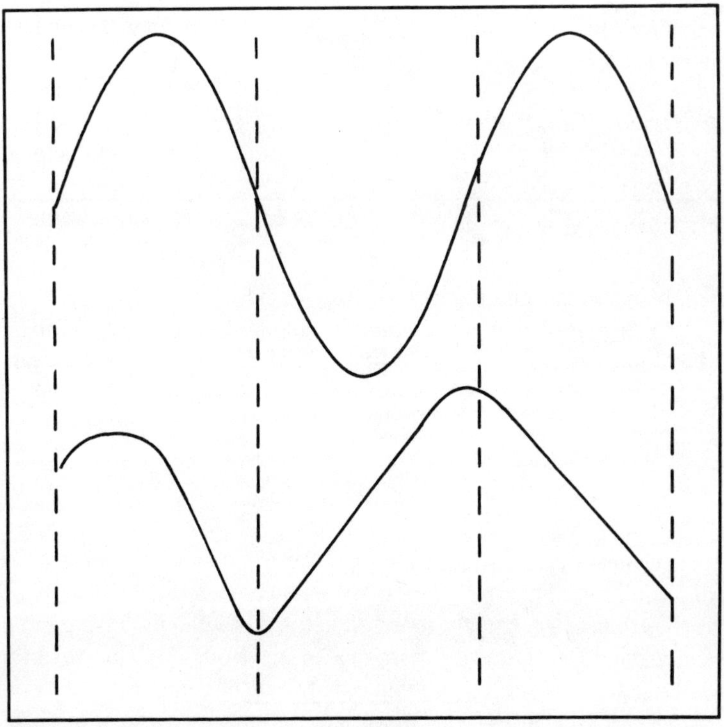

Fig. 5-7. Feeding a sine wave into an integrator causes a phase shift, that results in a cosine wave output.

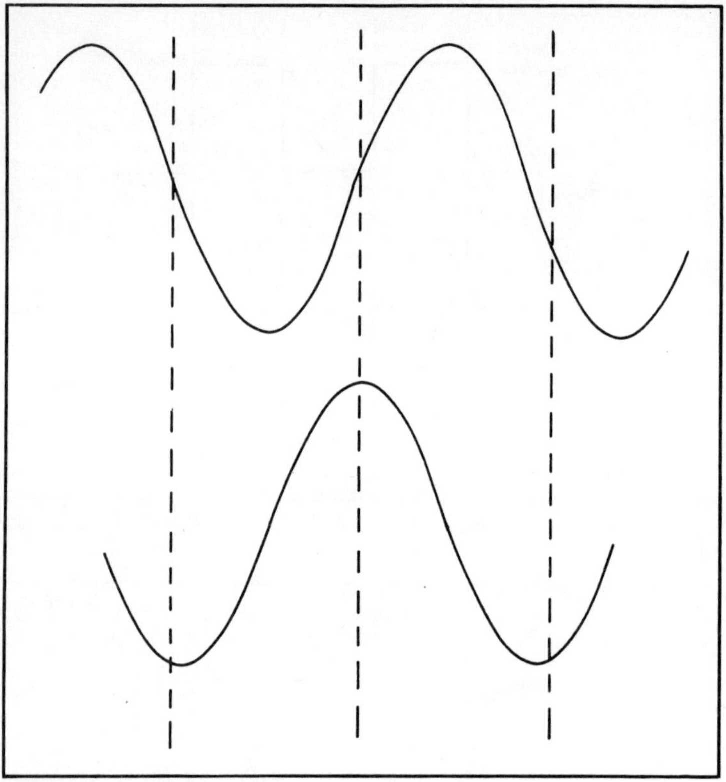

Fig. 5-8. An integrator will convert a cosine wave back into a sine wave.

where F_m is the minimum input frequency.

In practical design, a suitable standard capacitance will be selected, and the resistance value will be determined by the rearranged formula.

$$R = \frac{1}{2\pi F_m C}$$

Generally speaking, the higher the frequency an integrator is designed for, the lower the circuit gain can be.

A resistor is usually inserted between the noninverting input and ground to minimize errors due to input offsets. This resistor will usually have the same value as the input resistor.

An equivalent circuit for an integrator is shown in Fig. 5-9.

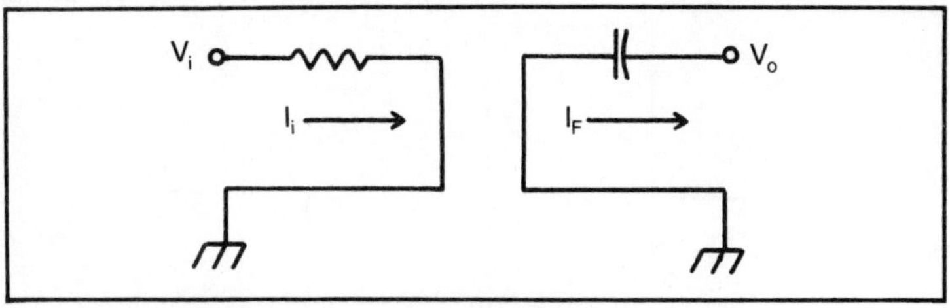

Fig. 5-9. The equivalent circuit for an ideal integrator.

DIFFERENTIATOR

If I reverse the positions of the capacitor and the resistor in the feedback network, as illustrated in Fig. 5-10, I have a new circuit type called a *differentiator*.

A differentiator, as may be expected, is just the opposite of an integrator. If a square wave is fed into an integrator, the output will be a triangle wave. Conversely, feeding a triangle wave into a differentiator will produce a square wave at the output.

If a square wave is fed into a differentiator circuit, the output will be a series of spikes, as shown in Fig. 5-11.

Since the input component is a capacitor, low frequency will be inhibited before they ever reach the input of the op amp. The result of this is that high frequencies will have higher gain than low frequencies.

Basically a differentiator responds only to differences or changes in the input signal. A continuous dc signal would effectively be ignored by the circuit.

The time constant of the RC feedback network is fairly critical, as in the integrator. The RC time constant should be set to approximately 10 percent of the cycle period (1/F) of the expected input wave.

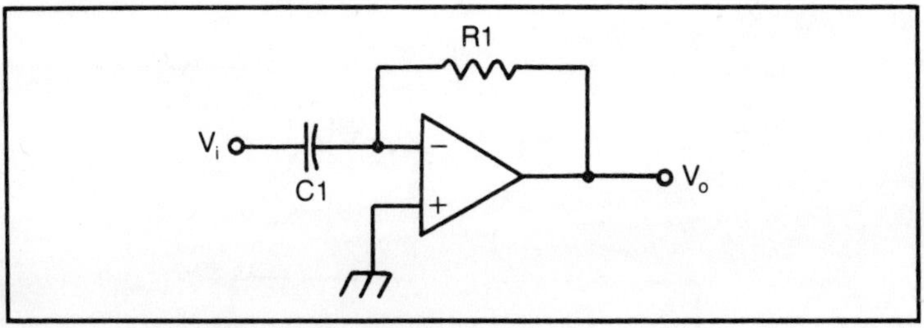

Fig. 5-10. If the input resistor of an inverting amplifier is replaced by a capacitor, the result is the basic differentiator circuit.

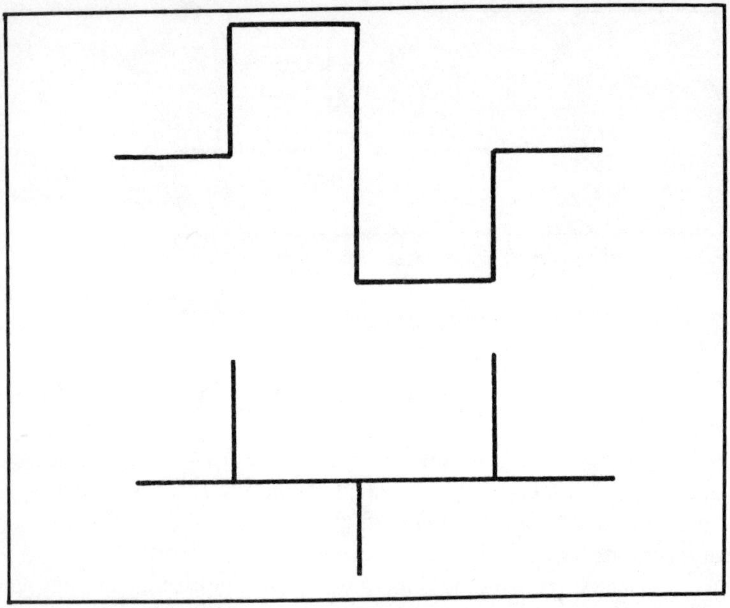

Fig. 5-11. If a square wave is the input signal for a differentiator, the output will be a dual polarity spike wave.

An equivalent circuit for a differentiator is shown in Fig. 5-12. The circuit shown in Fig. 5-10 is a simple idealized form of the differentiator. A more practical circuit is shown in Fig. 5-13. The added input resistor R2 has a fairly small value. Resistor R3 is used to control input offsets.

LOW-PASS FILTERS

An integrator tends to oppose high frequencies, but low frequencies are passed through to the output at full gain. It is a form of low-pass filter.

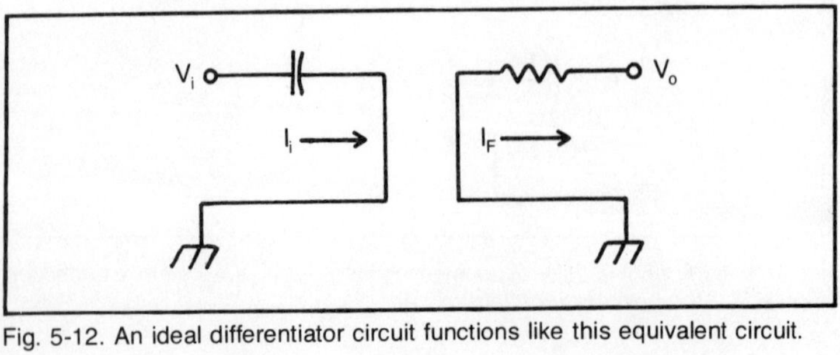

Fig. 5-12. An ideal differentiator circuit functions like this equivalent circuit.

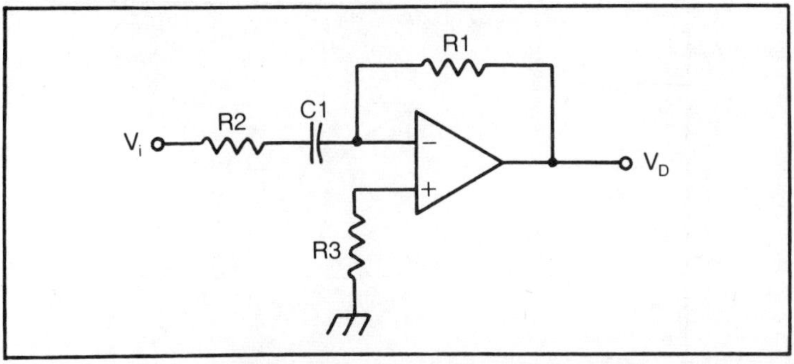

Fig. 5-13. A practical differentiator circuit requires a couple of extra resistors.

A simple passive (no gain—power drawn from the signal source) low-pass filter is shown in Fig. 5-14. Note that it is made up of just a resistor and a capacitor. Frequencies above a specific point (determined by the RC time constant) are shunted to ground through the capacitor. Lower frequencies are blocked by the capacitor, so they are present at the output.

When I graph frequency vs. output amplitude I get the graph shown in Fig. 5-15.

The cutoff frequency at which attenuation begins can be found according to the following formula:

$$F_c = \frac{159,000}{RC}$$

In designing a filter, I know the cutoff frequency and need to determine the appropriate component values. The input resistor is selected to match the square root of the product, of the source impedance, and the load impedance.

$$R = \sqrt{R_s R_l}$$

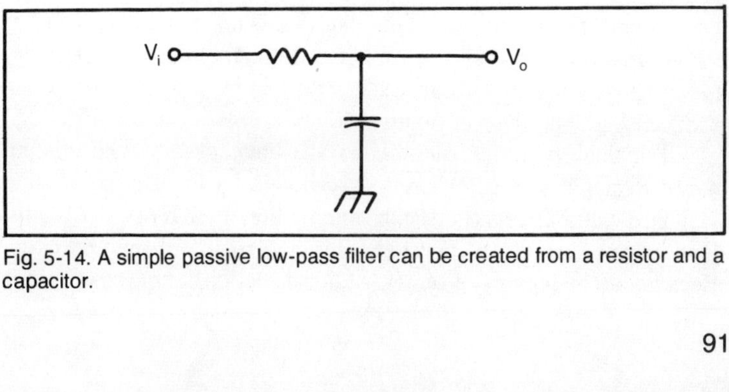

Fig. 5-14. A simple passive low-pass filter can be created from a resistor and a capacitor.

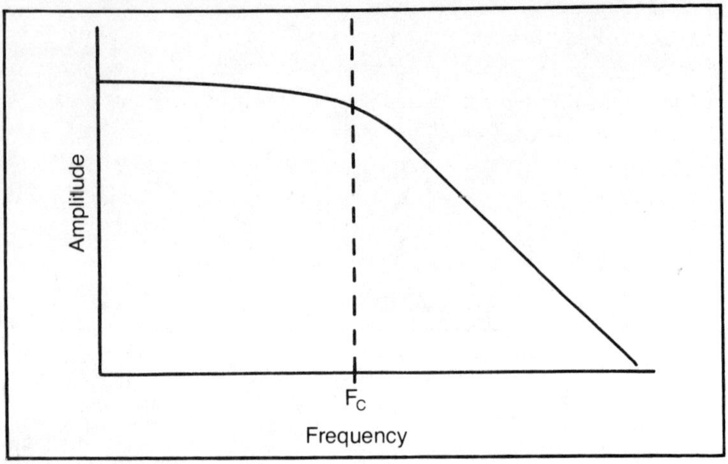

Fig. 5-15. Graphing the frequency response of the passive filter in Fig. 5-14 produces this result.

I can solve for C by rearranging the first equation.

$$C = \frac{159{,}000}{RF_c}$$

Let's try an example. I want to design a passive low-pass filter with a cutoff frequency of 1 kHz (1000 Hz). The impedance of the signal source is 10 K (10,000 ohms), and the load impedance is 600 ohms.

Now that I know the design requirements, I have to select appropriate component values for the filter. The first step is to determine the value of the resistor.

$$R = \sqrt{R_s R_l} = \sqrt{10{,}000 \times 600} = \sqrt{6{,}000{,}000}$$
$$= 2449.5 \text{ ohms}$$

I can round this off to 2500 ohms, but that is not a standard resistor value. I could work out a series/parallel combination of standard resistors to come up with the exact value, or I could use a trimpot to precisely set the desired value.

Fortunately, those techniques are unnecessary. The components in a passive filter are not critical. I can use the closest standard value. In this case I can select either a 2.2 K or a 2.7 K unit. Either 5 percent or 10 percent tolerance resistors may be used. Because of the leeway of resistor tolerances, either of these values may put us exactly at the calculated value, or reasonably close to it. For 5 percent resistors, a nominal value of 2.2 K could have an actual value anywhere between 2.09 K and 2.31 K. If this was a 10 percent tolerance device the range would be extended from 1.98 K to 2.42 K. If a 5 percent resistor is used, the actual value may be as low as 2565 ohms, or as high as 2835 ohms. If the resistor's tolerance is 10 percent the actual value will be somewhere between 2.43 K and 2.97 K. Any of these will be close enough.

I will select a 2.7 K resistor. Since nothing is critical in a passive filter, I can ignore tolerance error, and simply assume the nominal value is the actual value of the resistor. Now I can solve for the desired capacitance.

$$C = \frac{159{,}000}{RF_c} = \frac{159{,}000}{2700 \times 1000} = \frac{159{,}000}{2{,}700{,}000}$$

$$= 0.05888888$$

I can use a standard 0.47 μF capacitor.

Since I rounded off the values of both components, it is logical to assume that the cutoff frequency has been altered. Plug the selected component values back into the equation and recalculate the cutoff frequency (ignoring component tolerances).

$$F_c = \frac{159{,}000}{RC} = \frac{159{,}000}{2700 \times 0.047}$$

$$= \frac{159{,}000}{126.9} \cong 1253 \text{ Hz}$$

That's quite a large deviation from the nominal value of 1000 Hz! Surely, that can't be acceptable!

Actually, it's not at all unreasonable in a passive filter circuit, because the rolloff is so gradual that the difference will not be noticeable in practical applications.

An ideal filter would have an infinite rolloff as illustrated in Fig. 5-16. If the cutoff frequency was 1000 Hz, a 999 Hz input signal would be completely passed at full gain, but a 1001 Hz input signal would be completely blocked off from the output.

Real-world filters are nowhere near so precise. This is especially true of simple passive filters.

Refer back to the frequency response graph shown in Fig. 5-15.

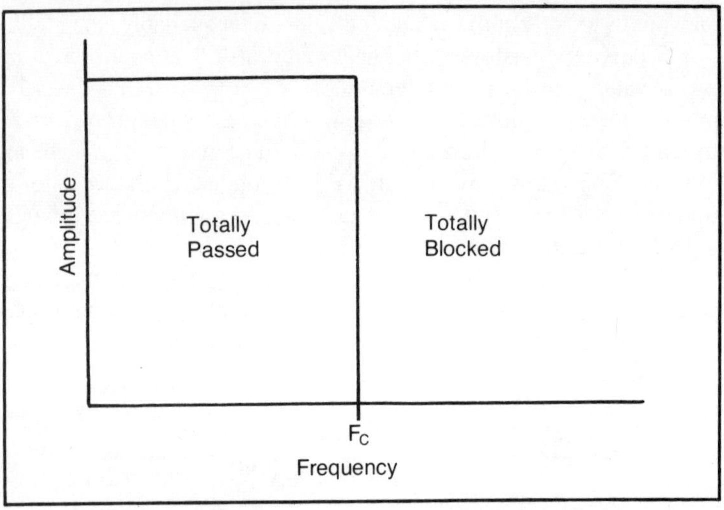

Fig. 5-16. An ideal low-pass filter would have an infinitely sharp rolloff slope.

Notice that there is no clearly defined point of division between passed and blocked frequencies. Instead, the output amplitude gradually tapers off. The cutoff frequency is defined as the point where the output amplitude drops to $1/\sqrt{2}$ of the maximum level. This is the same as multiplying the maximum amplitude by a factor of 0.707.

Filter response is usually defined in terms of decibels (dB). The cutoff frequency is the point where the output amplitude has dropped off −3 dB of the maximum amplitude.

The dB response drops off linearly as the frequency increases logarithmically. In a passive filter, the amplitude drops off another −3 dB for each doubling of frequency. Since a doubling of frequency produces a change of one octave, this is called a −3 dB per octave rolloff.

A −6 dB change is an attenuation change of 2. For a low-pass passive filter with a nominal cutoff frequency of 1000 Hz, an input signal of 2000 Hz would have half the output level of an input of 1000 Hz. Clearly, if the actual cutoff frequency (−3 dB point) is 1253 Hz, there won't be very much perceptible difference in the circuit's operation.

While a passive filter is sufficient for some noncritical applications, it leaves plenty to be desired. It's main limitations are as follows:

☐ No circuit gain, only attenuation. Even the passed frequencies are attenuated to some degree.

☐ Power is stolen from the signal source, rather than a power supply.

☐ The rolloff is very gradual.

☐ Poor isolation between the source and load circuits.

These limitations can be overcome by using active filter circuits. An active filter includes an amplifier to add some gain. An active filter also requires power supply connections.

Operational amplifiers can be the heart of many different active filter circuits to meet almost any required application.

Active filters are classified according to their *order*. This is a measurement of the rolloff characteristic of the filter. Each order represents −6 dB of rolloff per octave. The higher the order, the steeper the rolloff.

1st order	−6 dB/octave
2nd order	−12 dB/octave
3rd order	−18 dB/octave
4th order	−24 dB/octave
5th order	−30 dB/octave

Higher order filters are much better at discriminating between passed and blocked frequencies, but the circuits are more expensive and complex.

Another important aspect of filter design is damping. The *damping index* is a number that describes the circuit's tendency towards breaking into oscillation. Practical damping factors for active filter circuits range from 0 (an oscillator) to 2.

Selecting the proper damping factor inevitably involves compromise. Highly damped filters feature a smooth, even rolloff and good overshoot and transient response. Slightly damped filters, on the other hand, offer sharper rejection characteristics.

There are two basic types of filter response. (Actually there are many others, but these are the most commonly encountered.)

A Butterworth filter response, illustrated in Fig. 5-17, is highly damped and features the flattest passband. That is, frequencies below the cutoff frequency are evenly amplified. The higher the order of the filter, the steeper the rolloff will be.

Figure 5-18 shows a typical Chebyshev filter response. This type of filter is only slightly damped, resulting in a sharply defined cutoff frequency. Notice the shape of the passband (below the cutoff frequency). There is a slight dip within the passband. Frequencies

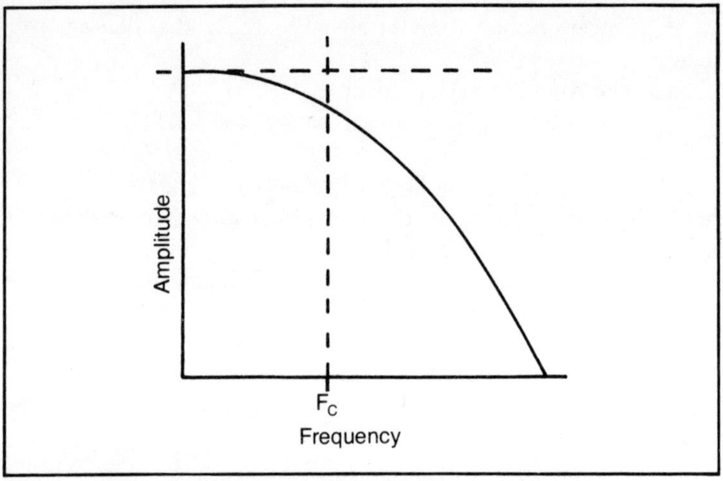

Fig. 5-17. Many active filters feature a Butterworth frequency response.

within this dip will have less than maximum gain. Again, the higher the order of the filter, the steeper the rolloff will be.

A simple first order low-pass active filter circuit is shown in Fig. 5-19. Notice that this is basically a noninverting amplifier circuit with a low-pass, passive RC filter added to the input. The passband gain is defined by R3 and R4 in the usual manner for noninverting amplifiers.

Rl and Cl determines the cutoff frequency of the filter, according to the following:

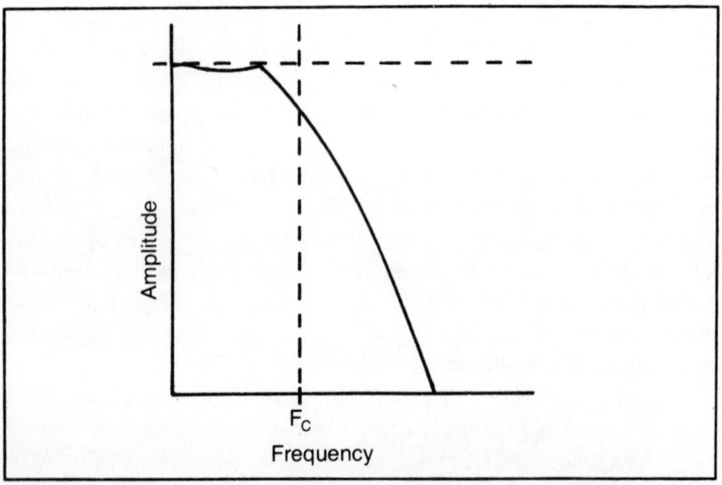

Fig. 5-18. Other active filters have a Chebyshev frequency response.

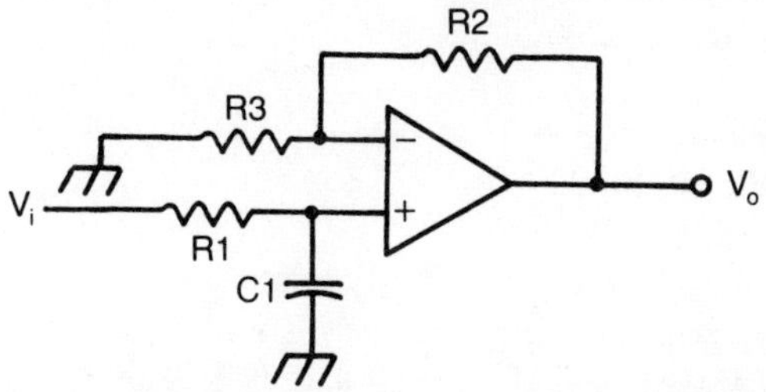

Fig. 5-19. One way to create a simple first order, active, low-pass filter is to place a passive, low-pass filter in the input path.

$$F_c = \frac{1}{2\pi\ R1C1}$$

The equation can be simplified in the following manner:

$$F_c = \frac{1}{6.28R1C1}$$

In a typical design, a suitable capacitor will be selected and the equation will be rearranged to solve for the unknown resistance.

$$R1 = \frac{1}{6.28F_c C1}$$

Resistors R1 and R3 will usually be given equal values, so for a specific gain, I can solve for the value of R2 using this formula:

$$R2 = R3(G - 1)$$

Let's try out a few examples. I want a low-pass filter with a cutoff frequency of 2500 Hz, and a passband gain of 4.

The first step is to select a capacitance value. This is done somewhat randomly. I'll try a 0.01 μF capacitor. Now I can solve for R1.

$$R1 = \frac{1}{6.28 \times 2500 \times 0.00000001} = \frac{1}{0.000157}$$

$$R1 \cong 6369 \text{ ohms}$$

I can round this off to 6.2 K (6200 ohms), which is a standard 5 percent value.

Next I need to determine the value for the feedback resistor (R2). R3 = R1 = 6200 ohms, so:

$$R2 = R3(G - 1) = 6200(4 - 1) = 6200 \times 3$$
$$R2 = 18{,}600 \text{ ohms} = 18.6 \text{ K}$$

As it turns out, 18 K is a standard resistor value, so I can use that. How much error has our rounding off introduced. Well, ignoring component tolerances, the actual gain is 3.9.

$$G = 1 + \frac{R2}{R3} = 1 + \frac{18{,}000}{6200} = 1 + 2.9 = 3.9$$

That is certainly close enough to the desired gain of 4.

Now, what about the cutoff frequency? Plugging in the selected component values I find:

$$F_c = \frac{1}{6.28R1C1} = \frac{1}{6.28 \times 6200 \times 0.00000001}$$

$$F_c = \frac{1}{0.0003894} = 2568 \text{ Hz}$$

This is also quite close to our desired cutoff frequency of 2500 Hz. Component tolerances could be expected to introduce this much error. The deviation from the specified values are insignificant.

I may want to design a similar filter without doing all of the work over. Let's say I now need a low-pass filter with the same passband gain, but a cutoff frequency of 5000 Hz. This is twice the nominal cutoff frequency of the original filter.

To change the cutoff frequency in a filter, determine the ratio of the old frequency to the new frequency.

$$F_r = \frac{F_o}{F_n}$$

Multiply this factor either by all of the frequency determining resistors, or all of the frequency determining capacitors, but not both.

In my example, I want to replace a 2500 Hz cutoff frequency with a 5000 Hz cutoff frequency. Therefore, the ratio is as follows:

$$F_r = \frac{2500}{5000} = \frac{1}{2}$$

Resistor R1 is the only frequency determining resistor in this simple circuit, so I only need to solve one equation:

$$R1_n = F_r R1_o = \frac{1}{2} \times 6200 = 3100 \text{ ohms}$$

This is not a standard resistance value, so I will substitute a standard 3.3 K resistor.

Plugging this altered resistor value into the formula, I find that the cutoff frequency of the revised filter is 4762 Hz.

$$F_c = \frac{1}{6.28 \times 3300 \times 0.00000001} = \frac{1}{0.00021}$$

$$= 4762 \text{ Hz}$$

This is reasonably close to our desired cutoff frequency of 5000 Hz.

The resistors in the feedback network (R2 and R3) may need to be changed along with R1. Capacitor C1 is left alone. Only resistors or only capacitors should be changed.

Let's try another example to illustrate revising a filter by changing frequency dependent capacitances. I will start off with my original 2500 Hz cutoff low-pass filter. To refresh your memory, the frequency determining components are R1 (6200 ohms) and C1 (0.01 μF). I now need to change the cutoff frequency to 7500 Hz. The ratio of the old and new frequencies is one third.

$$F_r = \frac{2500}{7500} = \frac{1}{3}$$

I use this ratio to determine the new capacitance value. All resistance values (including the feedback network) are left alone. The new filter should have a capacitor with a value of:

$$C1_n = F_r C1_o = \frac{1}{3} \times 0.01 = 0.0033\ \mu\text{F}$$

Fortunately, this is a standard capacitance value. Plugging it into the main equation, I can confirm the new cutoff frequency of the filter.

$$F_c = \frac{1}{6.28 \times 6200 \times 0.0000000033}$$

$$F_c = \frac{1}{0.0001285} = 7783 \text{ Hz}$$

Again, this should be close enough.

The cutoff frequency of a filter may be changed by changing all of the frequency determining capacitances, but not both.

Sometimes you may find it necessary to change the capacitor value without altering the cutoff frequency or other characteristics of the filter. This may be because a specific capacitor value is unavailable (it may be a nonstandard value), or it might be because of an impedance matching problem. In either case, a simple solution is readily available. Simply multiply all of the frequency determining resistances by the ratio of the old capacitance to the new capacitance.

$$R_n = R_o \times \frac{C_o}{C_n}$$

Let's say the 0.0033 μF capacitor called for in the last example is not available for some reason, but I do happen to have some 0.0047 μF units handy, I need to adjust the value of R1 accordingly.

$$R1_n = R1_o \times \frac{C_o}{C_n} = 6200 \times \frac{0.0033}{0.0047}$$

$$R1_n \cong 6200 \times 0.7 \cong 4353 \text{ ohms}$$

I can use a standard 4.2 K (4200 ohm) resistor. It should be close enough.

Now comes the big question. Will the filter circuit with these new component values still give us a cutoff frequency that is acceptably close to our nominal value of 7500 Hz? Let's check out the equation and find out.

$$F_c = \frac{1}{6.28R1C1} = \frac{1}{6.28 \times 4200 \times 0.0000000047}$$

$$F_c = \frac{1}{0.000124} \cong 8067 \text{ Hz}$$

The error turned out a little bit larger than in the earlier version, but it is still too close to really matter in a 1st order filter.

A variation on this circuit is shown in Fig. 5-20. The formula for finding the cutoff frequency is the same as for the active filter circuit of Fig. 5-19:

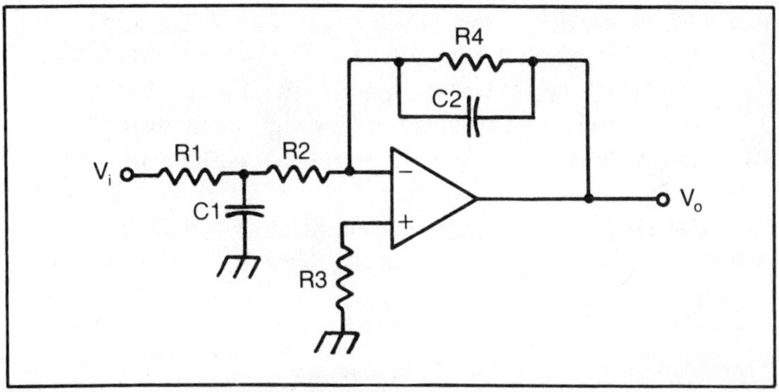

Fig. 5-20. An alternative first order, active, low-pass filter circuit is shown here.

$$F_c = \frac{1}{6.28R1C1}$$

Resistors R1, R2, and R4 all have identical values. The value of R3 is double this value. The two capacitors (C1 and C2) are also equal.

This circuit uses the inverting input rather than the noninverting input, so there will be a greater phase shift between the input and output signals. Remember that since capacitors are included in filter circuits, there will always be some degree of phase shifting in any filter circuits. Generally this won't cause any particular problems.

Both of the active filter circuits discussed so far are 1st order

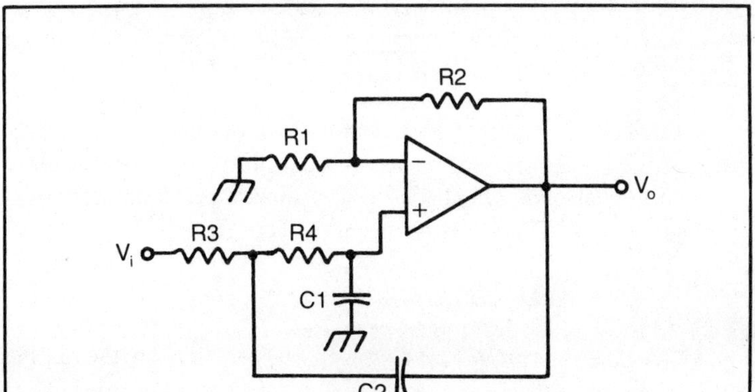

Fig. 5-21. This is a second order, active, low-pass VCVS filter circuit.

types. That is, the rolloff is −6 dB/octave. Many applications require a steeper rolloff slope.

A 2nd order active low-pass filter circuit is illustrated in Fig. 5-21. It's rolloff is −12 dB/octave. This type of circuit is a voltage-controlled voltage-source, or VCVS filter. Occasionally it may be called a Sallen and Key filter.

The formula for finding the cutoff frequency is a bit more complex than with the 1st order filters I've explained in the last few pages.

$$F_c = \frac{1}{2\pi\sqrt{R3R4C1C2}}$$

The factor 2π, of course, is a constant with a value of approximately 6.28.

Because of the complexity of solving for all of the frequency determining components, a variation called the *equal conponent VCVS* filter is generally used. The schematic is the same, but the following relationships between the components must be maintained:

$$R2 = 0.586R1$$
$$R3 = R4$$
$$C1 = C2$$

This allows a much simpler formula for the cutoff frequency.

$$F_c = \frac{1}{6.28\sqrt{R3R3C1C1}} = \frac{1}{6.28R3C1}$$

Does this revised equation look familiar? It certainly should! It's the same as the equation used for the 1st order filters discussed earlier.

The equal component VCVS filter has a peculiar limitation. The gain must be 1.586, or the circuit will not work. This is why R2 = 0.586R1.

If −12 dB/octave is still not a steep enough rolloff, higher order filters can be created by cascading 1st order and 2nd order filter sections, as illustrated in Fig. 5-22.

HIGH-PASS FILTERS

Let's return to the passive low-pass filter for a moment. It is

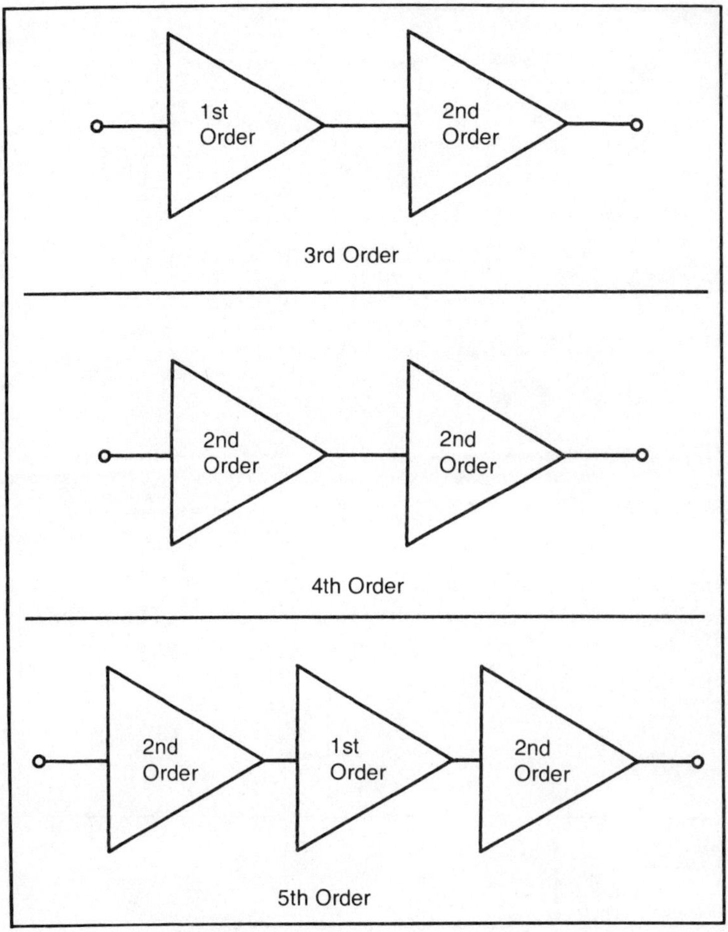

Fig. 5-22. Higher-order filters can be created by cascading low-order filter circuits.

shown again in Fig. 5-23A. The input signal passes through the resistor, and the capacitor shunts high frequencies to ground before they reach the output.

What would happen if I reverse the positions of these two components, as shown in Fig. 5-23B?

All of the signal must now pass through the capacitor. Since a capacitor opposes low frequencies, but passes high frequencies, the output will contain more high frequencies than low frequencies. A frequency response graph is shown in Fig. 5-24. Notice that it is a mirror image of the frequency response graph for a low-pass filter (refer back to Fig. 5-15).

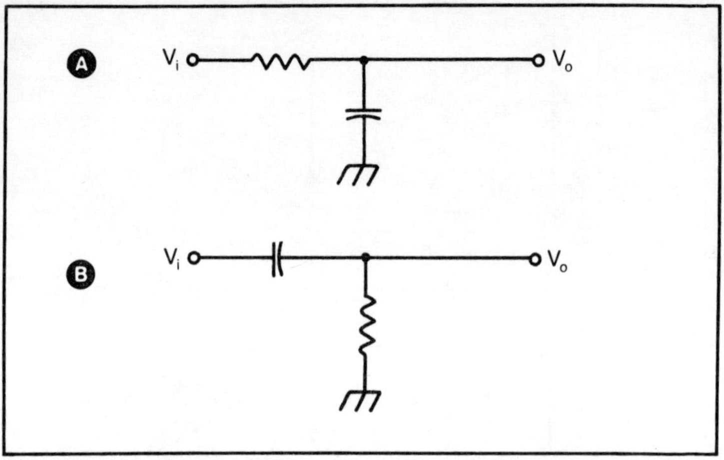

Fig. 5-23. A passive high-pass filter (B) is created by reversing the components of a low-pass filter (A).

Since high frequencies are passed and low frequencies are blocked, this is called a high-pass filter. It behaves just like a low-pass filter in all respects, except that the passed and blocked frequencies are reversed. The formula for finding the cutoff frequency remains the same.

$$F_c = \frac{159{,}000}{RC}$$

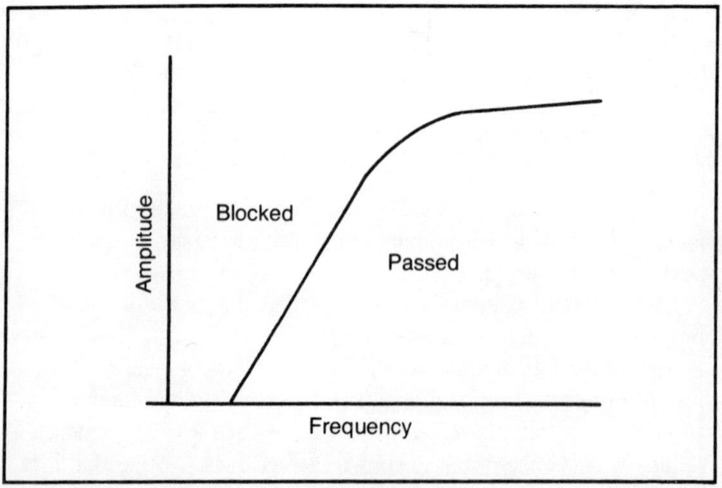

Fig. 5-24. The frequency response of a high-pass filter is essentially the mirror image of that of a low-pass filter.

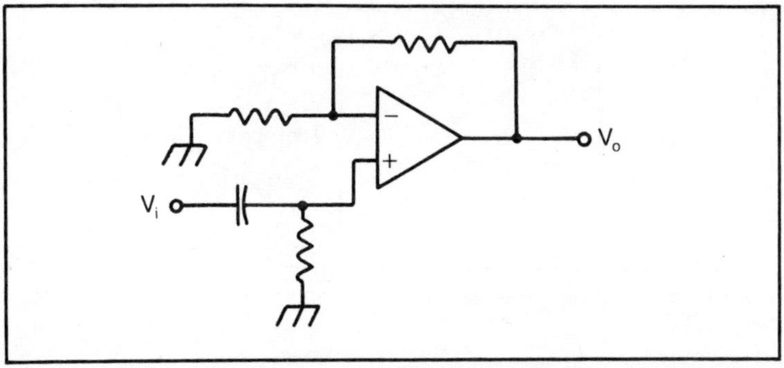

Fig. 5-25. Placing a passive high-pass filter in the input path of a noninverting amplifier creates a first order, active, high-pass filter.

A simple first order active high-pass filter is shown in Fig. 5-25. Figure 5-26 illustrates a second order VCVS active high-pass filter circuit. Both of these circuits function as mirror images of their low-pass counterparts.

BANDPASS FILTERS

Low-pass filters and high-pass filters can be combined to create two new filter types. The first of these is achieved by connecting a low-pass filter and a high-pass filter in series, as shown in Fig. 5-27.

The cutoff frequencies of these two filters must overlap. That is, the cutoff frequency of the low-pass filter must be higher than the cutoff frequency of the high-pass filter.

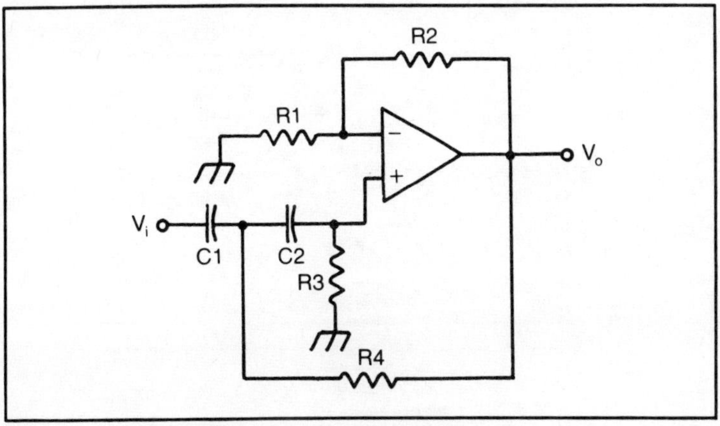

Fig. 5-26. The second order, active, high-pass VCVS filter circuit closely resembles the low-pass filter circuit shown in Fig. 5-21.

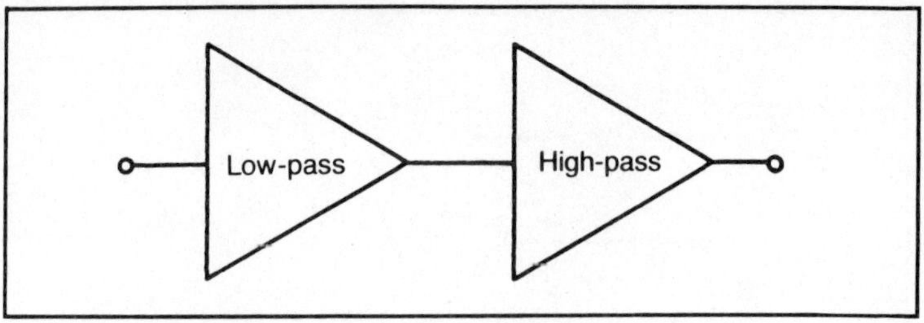

Fig. 5-27. Connecting a low-pass filter and a high-pass filter in series creates a band-pass filter.

Only those frequencies which are passed by both of the component filters will reach the output. Any frequency blocked by either of the component filters will be blocked by the composite filter.

The frequency response graph for this composite filter is shown in Fig. 5-28. Only a specific band of frequencies that is common to both of the component filters will be passed by the composite filter. This new type of filter is called a *bandpass filter* for obvious reasons.

Special bandpass filters rather than separate low-pass/high-pass series combinations are generally used in practical applications.

A bandpass filter has three important frequency points, as illustrated in Fig. 5-29. There are two cutoff frequencies—the lower cutoff frequency (F_l), and the upper cutoff frequency (F_h). The

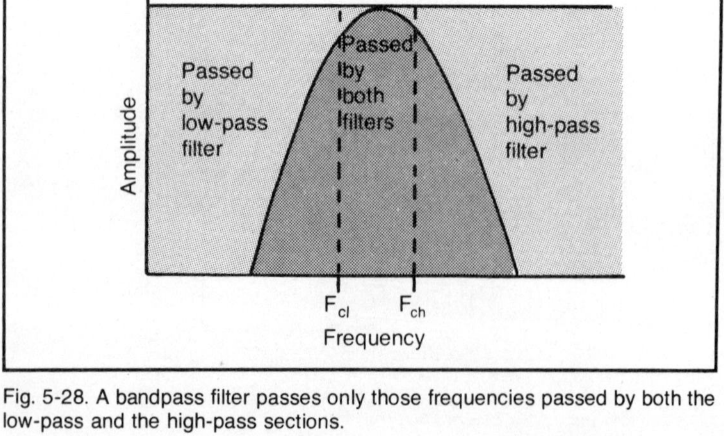

Fig. 5-28. A bandpass filter passes only those frequencies passed by both the low-pass and the high-pass sections.

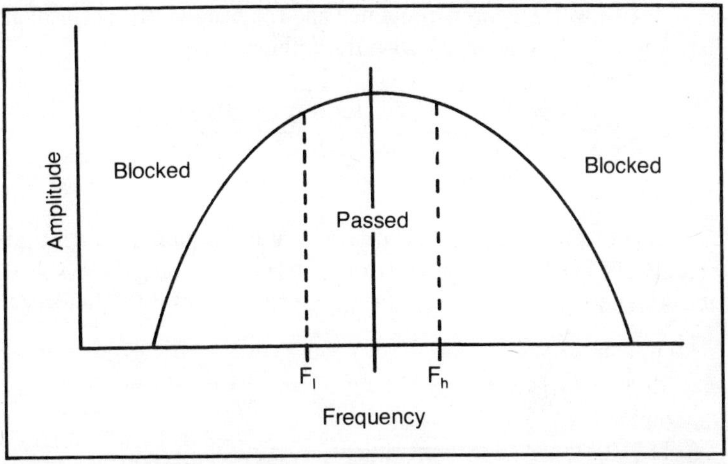

Fig. 5-29. A bandpass filter has two cutoff frequencies symmetrically arranged around the center frequency of the passband.

third significant frequency point is mid-way between the other two. It is the center frequency (F_c). As the name implies, this is the mid-point of the passed band of frequencies.

The center frequency is the geometric, rather than the algebraic, average of the two cutoff frequencies. It is equal to the square root of the product of the cutoff frequencies.

$$F_c = \sqrt{F_l F_h}$$

Let's say I have a bandpass filter with a lower cutoff frequency of 1000 Hz and an upper cutoff frequency of 2500 Hz. I can easily solve for the center frequency.

$$F_c = \sqrt{1000 \times 2500} = \sqrt{2{,}500{,}000}$$
$$F_c = 1581 \text{ Hz}$$

Another important specification for a bandpass filter is the *bandwidth* (BW). This is the size of the passband and is equal to the difference between the upper and lower cutoff frequencies:

$$BW = F_h - F_l$$

For our earlier example the bandwidth is 1500 Hz.

$$BW = 2500 - 1000 = 1500 \text{ Hz}$$

If I know the center frequency and the bandwidth, I can calculate the cutoff frequencies with the following equations:

$$F_l = \frac{\sqrt{BW^2 + (2F_c)^2} - BW}{2}$$

$$F_h = F_l + BW$$

In my example, I calculated a center frequency of approximately 1581 Hz and a bandwidth of 1500 Hz. Let's plug these values into the new equations to confirm the original cutoff frequencies.

$$F_l = \frac{\sqrt{1500^2 + (2 \times 1581)^2} - 1500}{2}$$

$$= \frac{\sqrt{2{,}250{,}000 + 10{,}000{,}000} - 1500}{2}$$

$$= \frac{\sqrt{12{,}250{,}000} - 1500}{2} = \frac{3500 - 1500}{2}$$

$$= \frac{2000}{2} = 1000 \text{ Hz}$$

Our original lower cutoff frequency was indeed 1000 Hz. Now, I check the upper cutoff frequency.

$$F_h = 1000 + 1500 = 2500 \text{ Hz}$$

Once again, I came up with my original value. The equations check out.

Filter frequency response graphs are plotted on a logarithmic scale. The bandpass response will appear to be symmetrical around the center frequency. That is, the center frequency will appear to be exactly in the middle of the two cutoff frequencies. This will not be true on a linear scale.

There is one final specification that is usually cited for bandpass filters. This is the *quality factor,* or Q.

Essentially, Q is a measurement of the sharpness of the rolloff as the frequency moves away from the center frequency in either direction.

One way to determine the Q of a bandpass filter circuit is to take the inverse of the damping value.

$$Q = 1/d$$

The ratio between the bandwidth and the center frequency also defines the Q of circuit.

$$Q = \frac{F_c}{BW}$$

Alternatively, it can be calculated from the upper and lower-cutoff frequencies by substituting earlier equations.

$$Q = \frac{F_c}{F_h - F_l} = \frac{\sqrt{F_h F_l}}{F_h - F_l}$$

Increasing the bandwidth will decrease the Q.

Returning to my example, I can easily find the Q by plugging in the center frequency and bandwidth values.

$$Q = \frac{F_c}{BW} = \frac{1581}{1500} \cong 1.05$$

The Q value for a practical bandpass filter circuit may be anywhere from less than 1 to several hundred.

Now that you are familiar with the basic specifications of a bandpass filter, it's time to look at some actual circuits.

A passive bandpass filter circuit can be created by adding a coil to a low-pass filter, as shown in Fig. 5-30.

The capacitor and coil form a parallel resonant circuit that is resonant at the center frequency of the filter. The center frequency can be found by using the following formula:

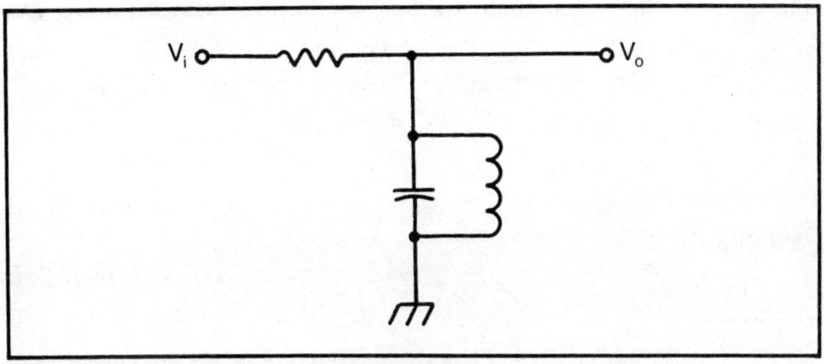

Fig. 5-30. A passive bandpass filter can be created by placing a coil across the capacitor in a passive low-pass filter.

$$F_c = \frac{159}{\sqrt{LC}}$$

The first step in designing a passive bandpass filter is to determine the bandwidth. Let's assume I need a circuit with the following parameters:

source impedance	10 K
load impedance	50 K
bandwidth	1200 Hz
center frequency	4000 Hz

First, I design a passive low-pass filter with a cutoff frequency equal to the desired bandwidth. This is 1200 Hz in our example.

The resistor should have a value equal to the square root of the product of the source and load impedances.

$$R = \sqrt{R_s R_l} = \sqrt{10,000 \times 50,000}$$
$$R = \sqrt{500,000,000} = 22,360 = 22,000 = 22\ K$$

The next step is to solve for C.

$$C = \frac{159,000}{RF_c} = \frac{159,000}{22,000 \times 1200} = \frac{159,000}{26,400,000}$$

$$C = 0.006$$

I can round this value off and use a 0.005 μF capacitor.

I now have a passive low-pass filter with a cutoff frequency of 1200 Hz. By adding a coil across the capacitor I can change this circuit into a passive bandpass filter with a bandwidth of 1200 Hz.

The coil's inductance value must be selected so that it and the capacitor will be resonant at the center frequency. The desired coil value can be found with this equation:

$$L = \frac{1}{C}\left(\frac{159}{F_c}\right)^2$$

Plugging in the values for our example I find:

$$L = \frac{1}{0.005}\left(\frac{159}{4000}\right)^2 = 200 \times (0.03975)^2$$

$$= 200 \times 0.0015801 = 0.3160125 \cong 0.3 \text{ mH}$$

Of course, a passive bandpass filter suffers from all of the disadvantages described earlier for the passive low-pass filter plus one—the coil itself. Coils tend to be relatively bulky and expensive. It is often difficult to find the needed value.

By using an op amp as the heart of an active bandpass filter you eliminate the need for coils.

A typical active bandpass filter circuit is shown in Fig. 5-31. This is called a multiple-feedback bandpass filter because there is more than one feedback path (through R2 and through C1 and C2).

This circuit is suitable for moderate amounts of gain and for Q values of about 15 to 20.

The formula for determining gain is a little more complicated than any that I have introduced in previous circuits.

$$G = \frac{R2}{R1\left(1 + \frac{C1}{C2}\right)}$$

This is the circuit's maximum gain and is valid only at (or close to) the filter's center frequency.

The equation for finding the center frequency is also rather complex, due to the relatively large number of frequency determining components. The following formula can be used to find the center frequency:

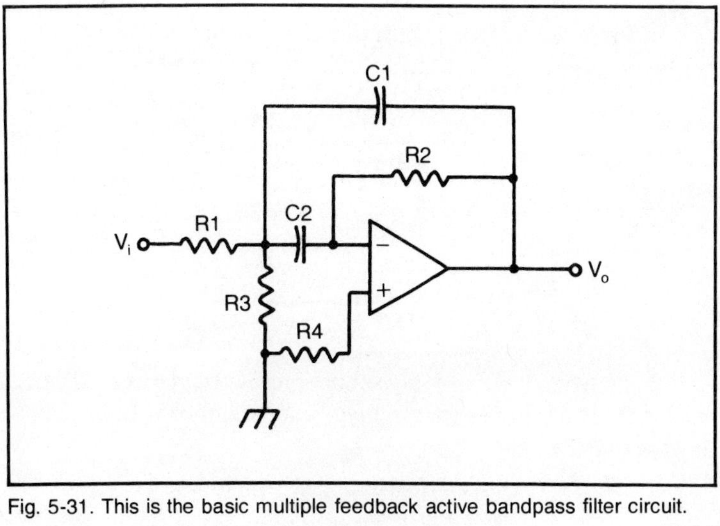

Fig. 5-31. This is the basic multiple feedback active bandpass filter circuit.

$$F_c = \frac{1}{2\pi}\sqrt{\frac{1}{R2C1C2}\left(\frac{1}{R1}+\frac{1}{R3}\right)}$$

The factor $1/2\pi$ is about 0.159.

Designing a multiple-feedback bandpass filter can be made considerably easier if the capacitor values are made equal (C1 = C2). This makes the equation much simpler.

$$F_c = 0.159 \times \sqrt{\frac{1}{R2 \times C} \times \left(\frac{1}{R1}+\frac{1}{R3}\right)}$$

Naturally, when designing filters you will probably know the desired center frequency and will need to calculate the appropriate component values. First, by making the capacitors of equal value I can simplify the gain equation.

$$G = \frac{R2}{R1\left(1+\frac{C}{C}\right)} = \frac{R2}{R1 \times (1+1)} = \frac{R2}{2R1}$$

Generally it is easiest to select a convenient value for C, then solve for the resistances with the desired values for Q, G (gain), and F_c.

The multiple-feedback bandpass filter circuit is best suited for relatively narrow bandwidths as compared to the center frequency. This holds the value of Q down.

Each of the frequency determining resistors can be solved for with the following equations:

$$R1 = \frac{Q}{2\pi F_c GC}$$

$$R2 = 2GR1$$

$$R3 = \frac{Q}{2\pi F_c C(2Q^2 - G)}$$

Let's test out these design equations by trying a typical design problem. I want to design an active bandpass filter with the following parameters:

$$F_c = 2000 \text{ Hz}$$

$$BW = 400\ Hz$$
$$G = 2.5$$

The first step is to determine the value of Q for the circuit.

$$Q = \frac{F_c}{BW} = \frac{2000}{400} = 5$$

Next I arbitrarily select a value of C. I will use 0.01 μF. Both capacitors (C1 and C2) will have the same value.

Plugging the known values into the equation for resistor R1, I get 15,924 ohms.

$$R1 = \frac{Q}{2\pi F_c GC} = \frac{5}{6.28 \times 2000 \times 2.5 \times 0.00000001}$$

$$= \frac{5}{0.000314} = 15{,}924 \text{ ohms}$$

I can use a 15 K resistor for R1. Next I calculate R2.

$$R2 = 2GR1 = 2 \times 2.5 \times 15{,}000 = 75{,}000 \text{ ohms}$$

Finally, I solve for resistor R3:

$$R3 = \frac{Q}{2\pi F_c C(2Q^2 - G)}$$

$$R3 = \frac{5}{6.28 \times 2000 \times 0.00000001 \times ((2 \times 5^2) - 2.5)}$$

$$R3 = \frac{5}{0.0001256 \times ((2 \times 25) - 2.5)}$$

$$R3 = \frac{5}{0.0001256 \times (50 - 2.5)} = \frac{5}{0.0001256 \times 47.5}$$

$$R3 = \frac{5}{0.005966} = 838 \text{ ohms}$$

A 820 ohms resistor will certainly be close enough.

Now let's go back and plug these component values into the main F_c equation and see if I actually come up with something reasonably close to the original desired center frequency (2000 Hz). There will probably be some error due to the rounding off of the component values, but the calculated center frequency should be approximately the nominal value.

$$F_c = 0.159 \sqrt{\frac{1}{R2C^{-2}} \times \left(\frac{1}{R1} + \frac{1}{R3}\right)}$$

$$F_c = 0.159 \sqrt{\frac{1}{75{,}000 \times (0.00000001^{-2}} \times \left(\frac{1}{15{,}000} + \frac{1}{820}\right)}$$

$$F_c = 0.159 \sqrt{\frac{1}{0.0000000000075} \times 0.0012862}$$

$$F_c = 0.159 \sqrt{133{,}333{,}333{,}333 \times 0.0012862}$$

$$F_c = 0.159 \sqrt{171{,}490{,}510} = 0.159 \times 13{,}095$$

$$F_c = 2082 \text{ Hz}$$

That is quite close to our design goal of a center frequency of 2000 Hz.

One very nice feature of the multiple-feedback bandpass filter is that the center frequency can be changed simply by changing the value of resistor R3. Simply multiply the original value of R3 by the square of the ratio of the old frequency to the new frequency.

$$R3_n = R3_o \times \left(\frac{F_{co}}{F_{cn}}\right)^2$$

For instance, if I want to adapt the filter circuit I've just designed to a center frequency of 1200 Hz, I would need to change R3 to 2278 ohms.

$$R3 = 820 \times \left(\frac{2000}{1200}\right)^2 = 820 \times 1.6667^2$$

$$R3 = 820 \times 2.7778 = 2278 \text{ ohms}$$

I can use a 2.2 K resistor as the new R3. Let's double-check the F_c equation to make sure that everything works out the way I want it to.

$$F_c = 0.159 \sqrt{\frac{1}{75{,}000 \times (0.00000001)^{-2}} \times \left(\frac{1}{15{,}000} + \frac{1}{2200}\right)}$$

$$F_c = 0.159 \sqrt{133{,}333{,}333{,}333 \times (0.0000667 + 0.004545)}$$

$$F_c = 0.159 \sqrt{133{,}333{,}333{,}333 \times 0.0005212}$$

$$F_c = 0.159 \sqrt{69494949} = 0.159 \times 8336$$

$$F_c = 1325 \text{ Hz}$$

That's still reasonably close to the desired value of 1200 Hz.

There is one important limitation of the multiple-feedback bandpass filter circuit of Fig. 5-31. If the gain is high, the Q must also be high.

$$Q > \sqrt{\frac{G}{2}}$$

If this algebraic statement is not true, resistor R3 will result in a negative value. Obviously, that can't be done.

A variation on this basic circuit is shown in Fig. 5-32. Resistor R3 has been eliminated altogether. Effectively, the circuit behaves in the same manner as the one in Fig. 5-31, except R3 is an open circuit. That is, its value is infinity. This results in some changes in the basic equations. Again, matters are greatly simplified by letting C1 = C2. The center frequency equation becomes:

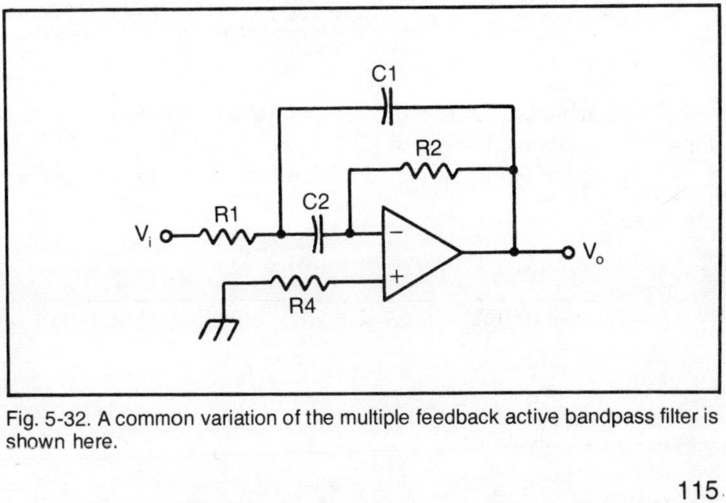

Fig. 5-32. A common variation of the multiple feedback active bandpass filter is shown here.

$$F_c = 0.159 \times \frac{1}{\sqrt{R1R2C^2}}$$

The two frequency determining resistors (R1 and R2) take their values from the following equations:

$$R1 = \frac{Q}{2\pi GCF_c}$$

$$R2 = \frac{G}{2\pi QCF_c}$$

The circuit gain is dependent on the Q and the values of R1 and R2.

$$G = Q \times \sqrt{\frac{R2}{R1}}$$

Let's try out an example. I want an active bandpass filter with the following characteristics:

$$F_c = 1800 \text{ Hz}$$
$$BW = 300 \text{ Hz}$$
$$G = 2.25$$

The Q, of course, is the ratio of the center frequency (F_c) to the bandwidth (BW).

$$Q = \frac{F_c}{BW} = \frac{1800}{300} = 6$$

I will arbitrarily select a value of 0.02 μF (0.00000002 farad) for C. Remember C1 = C2 = C.

Plugging in the known values I can find the required value for resistor R1.

$$R1 = \frac{Q}{2\pi GCF_c} = \frac{6}{6.28 \times 2.25 \times 0.00000002 \times 1800}$$

$$R1 = \frac{6}{0.0005087} = 11795 \text{ ohms}$$

A 12 K resistor will do just fine for R1. Next, solving for R2 I find the following:

$$R2 = \frac{G}{2\pi QCF_c} = \frac{2.25}{6.28 \times 6 \times 0.00000002 \times 1800}$$

$$= \frac{2.25}{0.0013565} = 1659 \text{ ohms}$$

I could probably get away with either a 1.5 K or a 1.8 K resistor here. 1.5 K is a little closer to the calculated value, and I rounded up for R1, so rounding down should help cancel out some of the round-off error.

I have two equations that I can use to double-check my selected resistor values. One is the gain formula.

$$G = Q \times \sqrt{\frac{R2}{R1}} = 6 \times \sqrt{\frac{1500}{12{,}000}}$$

$$G = 6 \times \sqrt{0.125} = 6 \times 0.35 = 2.12$$

That is acceptably close to my specified gain of 2.25. I am not likely to notice this much error in most practical applications.

Now let's double-check the center frequency (F_c) equation.

$$F_c = 0.159 \times \frac{1}{\sqrt{R1R2C^2}}$$

$$F_c = 0.159 \times \frac{1}{\sqrt{12{,}000 \times 1500 \times (0.00000002)^2}}$$

$$F_c = 0.159 \times \frac{1}{\sqrt{18{,}000{,}000 \times 0.0000000000000004}}$$

$$F_c = 0.159 \times \frac{1}{\sqrt{0.0000000072}} = \frac{1}{0.0000849}$$

$$F_c = 11{,}785 \text{ Hz}$$

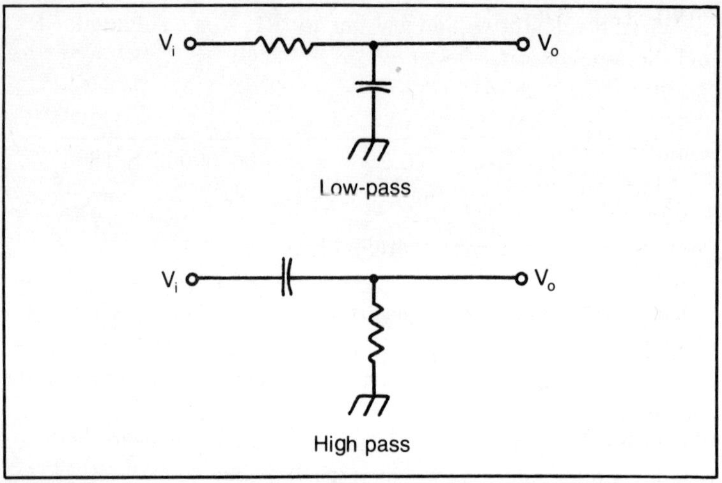

Fig. 5-33. A low-pass filter can be converted into a high-pass filter by simple inversion.

Yes, that is close enough to my specified center frequency of 12,000 Hz.

In critical applications it may be necessary to use precision resistors with values closer to the calculated values. Rounding off to standard resistance values will inevitably introduce errors.

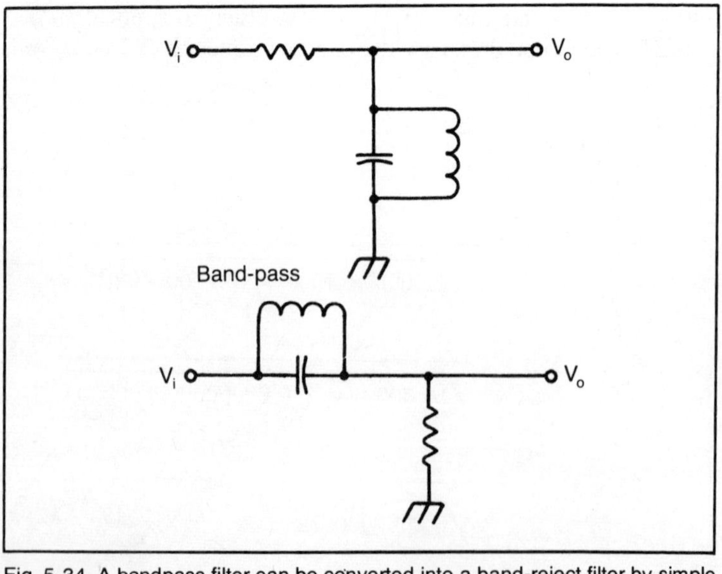

Fig. 5-34. A bandpass filter can be converted into a band-reject filter by simple inversion.

BAND-REJECT FILTERS

Remember how I created a passive high-pass filter by reversing the position of the components in a passive low-pass filter, as shown in Fig. 5-33? What do you suppose will happen if I reverse the components in a passive bandpass filter, as shown in Fig. 5-34?

I get a filter that is the mirror image of a bandpass filter. Where a bandpass filter rejects everything but a specific band of frequencies, as illustrated in the frequency response graph of Fig. 5-35, this new filter passes everything except a specific band of frequencies which is rejected. The frequency response graph for this type of filter is illustrated in Fig. 5-36.

This new filter type is called a *band-reject filter*, for obvious reasons. It is also sometimes called a notch filter because of the appearance of the frequency response graph.

The calculations for a passive band-reject filter are the same as those for a comparable passive bandpass filter. The center frequency is the mid-point of the rejected band. The bandwidth is the size of the rejected band.

A bandpass filter can be created by connecting a low-pass filter and a high-pass filter in series, as shown in Fig. 5-27. Similarly, a band-reject filter can be created by connecting a low-pass filter and a high-pass filter in parallel, as illustrated in Fig. 5-37.

Only those signals which are blocked by both of the component filters will be deleted from the output. Any frequency passed by

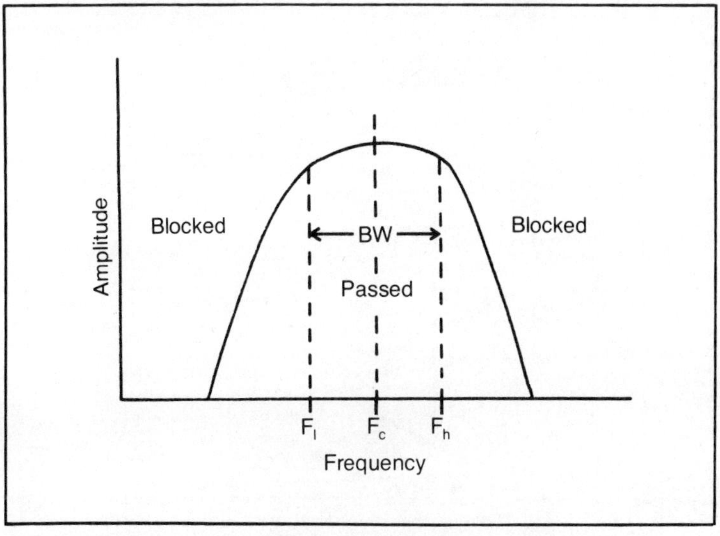

Fig. 5-35. A bandpass filter passes a midrange band of frequencies.

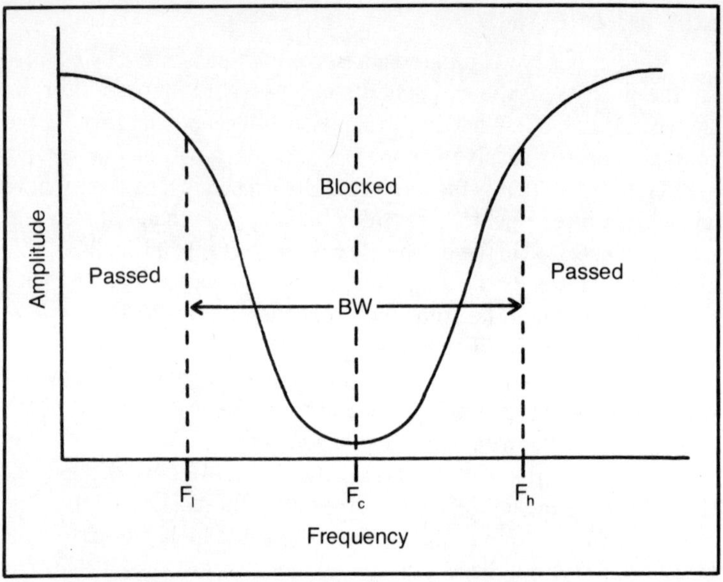

Fig. 5-36. A band-reject filter passes everything except a midrange band of frequencies.

either or both of the component filters will appear in the output signal.

Band-reject filters generally have narrow bandwidths, and are most frequently used to block noise or interference signals. For

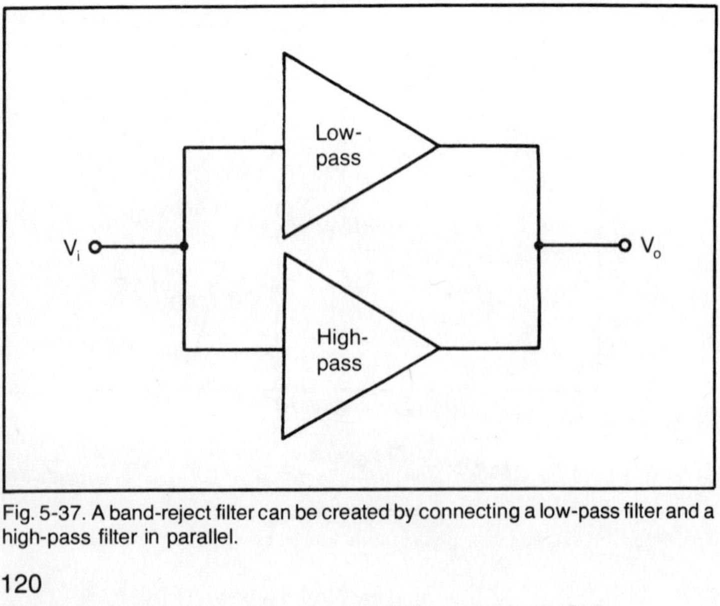

Fig. 5-37. A band-reject filter can be created by connecting a low-pass filter and a high-pass filter in parallel.

instance, ac power lines can generate strong signals at 60 Hz and its multiples (120 Hz, 180 Hz, 240 Hz, etc.). One or more of these frequencies could cause interference with a desired signal at a nearby frequency. A band-reject filter tuned to the ac interference frequency may be used to clean up the output signal.

Of course, if you have a strong interfering signal at 60 Hz, and your desired signal is 59 Hz, it may be very difficult to come up with a band-reject filter with a narrow enough bandwidth to successfully suppress the unwanted signal, while passing the desired signal.

There is another type of passive band-reject filter circuit that does not use bulky or expensive coils at the price of a somewhat larger component count. This type of circuit, shown in Fig. 5-38, is called a *twin-T filter,* because of the physical arrangement of the components. Looking at the schematic, you should be able to see that resistors R1 and R2 and capacitor C3 make up a T-shape. Similarly, the second T is comprised of C1, C2, and R3.

Solving for the component values is not particularly difficult, because certain simple rules must be followed for the circuit to work. They are as follows:

$$R1 = R2 \qquad C1 = C2$$

$$C3 = 2 \times C1$$

$$R3 = \frac{R1}{2}$$

The center frequency (sometimes called the null frequency) is determined with the following formula:

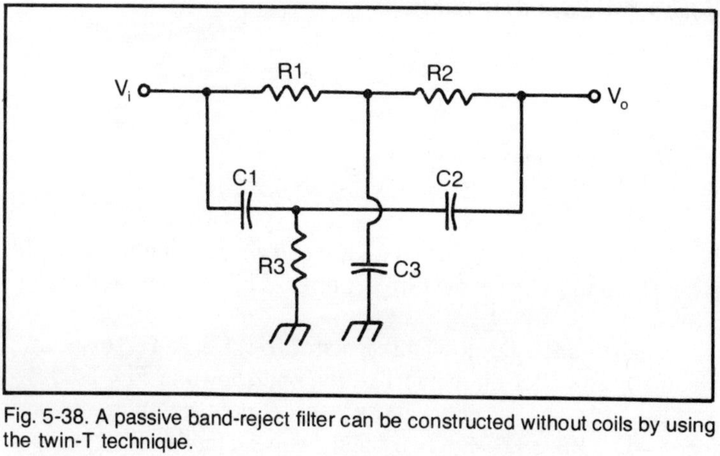

Fig. 5-38. A passive band-reject filter can be constructed without coils by using the twin-T technique.

$$F_c = \frac{1}{2\pi R1C1}$$

If the component values are perfectly matched, the center frequency should theoretically be completely removed from the output signal. In the real world, however, there will inevitably be some degree of mismatch. Nevertheless, this passive circuit is capable of very good rejection, and the center frequency will be greatly attenuated at the output.

Let's try out an example. I want a band-reject filter to suppress a 120 Hz interference signal. I can select a likely value for C1, and rearrange the F_c equation to solve for R1:

$$R1 = \frac{1}{2\pi C1F_c}$$

I will use a 0.1 μF capacitor for C1. This means R1 should have a value of 13,270 ohms.

$$R1 = \frac{1}{6.28 \times 120 \times 0.0000001} = \frac{1}{0.00007536}$$

$$R1 = 13{,}270 \text{ ohms}$$

I can round that off to 12 K.

Now it's a simple matter to find the other necessary component values:

$$R2 = R1 = 12 \text{ K}$$

$$R3 = \frac{R1}{2} = \frac{12{,}000}{2} \times 6000 \cong 6.2 \text{ K}$$

$$C2 = C1 = 0.1\ \mu F$$
$$C3 = 2 \times C1 = 2 \times 0.1 = 0.2\ \mu F$$

The circuit can be improved a great deal by adding an op amp to convert it into an active band-reject filter. This is illustrated in Fig. 5-39.

Resistors R1, R2, R3, and capacitors C1, C2, and C3 make up a passive band-reject filter just like the one shown in Fig. 5-38.

Besides the op amp itself, three new components are added—R4, R5, and C4. Their values depend on the other compo-

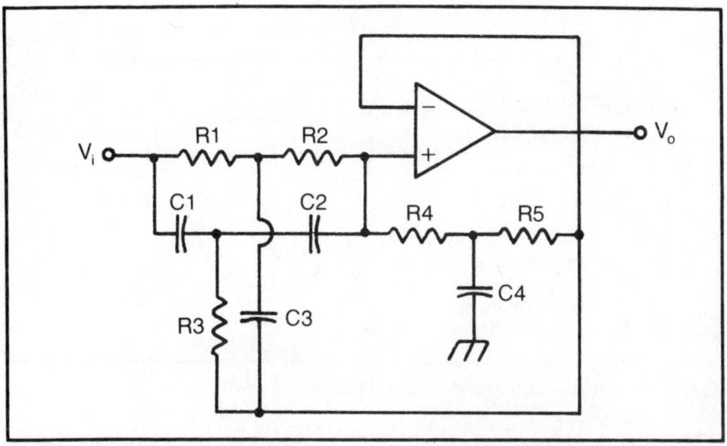

Fig. 5-39. Combining a passive, twin-T, band-reject filter with an op amp creates an active band-reject filter.

nent values, and the circuit Q, according to the following equations:

$$R4 = R5 = 2QR1$$

$$C4 = \frac{C1}{Q}$$

Let's continue with the sample filter design I started with the passive version. Remember, I have the following values:

$$F_c = 120 \text{ Hz} \qquad C1 = C2 = 0.1\ \mu F$$
$$R1 = R2 = 12\text{ K} \qquad C3 = 0.2\ \mu F$$
$$R3 = 6.2\text{ K}$$

Before I can go on, I need to determine how much bandwidth I want my circuit to have. In most applications a band-reject filter will call for a fairly narrow bandwidth. Let's say I want a 20 Hz bandwidth. I am now able to solve for Q.

$$Q = \frac{F_c}{BW} = \frac{120}{20} = 6$$

Now I have all of the information I need to determine the values for R4, R5, and C4. First, I will solve for R4.

$$R4 = 2QR1 = 2 \times 6 \times 12{,}000 = 144{,}000 \text{ ohms}$$

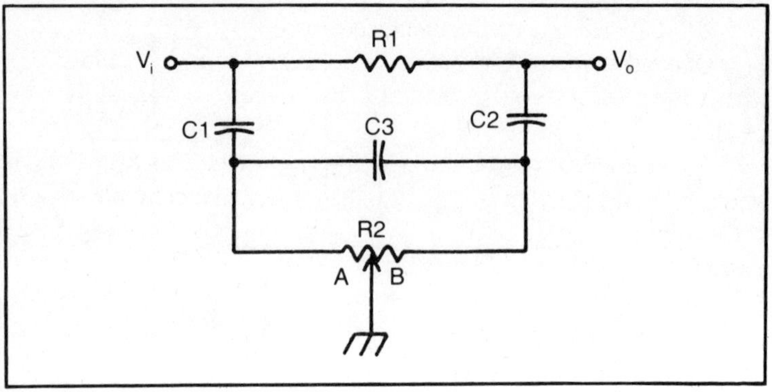

Fig. 5-40. A passive, bridged differentiator is shown here.

The closest standard resistor value is 150 K. Resistor R5 should also be equal to 150 K, of course.

Finally, I find the value for capacitor C4.

$$C4 = \frac{C1}{Q} = \frac{0.0000001}{6} = 0.00000001667 \text{ farad}$$

$$C4 = 0.01667 \ \mu F$$

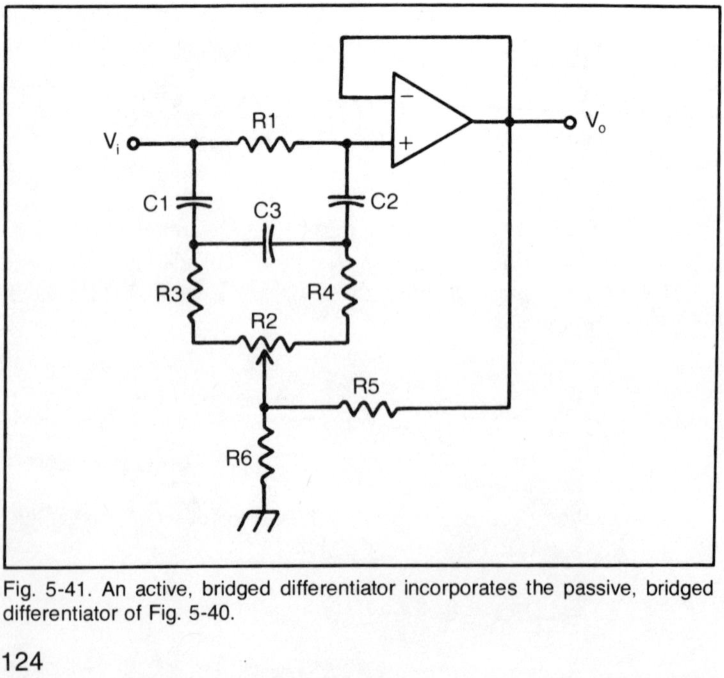

Fig. 5-41. An active, bridged differentiator incorporates the passive, bridged differentiator of Fig. 5-40.

A 0.015 μF capacitor should be close enough.

Of course, for critical applications, high tolerance components with values as close as possible to the calculated values may be used.

An interesting variation on the basic twin-T passive band-reject filter is shown in Fig. 5-40. Close tolerance components are not essential because potentiometer R2 can adjust the center frequency over a fairly wide range. This circuit is called an *abridged differentiator.*

The three capacitors have identical values. Resistor R1 should be about 6 times the total resistance of the potentiometer. The cutoff frequency can be found with the following formula:

$$F_c = \frac{1}{2\pi C\sqrt{3R2_a R2_b}}$$

A bridged differentiator circuit can be used as the input network for an active band-reject filter, as illustrated in Fig. 5-41.

6

Signal Generators

In the previous chapters, while discussing amplifier and filter circuits, the problem of parasitic oscillations was touched on. If the feedback signal is in phase with the original input signal, an op amp can break into oscillation and start generating a spurious signal of its own. Obviously, this is undesirable in an amplifier or a filter. There are occasions, however, when you want to generate a specific signal. The operational amplifier is also capable of controlled oscillation. In this chapter, I want to introduce a new class of op amp applications—signal generation.

SINE WAVE OSCILLATORS

The simplest of all ac signals is the sine wave, shown in Fig. 6-1. This is a very pure signal with no harmonic content at all. A sine wave consists only of a single fundamental frequency.

Actually it is difficult to generate a truly pure, distortion free sine wave. Fortunately, I can come quite close to the ideal with an oscillator circuit built around an op amp.

A typical op amp sine wave oscillator circuit is shown in Fig. 6-2. The feedback network is a twin-T circuit like the one used in the band-reject filter in Chapter 5. Resistors R1 and R2 along with capacitor C1 form one T. The other T is made up of C2, C3, R3, and R4. It is upside down in the schematic.

The oscillation frequency is determined by this formula provided that the following relationships are true.

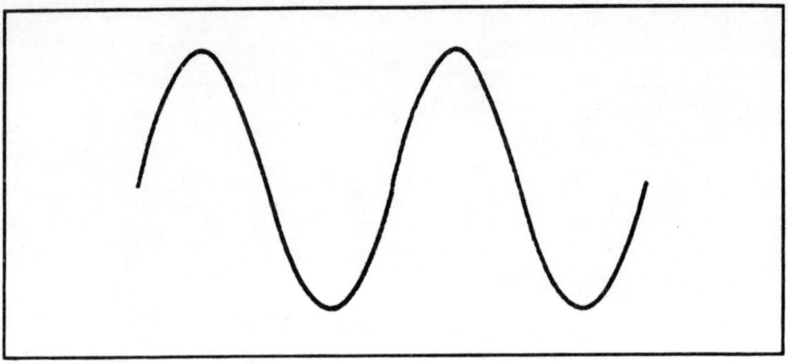

Fig. 6-1. The simplest ac signal is the sine wave.

$$F = \frac{1}{2\pi \, R1C2}$$

$$R2 = R1 \qquad C3 = C2$$

$$R3 = \frac{R1}{4} \qquad C1 = 2C2$$

$$R4 \cong \frac{R1}{2}$$

The twin-T feedback network is detuned slightly by adjusting the value of R4. This will usually be a miniature trimmer potentiometer. The potentiometer is set for maximum resistance, then slowly decreased in value until the circuit begins to oscillate. If this resistance is set too low, the sine wave will be distorted at the output.

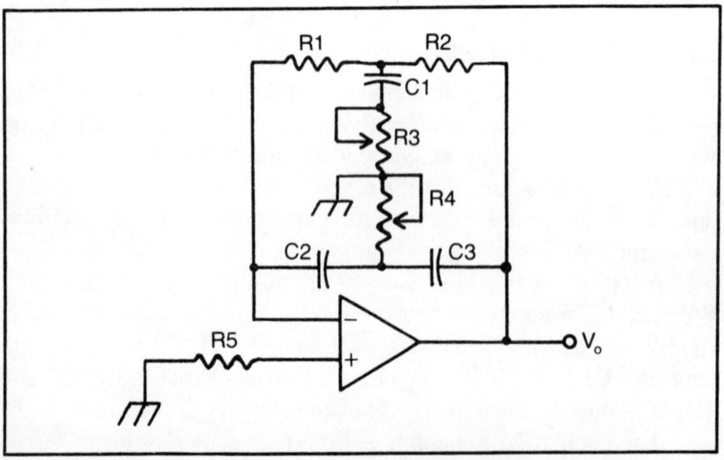

Fig. 6-2. Here is the circuit for a twin-T sine wave oscillator.

This circuit works because the twin-T network phase shifts the output signal by 180° before feeding it back into the inverting input, which by definition introduces another 180° of phase shift for a total shift of 360°—that is, positive feedback.

Internal noise within the IC itself generates enough of a signal to start the oscillation process when power is applied to the circuit.

Let's work through a typical design example. I will design a sine wave oscillator with a frequency of 600 Hz.

The first step is to select a reasonable value for the capacitor. I'll try a 0.01 μF capacitor for C2 and C3 (remember, these two components must have identical values).

Now, by rearranging the frequency equation, I can solve for the value of R1.

$$R1 = \frac{1}{2\pi FC2} = \frac{1}{6.28 \times 600 \times 0.00000001}$$

$$R1 = \frac{1}{0.0000377} = 26{,}526 \text{ ohms}$$

A 27 K resistor should be close enough. Now it's no problem at all to find the values for the other components in the circuit.

$$R2 = R1 = 27 \text{ K}$$

$$R3 = \frac{R1}{4} = \frac{27{,}000}{4} = 675 \cong 680 \text{ ohms}$$

$$R4 = \frac{R1}{2} = \frac{27{,}000}{2} = 13{,}500 \cong 15 \text{ K to } 25 \text{ K trimpot}$$

$$C1 = 2C2 = 2 \times 0.01 = 0.02\ \mu\text{F}$$

$$C3 = C2 = 0.01\ \mu\text{F}$$

The major disadvantage of this circuit is that the frequency and purity of the output waveform is dependent on the interlocking values of all of the components. There is no easy way to make the output frequency variable.

Another popular op amp sine wave oscillator circuit is illustrated in Fig. 6-3. This is called a quadrature oscillator.

A quadrature oscillator has two outputs, labeled sine and cosine. A cosine wave is a sine wave that has been phase shifted 90°, as shown in Fig. 6-4.

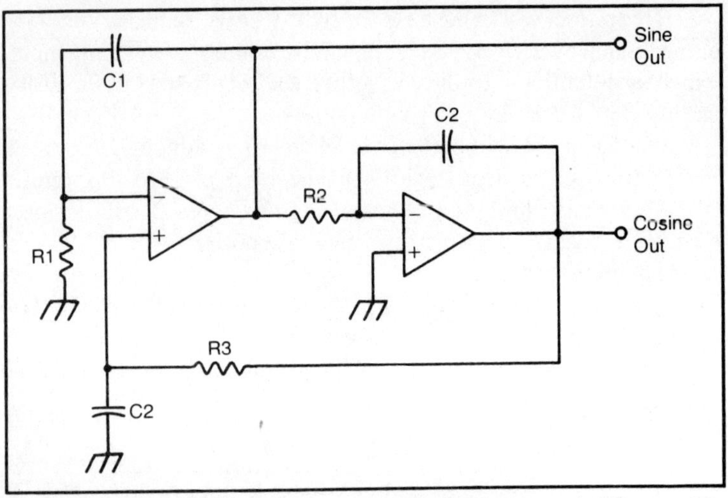

Fig. 6-3. A quadrature oscillator produces both sine waves and cosine waves.

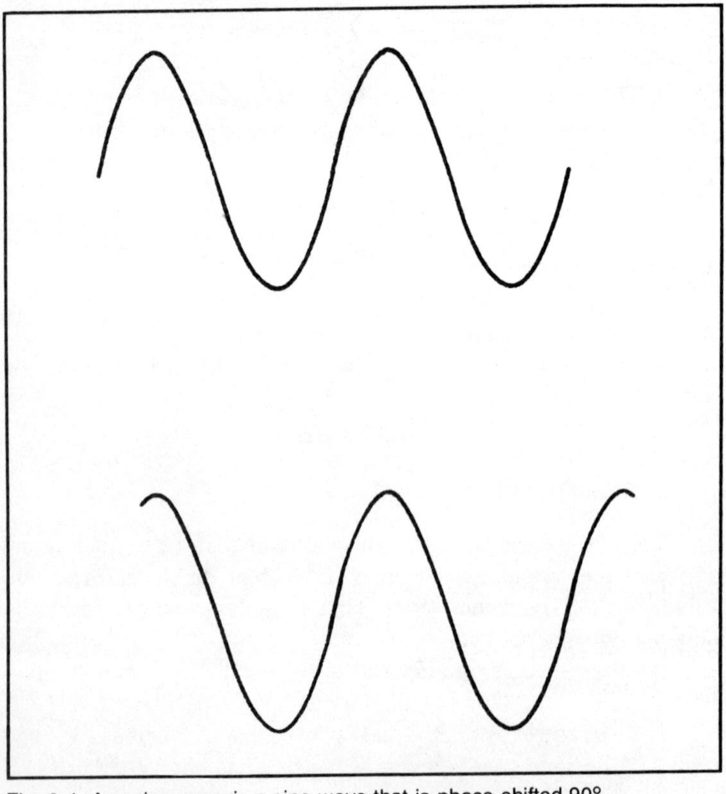
Fig. 6-4. A cosine wave is a sine wave that is phase shifted 90°.

Notice that two op amps are required for this circuit. Only six additional passive components are required. All three capacitors are given identical values. Resistors R2 and R3 are also equal. Resistor R1 has a value that is slightly less than R2 to make sure that the oscillations will begin as soon as power is applied. The exact value of R1 is not particularly critical, as long as it is somewhat smaller than R2.

This means I really only need to select two component values—C (C1, C2, and C3) and R (R2, R3).

The frequency equation is the same as the one used for the twin-T sine wave oscillator discussed in the last few pages. Once again the best design approach would be to select a likely value for C, then rearrange the frequency equation to solve for R.

$$R = \frac{1}{2\pi FC}$$

As an example, I will design a quadrature oscillator with an output frequency of 1000 Hz. I will assign a value of 0.047 μF to C. Therefore, the resistance value should be 3388 ohms.

$$R = \frac{1}{6.28 \times 1000 \times 0.000000047} = \frac{1}{0.0002952} = 3388 \text{ ohms}$$

This is very close to the standard resistor value of 3.3 K. In most cases, R1 will be made the next lower standard value, or 2.7 K in this case. This value is not at all critical.

Besides the basic simplicity and elegance of its design, the quadrature oscillator also offers the advantage of a potentially variable output frequency. This can be accomplished by replacing resistors R2 and R3 with a dual potentiometer. The shafts are ganged together so both resistances will always be equal. Fixed resistors should be used in series with the potentiometers so that the value of R will never drop below that of R1.

The sine wave and cosine wave outputs of this circuit will always have the same frequency. The only difference between the two outputs will be their relative phase.

SQUARE WAVE GENERATORS

Op amp circuits can also generate other waveforms. One of the most popular is the square wave, shown in Fig. 6-5. This waveform consists of two constant signal levels, theoretically switching instantaneously from one level to the other. In a square wave each

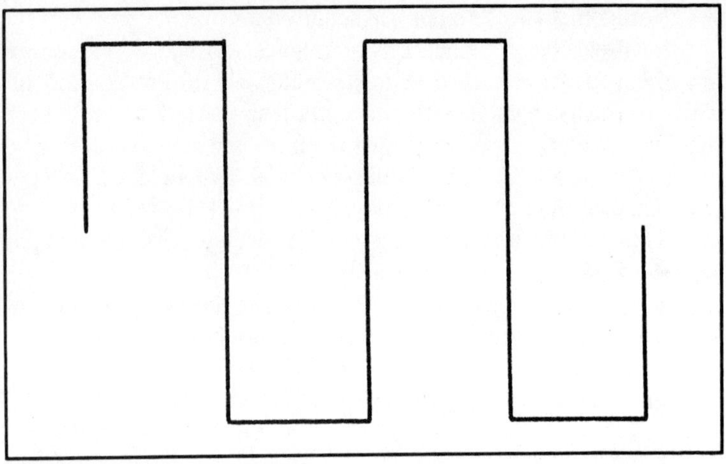

Fig. 6-5. An ideal square wave switches instantly between two voltage states.

level is held for exactly one half of the cycle. They are symmetrical, or equal in duration.

A square wave is a more complex signal than a sine wave. This waveform is made up of many frequencies. It can be synthesized by combining sine waves of the correct frequencies and relative amplitudes.

The lowest frequency component in a square wave is called the fundamental. This is the strongest component in the signal, and it defines the frequency of the signal as a whole and the perceived pitch if the waveform is fed through a loudspeaker.

Since the other frequency components in the signal are higher than the fundamental they are called overtones. If the overtones are integer multiples of the fundamental frequency they are called harmonics.

All of a square wave's overtones are harmonics. This signal contains only the odd harmonics (3 times the fundamental, 5 times the fundamental, 7 times the fundamental, and so forth). No multiples of two are included in the signal.

The harmonic content of a square wave is fairly strong. The amplitude of the harmonics decreases as they get further from the fundamental frequency (higher).

A 500 Hz square wave includes the following frequency components. Each frequency component may be thought of as a sine wave at the specified frequency.

500 Hz (fundamental)

1500 Hz	(third harmonic — 3 × 500)
2500 Hz	(fifth harmonic — 5 × 500)
3500 Hz	(seventh harmonic — 7 × 500)
4500 Hz	(ninth harmonic — 9 × 500)
5500 Hz	(eleventh harmonic — 11 × 500)
6500 Hz	(thirteenth harmonic — 13 × 500)
7500 Hz	(fifteenth harmonic — 15 × 500)

Higher harmonics (seventeenth, nineteenth, twenty-first, and so on) will also be present, but they will be relatively weak and can be ignored in many applications.

A square wave generator can be created by forcing an op amp into positive saturation, then into negative saturation, then into positive saturation, and so forth. With the right feedback network, the switching will be extremely rapid. A slight slurring, as illustrated in Fig. 6-6, will be introduced in practical devices. This is due to the slew rate of the op amp IC. The slew rate defines how fast the output can respond to changes in the input signal.

The slew shown in Fig. 6-6 is greatly exaggerated for clarity. In most cases, the actual slew rate will be very short, and probably unnoticeable in many applications. When the shape of the output square wave is critical, a high quality op amp IC with a very short slew rate must be used.

Figure 6-7 shows the circuit for what is probably the simplest possible square wave generator. Besides the op amp itself, only three resistors and one capacitor are required.

Resistor R1 and capacitor C1 define the time constant of the circuit, but the output frequency is also affected by the positive

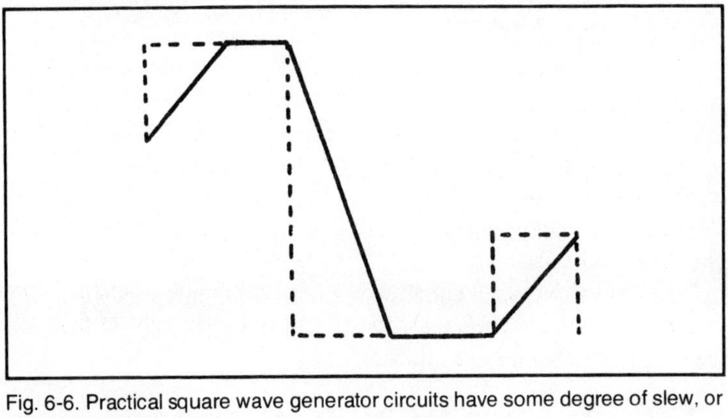

Fig. 6-6. Practical square wave generator circuits have some degree of slew, or switching time.

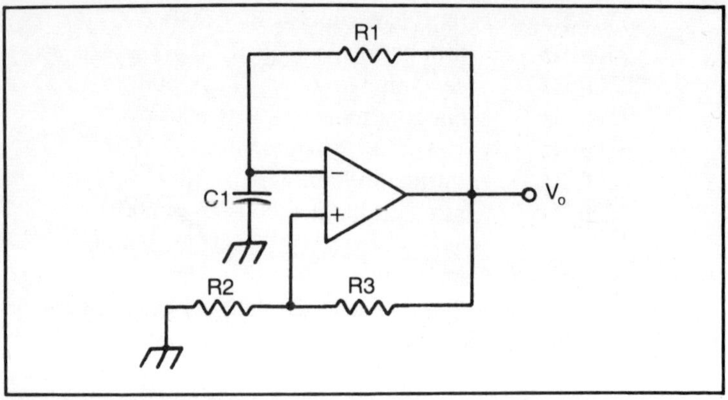

Fig. 6-7. This simple circuit will generate very good square waves.

feedback network (R3/R2). The general equations are rather complex, but I can simplify them for specific R3/R2 ratios.

$$\text{If } \frac{R3}{R2} = 1.1 \text{ then } F = \frac{0.5}{R1C1}$$

$$\text{If } \frac{R3}{R2} = 10 \text{ then } F = \frac{5}{R1C1}$$

For a beginning designer it would probably be a good idea to stick to one of these R3/R2 ratios to keep the design equations simple.

Let's try an example. I want a square wave generator with an output frequency of 1200 Hz.

First I'll select an R3/R2 ratio of 10. I can arbitrarily select a value for R2 and then find R3.

$$R3 = R2 \times \frac{R3}{R2} = R2 \times 10$$

If I use a 22 K resistor for R2, then R3 must be equal to 220 K.

$$R3 = 22{,}000 \times 10 = 220{,}000 \text{ ohms} = 220 \text{ K}$$

Now I know which equation to use for the output frequency.

$$F = \frac{5}{R1C1}$$

I know I want a value of 1200 Hz for F, and I need to find the necessary values for R1 and C1. I select a likely value for C1, and then rearrange the equation to solve for R1.

$$R1 = \frac{5}{FC1}$$

If I assign a value of 0.22 μF to C1, then R1 should become 18,940 ohms.

$$R1 = \frac{5}{1200 \times 0.00000022} = \frac{5}{0.000264} = 18{,}940 \text{ ohms}$$

A standard 18 K resistor should be close enough.

If I want to determine what the actual output frequency will be, I plug in the derived component values and use the frequency equation.

$$F = \frac{5}{R1C1} = \frac{5}{18{,}000 \times 0.00000022}$$

$$F = \frac{5}{0.00396} = 1262.6$$

This circuit can be made even more useful by adding a potentiometer, as shown in Fig. 6-8. This allows the output frequency to be manually changed.

The value of R1 is equal to the series combination of fixed resistor $R1_a$, and the adjusted value of potentiometer $R1_b$.

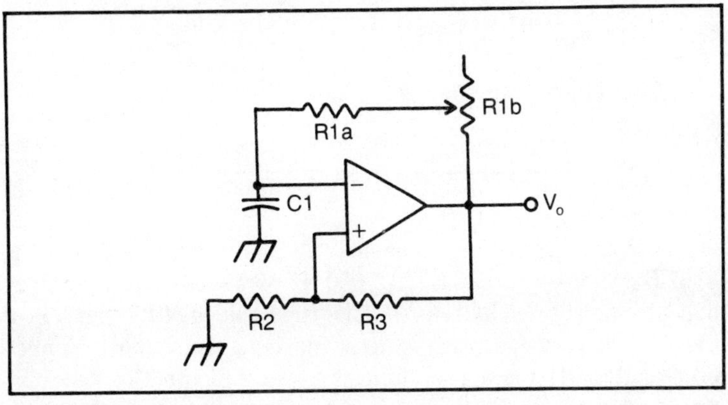

Fig. 6-8. Adapting the basic circuit of Fig. 6-7 permits manual control of the output frequency.

$$R1 = R1_a + R1_b$$

The fixed resistor is included to prevent the value of R1 from becoming zero. The fixed value of $R1_a$ and the maximum value of $R1_b$ sets the range of the output frequencies. Let's assume I have the following component values:

$$\begin{aligned} R1_a &= 10\text{ K} \\ R1_b &= 50\text{ K potentiometer} \\ R2 &= 22\text{ K} \\ R3 &= 220\text{ K} \\ C1 &= 0.1\ \mu\text{F} \end{aligned}$$

Feedback resistors R2 and R3 have the same values (and therefore the same ratio) as in the previous example.

At the minimum setting of $R1_b$, it is essentially 0 ohms, making the total value of R1 simply equal to $R1_a$.

$$R1 = R1_a + R1_b = 10{,}000 + 0 = 10{,}000$$

At this setting, the output frequency should be 5,000 Hz.

$$F = \frac{5}{R1\ C1} \times \frac{5}{10{,}000 \times 0.0000001} = \frac{5}{0.001} = 5{,}000\text{ Hz}$$

Moving potentiometer $R1_b$ to its maximum resistance setting, changes R1 to 60,000 ohms.

$$R1 = 10{,}000 + 50{,}000 = 60{,}000$$

This changes the output frequency to 833 Hz.

$$F = \frac{5}{60{,}000 \times 0.0000001} = \frac{5}{0.006} = 833\text{ Hz}$$

For the specified component values, the frequency range is 4167 Hz wide—from 833 Hz to 5,000 Hz. Wider or narrower ranges may also be created by selecting the right values of $R1_a$ and $R1_b$. If the range is very wide (that is, if the maximum value of $R1_b$ is much greater than $R1_a$) it will be difficult to set the output to a specific desired frequency. For instance, if I replaced the potentiometer in the earlier example with a 5 megohm unit, the output frequency will range from 10 to 5,000 Hz, but the potentiometer shaft will not have any greater range of movement.

A switch selectable range feature can be added by setting up different values for C1, as shown in Fig. 6-9.

As a rule of thumb, large capacitances and/or large resistances are used for low frequencies. This circuit is capable of generating very low frequencies.

The circuit works by forcing the op amp to switch between positive saturation (maximum positive voltage output) and negative saturation (maximum negative voltage output).

Actually the best square wave can be achieved if R2 has the same value as R3. When the op amp is in positive saturation, the noninverting input is held at one half of the positive saturation voltage by the feedback loop of R2 and R3, which act as a simple voltage divider. Meanwhile, capacitor C1 is being charged through $R1_a$ and $R1_b$.

At some point, the charge on C1 will exceed one half of the saturation voltage. Now the inverting input is more positive than the noninverting input. This forces the op amp to swing into negative saturation.

The noninverting input is now held at one half of the negative saturation voltage by R2 and R3. This condition holds true until capacitor C1 discharges to less than half of the negative saturation

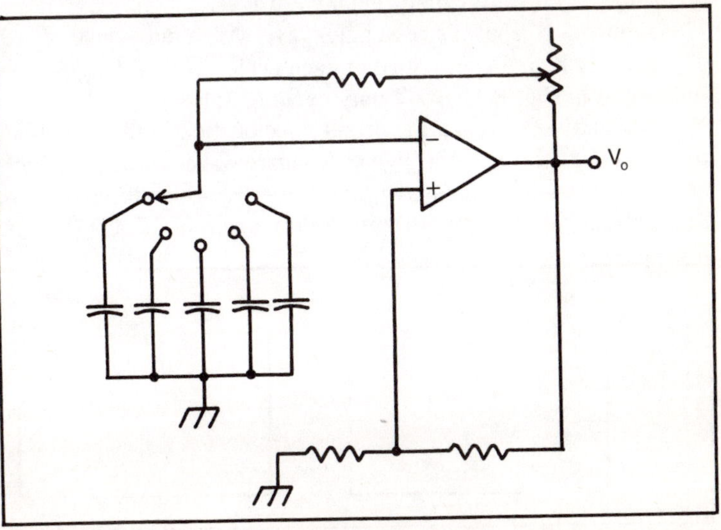

Fig. 6-9. Selecting different capacitance values changes the frequency range of the circuit.

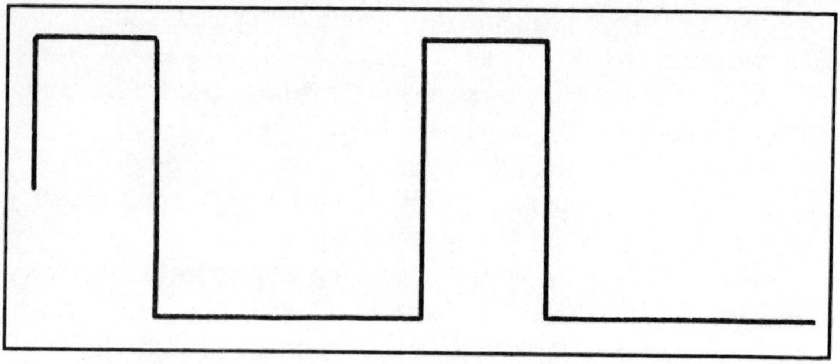

Fig. 6-10. A rectangle wave with a 1:3 duty cycle is at the high-voltage level for one third of each complete cycle.

voltage. This makes the noninverting input more positive (less negative) than the inverting input. The op amp switches back into positive saturation and the cycle repeats.

RECTANGLE WAVE AND PULSE GENERATORS

A square wave is perfectly symmetrical. It is in its high level state for one half of each cycle. Refer back to Fig. 6-5.

The ratio of high level time to total cycle time is called by the *duty cycle* of the signal. The duty cycle of a square wave is 1:2 by definition.

Waveforms that are closely related to the square wave, but other duty cycles are called rectangle waves.

Figure 6-10 shows a rectangle-wave with a duty cycle of 1:3. The level is high for one third of each cycle. The rectangle-wave illustrated in Fig. 6-11 has a duty cycle of 1:4.

The duty cycle is an important specification in that it defines the harmonic content of the signal. A square wave, with a duty cycle of 1:2 contains all harmonics except those that are multiples of two. Similarly, a rectangle wave with a duty cycle of 1:3 contains all

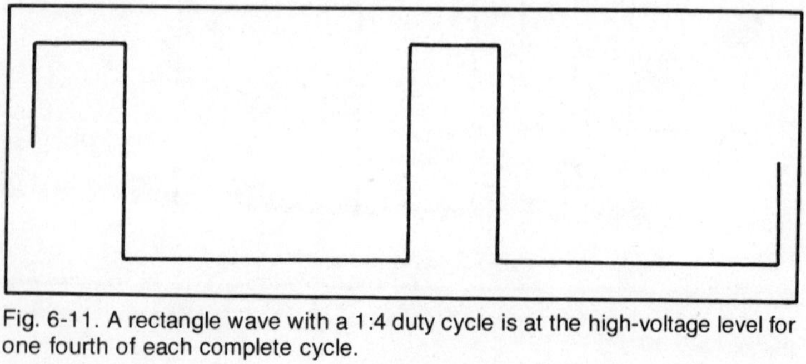

Fig. 6-11. A rectangle wave with a 1:4 duty cycle is at the high-voltage level for one fourth of each complete cycle.

Table 6-1. Harmonic Content for Various Rectangle Waves.

Duty Cycle	1:2 (square-wave)	1:3	1:4	1:5
	fundamental	fundamental	fundamental	fundamental
	— —	second	second	second
	third	— —	third	third
	— —	fourth	— —	fourth
	fifth	fifth	fifth	— —
	— —	— —	sixth	sixth
	seventh	seventh	seventh	seventh
	— —	eighth	— —	eighth
	ninth	— —	ninth	ninth
	— —	tenth	tenth	tenth
	eleventh	eleventh	eleventh	eleventh
	— —	— —	— —	twelth
	thirteenth	thirteenth	thirteenth	thirteenth
	— —	fourteenth	fourteenth	fourteenth
	fifteenth	— —	fifteenth	— —
	— —	sixteenth	— —	sixteenth
	seventeenth	seventeenth	seventeenth	seventeenth
	— —	— —	eighteenth	eighteenth
	nineteenth	nineteenth	nineteenth	nineteenth
	— —	twentieth	— —	— —

harmonics except for those that are multiples of three. A rectangle wave with a duty cycle of 1:4 is missing all harmonics that are multiples of four, and so on. The harmonic content of various rectangle waves is summarized in Table 6-1.

The square wave generator of Fig. 6-7 through Fig. 6-9 can be converted into a variable duty cycle rectangle-wave generator by adding a few components. The revised circuit is shown in Fig. 6-12.

On negative half-cycles, diode D1 blocks the flow of current through R5 and R6. The time constant is comprised of R1, R2, and

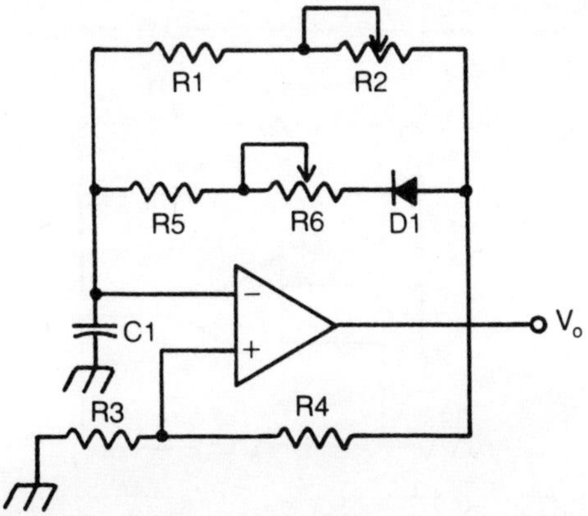

Fig. 6-12. The square wave generator circuit of Fig. 6-8 can easily be converted to a rectangle wave generator.

C1. On positive half cycles the diode conducts and the time constant is defined by C1 and the parallel combination of R1/R2 and R5/R6.

$$t_{c+} = C1 \times \frac{(R1 + R2) \times (R5 + R6)}{(R1 + R2) + (R5 + R6)}$$

Since the time constant will be different for the positive and negative portions of the cycle, the duty cycle will be something other than 1:2. The output waveform becomes asymmetric.

A simplified version of this circuit is shown in Fig. 6-13. On positive half-cycles, D1 is conducting, and D2 is blocking the feedback path. The positive time constant is simply $t_1 = C1R1$.

On negative half-cycles the situation is reversed. Diode D1 blocks current flow, and D2 conducts through R2, making the negative time constant equal to $t_2 = C1R2$.

The length of the total cycle is simply the sum of the two half-cycle time constants.

$$t_T = t_1 + t_2$$

The output frequency is the reciprocal of the total time constant of the complete cycle.

$$F = \frac{1}{t_T}$$

The duty cycle is the ratio between the positive time constant (t_1) and the total cycle time (t_T). This can be simplified by factoring.

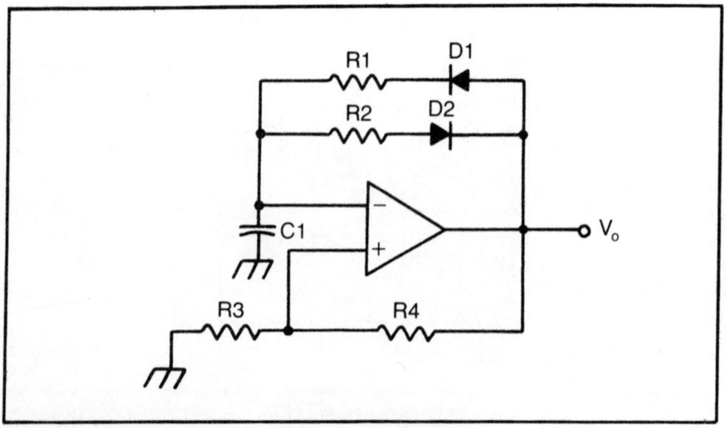

Fig. 6-13. A simple rectangle wave generator circuit.

$$t_1 : t_T =$$

$$C1R1 : C1R1 + C1R2 =$$

$$\frac{C1R1}{C1} : \frac{C1R1}{C1} + \frac{C1R2}{C1} =$$

$$R1 : R1 + R2$$

If R1 is equal to R2 the duty cycle becomes 1:2, or a square wave.

$$R1 : R1 + R1 =$$

$$R1 : 2R1 =$$

$$\frac{R1}{R1} : \frac{2R1}{R1} =$$

$$1 : 2$$

For this case, why not just use the circuit of Fig. 6-7 and forget about the diodes and separate resistors?

For nonsquare rectangle waves, R1 and R2 must have different values. Let's say C1 is a 0.1 μF capacitor, R1 is a 10 K resistor, and R2 is a 22 K resistor.

$$t_1 = C1R1 = 0.0000001 \times 10{,}000 = 0.001 \text{ second}$$
$$t_2 = C1R2 = 0.0000001 \times 22{,}000 = 0.0022 \text{ second}$$
$$t_T = t_1 + t_2 = 0.001 + 0.0022 = 0.0032 \text{ second}$$

$$F = \frac{1}{t_T} = \frac{1}{0.0032} = 312.5 \text{ Hz}$$

I can rearrange these equations to come up with a single formula for determining the frequency.

$$F = \frac{1}{t_T}$$

$$F = \frac{1}{t_1 + t_2}$$

$$F = \frac{1}{C1R1 + C1R2}$$

$$F = \frac{1}{C1(R1 + R2)}$$

I can confirm this new equation by plugging in the same component values I just used with the original equations.

$$F = \frac{1}{0.0000001(10{,}000 + 22{,}000)} = \frac{1}{0.0000001 \times 32{,}000}$$

$$= \frac{1}{0.0032} = 312.5 \text{ Hz}$$

I get the same answer with either method.

Now, what about the duty cycle? Earlier I concluded that I could define the duty cycle ratio with just the two resistor values (R1 and R2).

$$R1 : R1 + R2 =$$

$$10{,}000 : 10{,}000 + 22{,}000 =$$

$$10{,}000 : 32{,}000 =$$

$$\frac{10{,}000}{10{,}000} : \frac{32{,}000}{10{,}000} =$$

$$1 : 3.2$$

I may round this off to 1:3.

In designing a rectangle wave generator for a specific output frequency, I select a reasonable value for C1, then rearrange the frequency equation to solve for the sum of R1 and R2.

$$R1 + R2 = \frac{1}{FC1}$$

To separate the values of R1 and R2, I need to decide on the waveform's duty cycle. This ratio will always be in the form of 1:X. The total cycle time is by definition equal to the following equation:

$$t_T = Xt_1$$

By the same token, the total resistance is equal to this equation:

$$R_t = XR1$$

I can rewrite this equation as;

$$R1 = \frac{R_t}{X}$$

Once I know the value of R1, it is a simple matter to find the value of R2.

$$R2 = R_t - R1$$

Let's try out a typical problem. I need to design a rectangle wave generator with an output frequency of 1400 Hz, and a duty cycle of 1:4.

I select a value for C1. I will use a 0.033 μF capacitor. This makes the value of R_t (R1 + R2) equal to 21,645 ohms.

$$R_t = \frac{1}{FC1} = \frac{1}{1{,}400 \times 0.000000033}$$

$$R_t = \frac{1}{0.0000462} = 21{,}645 \text{ ohms}$$

Now I can solve for R1.

$$R1 = \frac{R_t}{X} = \frac{21{,}645}{4} = 5411 \text{ ohms}$$

Let's round that off to a standard value of 5.6 K. This gives R2 a value of 16,045.

$$R2 = R_t - R1 = 21{,}645 - 5600 = 16{,}045 \text{ ohms}$$

The nearest standard value to this is 15 K.

Let's double-check the duty cycle with the selected standard values.

$$R1 : R1 + R2 =$$

$$5600 : 5600 + 15{,}000 =$$

$$5600 : 20{,}600 =$$

$$\frac{5{,}600}{5{,}600} : \frac{20{,}600}{5600} =$$

$$1 : 3.68$$

This is close to 1:4. The error is due to rounding off the resistor values. It should not matter too much in most practical applications.

The final double-check is the output frequency. I can expect some degree of error due to the rounding off of resistor values. For critical applications that require precision, low-tolerance components with values as close as possible to the calculated values should be used.

For my example, the actual output frequency works out to 1471 Hz.

$$F = \frac{1}{C1(R1 + R2)} = \frac{1}{0.000000033 \times (5600 + 15{,}000)}$$

$$F = \frac{1}{0.000000033 \times 20600} = \frac{1}{0.006798} = 1471 \text{ Hz}$$

That isn't too far off from our nominal design value of 1400 Hz.

TRIANGLE WAVE GENERATORS

Another common waveform is the triangle wave, shown in Fig. 6-14. Compare this waveform with the sine wave of Fig. 6-1.

This type of signal contains the same harmonic content as a square wave—the fundamental, and all odd harmonics. All harmonics that are multiples of two are missing. The difference between the triangle wave and the square wave is the relative amplitudes of the harmonics. In a square wave, the harmonics are quite strong when compared to the fundamental. In a triangle wave, the fundamental is very strong, but the harmonics are very weak. Their amplitude drops off sharply as the harmonics get further from the fundamental frequency.

Since a triangle wave is so similar to a square wave, it is not

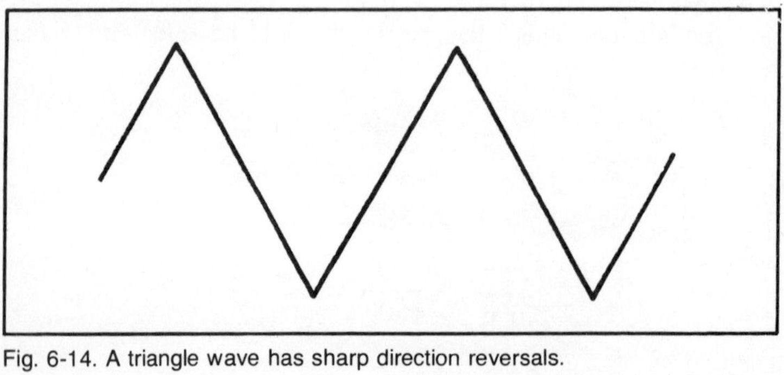

Fig. 6-14. A triangle wave has sharp direction reversals.

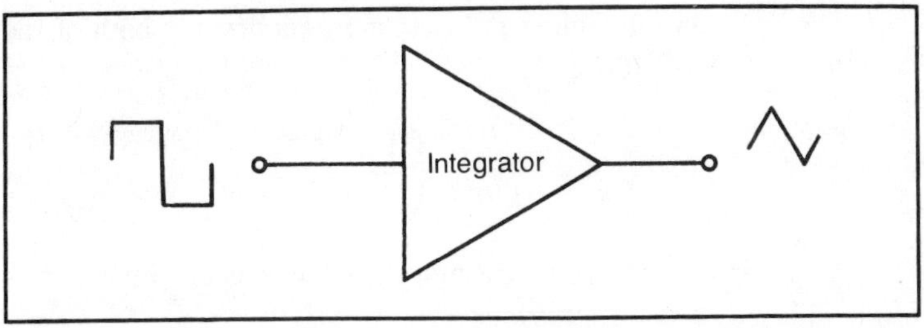

Fig. 6-15. A triangle wave can be created by passing a square wave through an integrator.

difficult to convert one into the other. Passing a square wave signal through an integrator (or low-pass filter) will produce a triangle wave signal at the output. This is illustrated in Fig. 6-15.

Most triangle wave generators built around op amps use this approach. A typical circuit is shown in Fig. 6-16.

Op amp A is wired as a comparator (discussed in a later chapter). It generates a square wave signal. This signal is passed through op amp B that is connected as an integrator. It converts the square wave into the triangle wave.

Notice that both waveforms can be tapped off from the circuit simultaneously. Two outputs are provided. Of course, both waveforms will have exactly the same frequency.

The square wave will swing back and forth between the op amp's positive and negative saturation voltage. The amplitude of the triangle wave is controlled by the ratio of R2/R1. These components also determine the output frequency, making adjustment a little tricky.

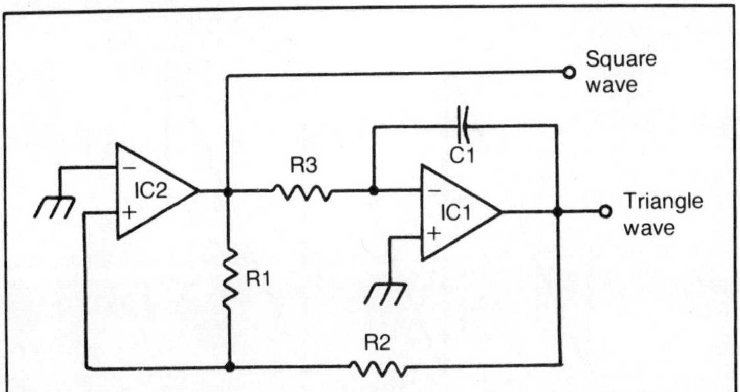

Fig. 6-16. Many triangle wave generator circuits also feature square wave outputs.

The formula for finding the output frequency for both of the waveforms is as follows:

$$F = \frac{1}{4R3C1}\left(\frac{R1}{R2}\right)$$

A circuit that can generate more than one basic waveform is often called a *function generator.*

SAWTOOTH WAVE GENERATORS

Back in Chapter 5 it was mentioned that feeding a constant dc voltage into an integrator produces a ramp-wave at the output. This waveform, which is also known as a sawtooth wave is shown in Fig. 6-17.

The simple integrator discussed in Chapter 5 needs to be manually reset to begin a new cycle. In a practical sawtooth wave generator, of course, there must be some means of resetting the output cycle at a regular rate corresponding to the desired output frequency.

The basic square wave/rectangle wave generators described earlier in this chapter can also generate a sawtooth wave by tapping off an output from the capacitor. The capacitor is charged on the positive portion of the cycle and discharged during the negative portion of the cycle, producing a modified sawtooth wave, as illustrated in Fig. 6-18. The circuit with the sawtooth-output is shown in Fig. 6-19.

The shorter the discharge time in comparison with the charge

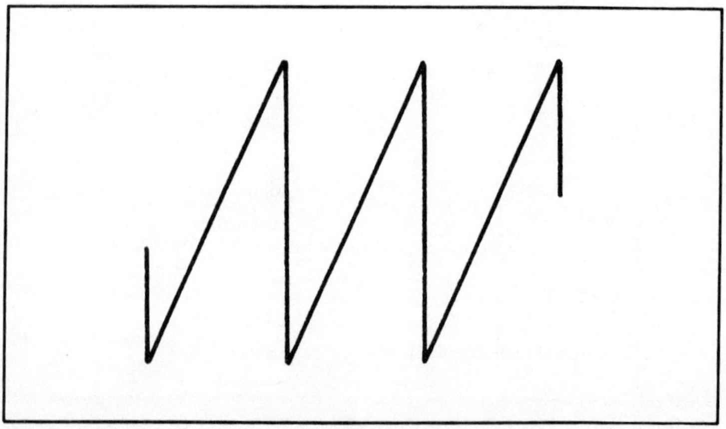

Fig. 6-17. This is an ascending sawtooth wave.

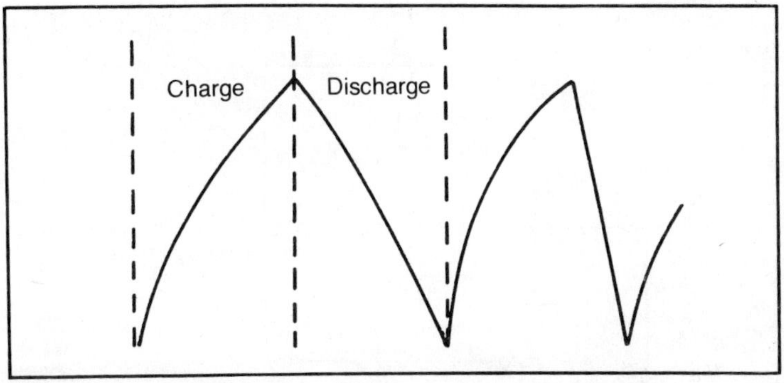

Fig. 6-18. This modified ascending sawtooth wave is generated by the circuit of Fig. 6-19.

time, the more linear the generated ramp signal will be. It is not really possible to achieve the maximum linearity of the signal shown in Fig. 6-17 with this circuit, but it is possible to come reasonably close.

The sawtooth wave generated in this manner will have the same frequency as the original rectangle wave output signal.

Sawtooth waves contain a great many strong harmonics. All integer harmonics (second, third, fourth, fifth, sixth, seventh, eighth, ninth, tenth, and so on) are present in a sawtooth wave. None are skipped like they are with the rectangle waves and triangle waves discussed earlier.

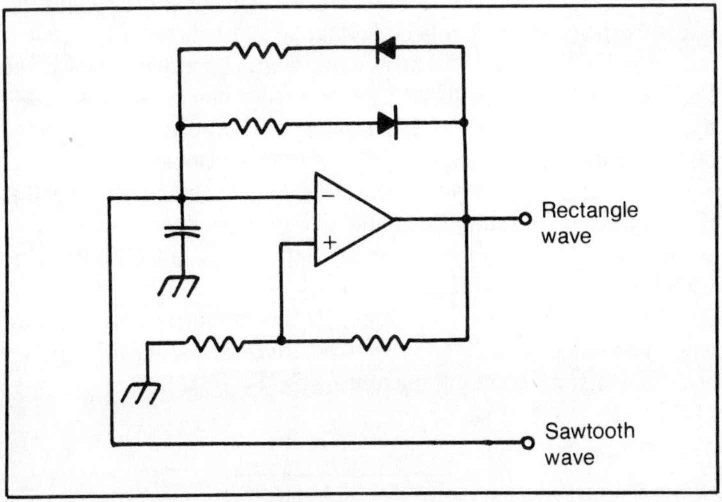

Fig. 6-19. A fair approximation of a sawtooth wave can be tapped off of the capacitor in a rectangle wave generator circuit.

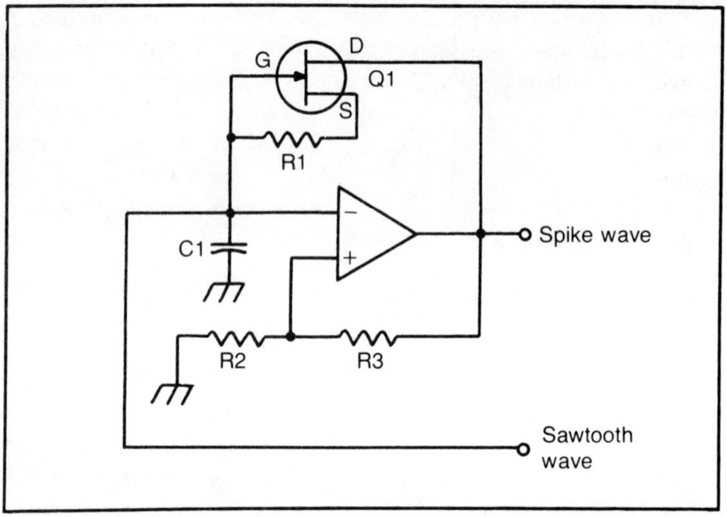

Fig. 6-20. Placing a FET in the feedback path produces better sawtooth wave signals.

An improved sawtooth wave generator circuit is shown in Fig. 6-20. This circuit produces a much more linear sawtooth signal than the circuit of Fig. 6-19.

The feedback resistors and diodes are replaced by Q1, a FET (Field-Effect Transistor). Resistor R1 biases the FET to pinch-off.

Capacitor C1 is charged linearly because the FET acts as a constant current source. When the voltage across the capacitor exceeds the positive feedback voltage at the noninverting input of the op amp, the device will switch to a negative output. The FET in the inverting-input feedback path now acts like a forward biased diode, allowing capacitor C1 to rapidly discharge and to switch the op amp back into its positive saturation condition.

This circuit also produces an odd negative spike wave signal at the output of the op amp. Each time the op amp switches to negative saturation a brief negative spike appears at its output. This waveform is shown in Fig. 6-21.

The output frequency formula for this circuit is a bit awkward since you must know the op amp's positive saturation voltage (V_s) and the constant pinch-off current of the FET (I_{ds}).

$$F = \frac{I_{ds}}{2V_s C1 \times 1 + \left(\frac{R3}{R2}\right)}$$

An even more linear sawtooth wave can be generated by a circuit containing more than a single op amp. A typical circuit is shown in Fig. 6-22. Unlike the other sawtooth generator circuits discussed in this section, this one is not a variation on the basic rectangle wave generator. It is strictly a sawtooth wave generator circuit.

For this circuit to function properly, the following component value relationships must be true:

$$R1 = R2$$

$$R3 = \frac{R1}{2}$$

$$R4 = R5$$

$$R6 = \frac{R4}{2}$$

$$R8 = R9$$

IC3 is a linear integrator. IC1 is a comparator, and IC2 is an inverter. The positive feedback loop through R1 around IC1 and IC2 latches these device until the signal from IC3 (through R2) fires the comparator.

When the output of IC2 is negative, it is fed into the inverting input of IC3 through R8. Diode D3 blocks the current flow through R7. This causes IC3 to integrate positively at a rate defined as t_1. This time constant is defined by the values of capacitor C1 and resistor R8.

$$t_1 = \frac{1}{C1R8}$$

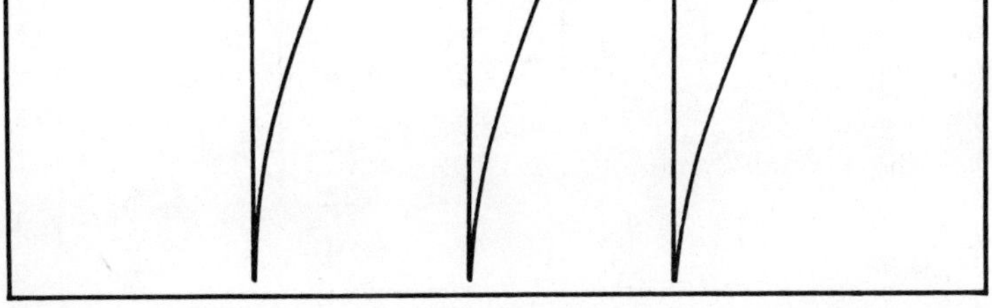

Fig. 6-21. The output of the op amp in the circuit of Fig. 6-20 is a train of negative spikes.

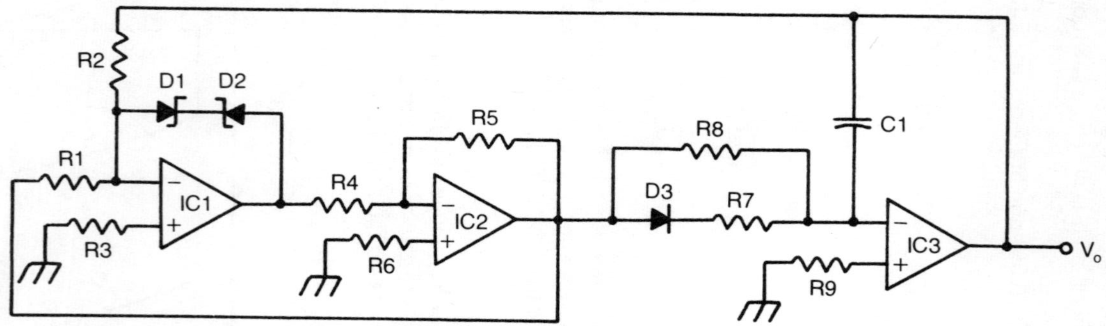

Fig. 6-22. Very clean sawtooth waves can be generated by this multiple op amp circuit.

When the output voltage equals the latching voltage set by zener diodes D1 and D2, the comparator (IC1) reverses its output state. This makes the output of IC2 positive. Diode D1 conducts, and the signal is fed into the inverting input of IC3 through the parallel combination of R7 and R8. This discharges capacitor C1 at a rate labelled t_2. This time constant is controlled by the values of capacitor C1 and resistors R7 and R8.

$$t_2 = \frac{1}{C1 \times \left(\frac{R7 \times R8}{R7 + R8}\right)}$$

If the value of R7 is much smaller than that of R8, the discharge rate can be made very steep, improving the linearity of the output waveform.

When the output voltage equals the inverse of the voltage set by the zener diodes ($V_o = -V_z$), comparator A returns to its original state, and the cycle repeats.

Obviously the complete cycle time equals the sum of the charging time (t_1) and the discharging time (t_2).

$$t_T = t_1 + t_2 = \frac{1}{C1R8} + \frac{1}{C1 \times \left(\frac{R7 \times R8}{R7 + R8}\right)}$$

The output frequency equals the reciprocal of the cycle time.

$$F = \frac{1}{t_T} = \frac{1}{\frac{1}{C1R8} + \frac{1}{C1 \times \left(\frac{R7 \times R8}{R7 + R8}\right)}}$$

$$F = \frac{C1}{\frac{1}{R8} + \frac{1}{\left(\frac{R7 \times R8}{R7 + R8}\right)}} = \frac{C1}{\frac{1}{R8} + \frac{R7 + R8}{R7 \times R8}}$$

$$F = \frac{C1R8}{\frac{(R7 + R8) + 1}{R7}} = \frac{C1R7R8}{2R7 + R8}$$

All of that may look terribly complicated, but I have shown how the formula was derived from earlier equations. In practical use, you will only have to work with the simplified final version of the equation.

$$F = \frac{C1R7R8}{2R7 + R8}$$

A rectangle wave signal made up of positive pulses can also be tapped off at the output of op amp B.

The main output of this circuit is called an ascending sawtooth wave because the voltage starts at a minimum level, then builds smoothly up to a maximum level, then quickly returns to the minimum level and starts over.

In some applications, I may need to reverse this. I might need a signal that starts at its maximum level, smoothly drops down to the minimum level, then quickly returns to the maximum level to begin the next cycle. This type of waveform is called a descending sawtooth wave, and is illustrated in Fig. 6-23.

One way to generate a descending sawtooth wave is to pass the output signal of an ascending sawtooth wave generator through an inverting voltage follower, as illustrated in Fig. 6-24.

The circuit of Fig. 6-25 offers an even simpler method. Descending sawtooth waves can be generated directly by reversing the polarity of diode D3, as shown in Fig. 6-25. This reverses time constants t_1 and t_2.

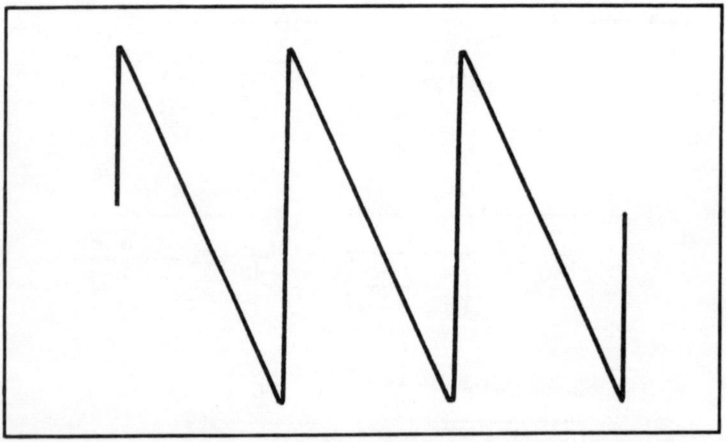

Fig. 6-23. A descending sawtooth wave is the mirror image of the ascending sawtooth wave shown in Fig. 6-17.

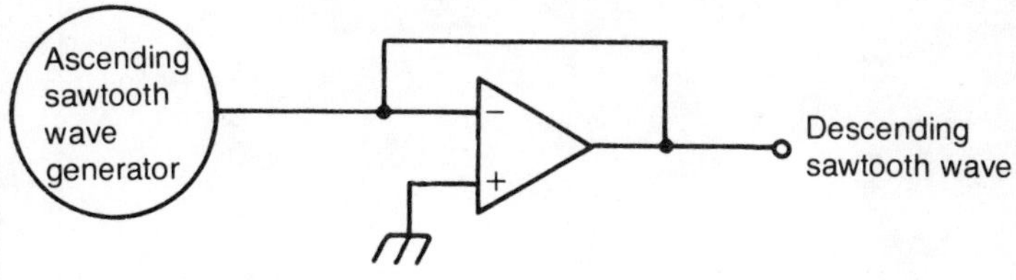

Fig. 6-24. A descending sawtooth wave may be derived from an ascending sawtooth wave signal by passing it through an inverting voltage follower.

Remember that to create a linear sawtooth wave with this circuit, resistor R7 must have a much smaller value than resistor R8. If R7 is much larger than R8, a signal that is similar to a triangle wave (discussed earlier in this chapter) will be generated. A number of unusual waveforms can be created by experimenting with the relative values of R7 and R8. Bear in mind that changing either or both of these resistances will have an effect on the output frequency.

STAIRCASE WAVE GENERATORS

A fairly novel waveform that can be created by an op amp is the staircase wave, which is shown in Fig. 6-26. The circuit illustrated in Fig. 6-27 can derive a staircase wave from a square wave.

Assuming capacitor C2 is fully discharged, the initial output will be at the positive saturation voltage. For each positive pulse received at the input, the output voltage will drop a step. The amplitude of each can be found with the following equation:

$$V_s = -\frac{C1 \times (V_i - 2V_d)}{C2}$$

V_s is the step voltage, V_i is the peak-to-peak voltage of the input pulse, and V_d is the voltage drop across each of the diodes. These diodes should be very closely matched so that $V_{D1} = V_{D2}$.

At some point the op amp will reach its negative saturation point. Of course, there can be no further steps down in output voltage after that.

The output level can be reset back to the original positive saturation voltage by momentarily closing switch S1, allowing C2 to completely discharge. In practical circuits, this switch will be re-

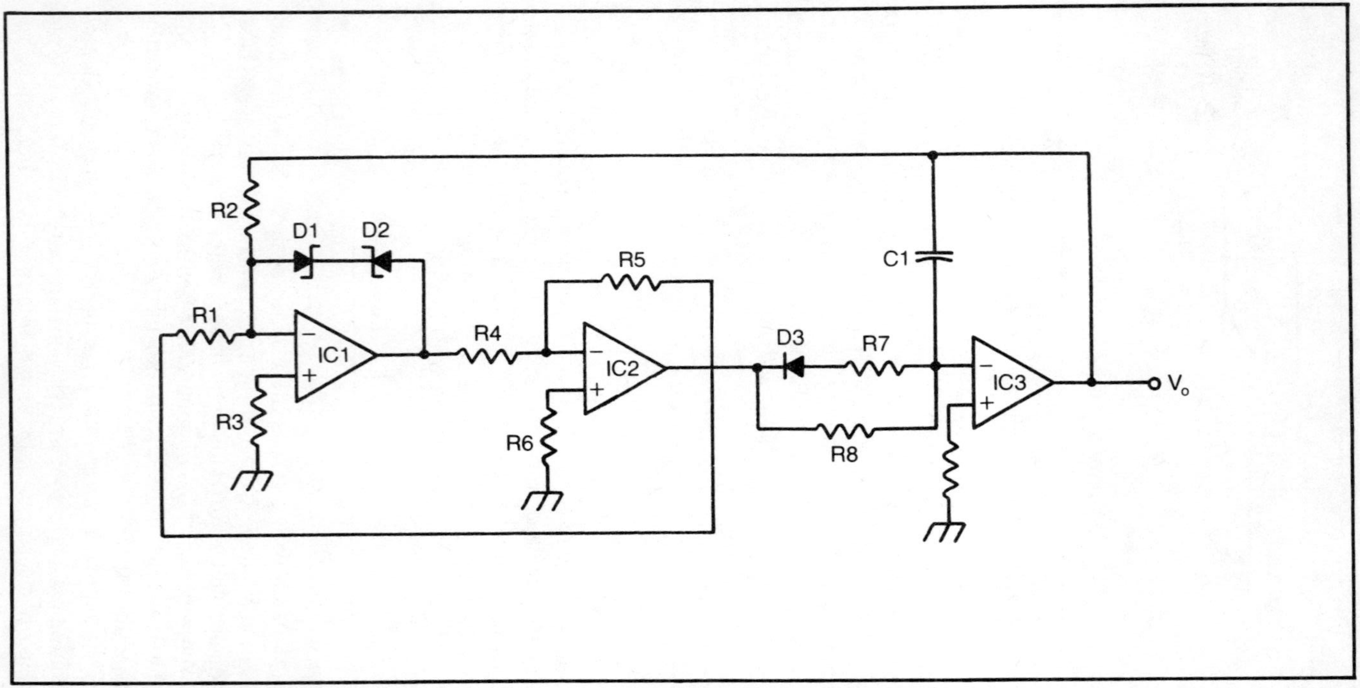

Fig. 6-25. The circuit of Fig. 6-22 can be modified to generate descending sawtooth waves directly.

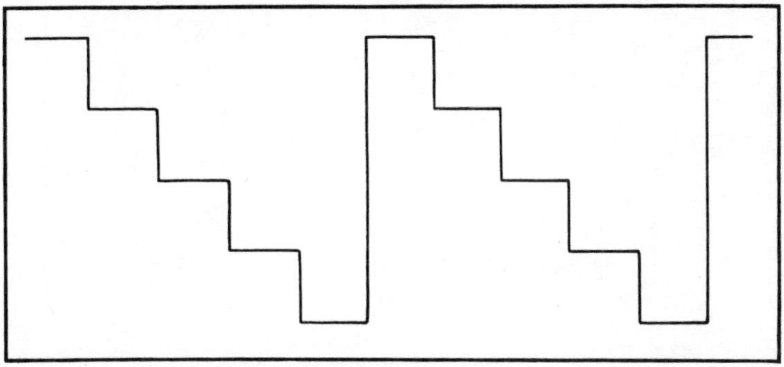

Fig. 6-26. Staircase waves are sometimes useful.

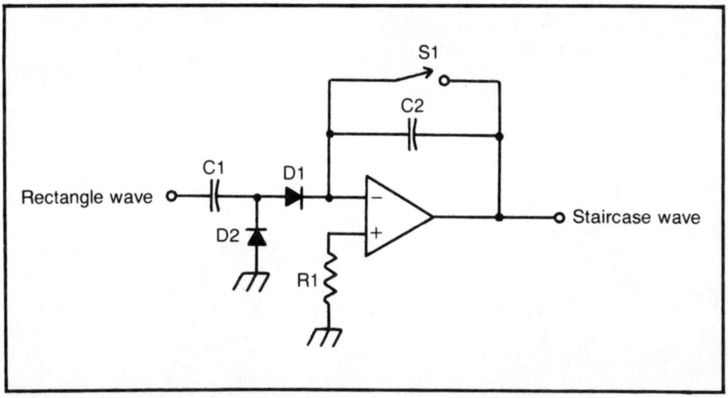

Fig. 6-27. This circuit will derive a staircase wave output from a rectangle wave input.

placed by circuitry that will automatically perform the switching operation at a rate equal to the frequency of the output signal.

The number of steps in the staircase wave will depend on the ratio of the input frequency to the reset frequency (output frequency). The input frequency must be higher than the output frequency. If the input frequency is four times the reset frequency, the output waveform will consist of four steps per cycle. If the input frequency is five times the reset frequency, the staircase wave at the output will be made up of five steps per cycle. If the input frequency is not an exact integer multiple of the reset frequency, the output waveform may not always contain the same number of steps. The first and last steps in the cycle may be shorter in duration than the other steps.

7

Pulse Circuits

In the last chapter, I concentrated a lot on rectangle wave generator circuits. These circuits are so important they have a special name—multivibrators.

A *multivibrator* has two output states and (ideally) switches instantly between its maximum and minimum output voltage conditions. For op amp circuits this is usually the positive and negative saturation voltages for the op amp device being used.

A rectangle wave generator as discussed in the last chapter belongs to one of the three major classes of multivibrators. Specifically, these circuits are known as *astable* multivibrators. The other two classes are *bistable* multivibrators and *monostable* multivibrators. These will be discussed shortly.

In an astable multivibrator, neither of the output states is stable. They continuously switch back and forth. In a bistable multivibrator, either of the output states may be stable and can be held indefinitely. A monostable multivibrator has only one stable state.

BISTABLE MULTIVIBRATORS

A bistable multivibrator has two stable output states. In a sense, it can be called a circuit with a memory. It remembers its previous output state until an input signal called a trigger pulse or (sometimes) simply a trigger, tell it to change.

Bistable multivibrators are also known as flip-flops. One trigger pulse causes the output to flip to its maximum voltage, and the next trigger pulse tells the output to flop back to its minimum voltage. Ideally the switching should be instantaneous, but practical op amps have finite slew rates. In most cases, the switching will be so rapid that the difference won't really be noticeable.

There are two primary ways to feed the trigger pulse into a bistable multivibrator circuit. The input signal may be dc coupled (direct coupled), or ac coupled (capacitively coupled).

Dc-Coupled Bistables

A typical dc-coupled bistable multivibrator circuit is shown in Fig. 7-1. Notice that the trigger pulse signal is fed directly into the noninverting input of the op amp through resistor R1.

Resistor R3 provides a positive feedback path to the noninverting input. Typical values for R1 and R2 are about 10 K, and R3 would typically be about 100 K.

Because of the positive feedback loop, if the output voltage deviates even slightly from true 0 (as it almost invariably will with practical devices), the deviation will be amplified repeatedly until the op amp reaches its saturation voltage. The polarity will depend on the polarity of the original deviation voltage.

Let's assume that the op amp starts out in positive saturation. If a positive trigger pulse is fed into the input, nothing will happen. The output will remain in positive saturation. If the trigger pulse is negative, however, the noninverting input will go negative, driving the output into negative saturation. This state will be held even when the trigger pulse is removed.

If a second negative trigger pulse is fed into the circuit, there will be no change in the output signal. The op amp will remain in negative saturation.

A positive trigger pulse at the input, will reverse the output again, returning the output to the positive saturation level.

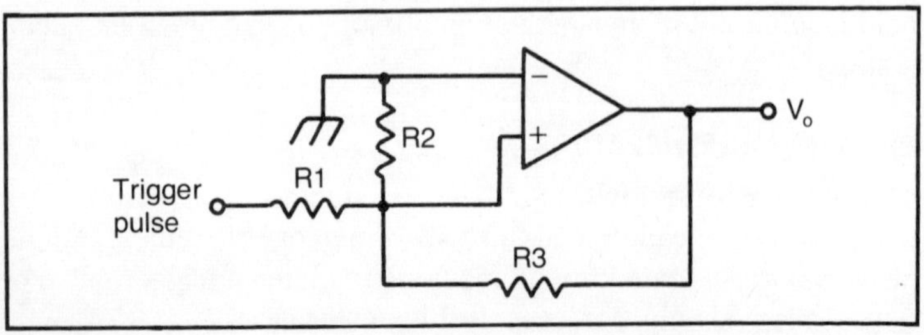

Fig. 7-1. A dc-coupled bistable multivibrator.

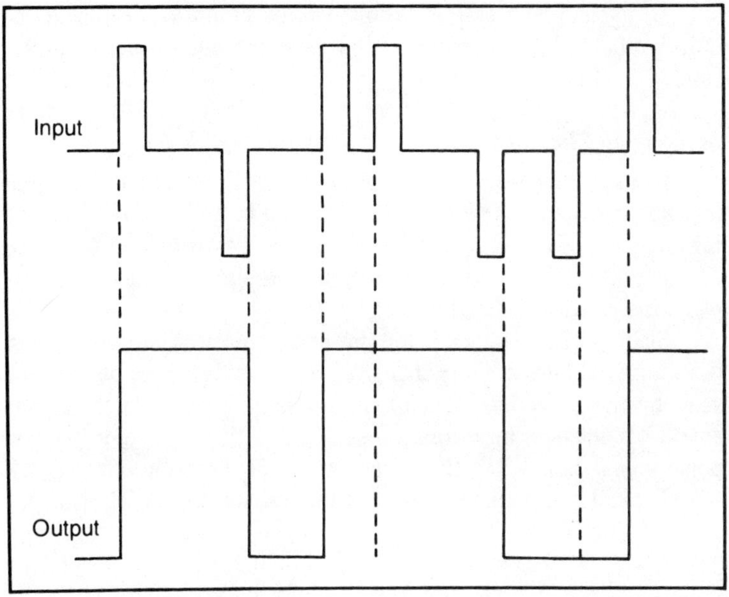

Fig. 7-2. The action of the circuit in Fig. 7-1 is shown by these typical input and output signals.

No matter what state the output is in, that state will be held constant until the opposite trigger pulse is applied to the input. Some typical input and output signals for this circuit are illustrated in Fig. 7-2.

A variation on the basic dc-coupled bistable multivibrator circuit is given in Fig. 7-3. The zener diodes limit output voltage swings below the absolute saturation voltages.

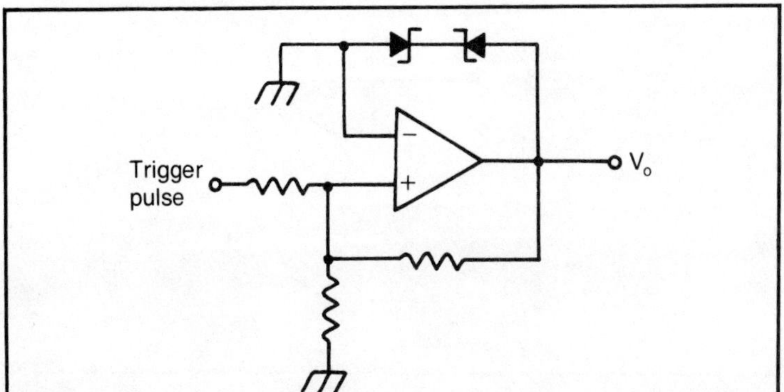

Fig. 7-3. This variation of the circuit of Fig. 7-1 offers improved performance.

A dc-coupled bistable circuit's major limitation is that it requires different types of signals to change the output in each direction.

Ac-Coupled Bistables

In an ac-coupled bistable multivibrator circuit, the input signal is passed through a capacitor (C1) before it is applied to the inverting input of the op amp. Positive feedback is returned from the output to the noninverting input through resistor R3 and capacitor C2. This circuit is shown in Fig. 7-4.

Since the input signal is fed through a capacitor, any continuing dc voltage is blocked. Only changes in the signal (trigger pulses) are seen by the inverting input of the op amp. While this might not sound particularly important, the result is that the same trigger pulse signal can switch the output in either direction.

Typical component values for this circuit are as follows:

R1 = 1 K	C1 = 0.1 μF
R2 = 100 K	C2 = 0.1 μF
R3 = 1 MΩ	

Resistors R2 and R3 form a voltage divider to feed back a portion of the output signal to the noninverting input.

Let's begin examining the action of the circuit in Fig. 7-4 by assuming that the op amp is in positive saturation. It will remain in positive saturation until a positive trigger pulse is received at the input. With the component values specified, the input pulse should be at least +2 volts.

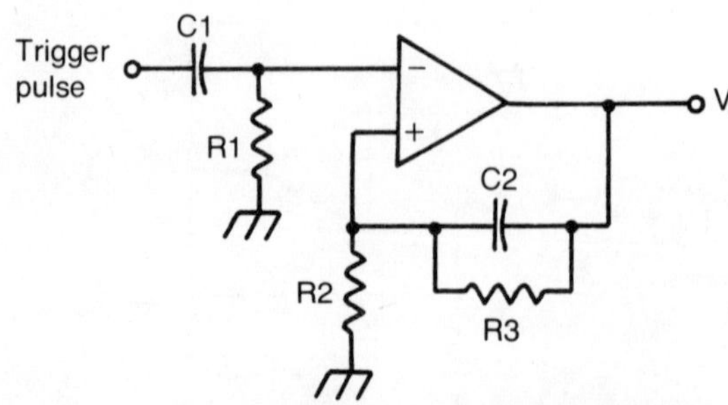

Fig. 7-4. An ac-coupled bistable multivibrator requires only single-polarity input trigger pulses.

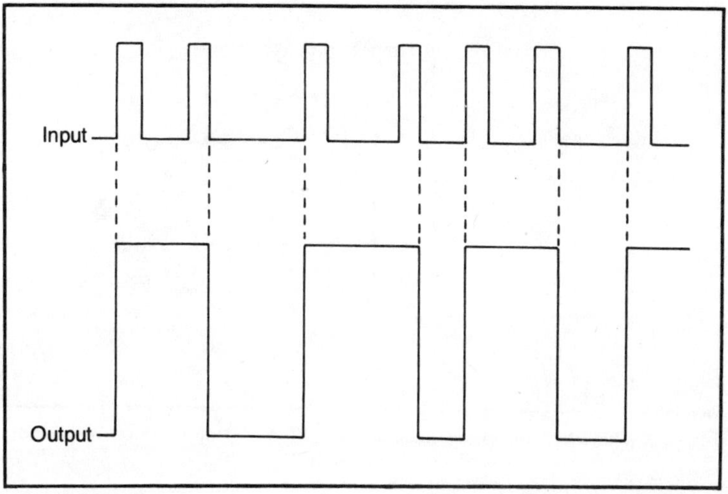

Fig. 7-5. Typical input and output signals for the ac-coupled bistable multivibrator in Fig. 7-4 are shown here.

The leading edge of this trigger pulse, coupled through C1, makes the inverting input of the op amp more positive than the noninverting input. This causes the circuit to swing into negative saturation.

Again this condition will be held until another positive trigger pulse is fed into the input. The trailing edge of this pulse makes the inverting input more negative (less positive) than the noninverting input. This forces the op amp back into positive saturation until the leading edge of the next trigger pulse.

Typical input and output signals for an ac-coupled bistable multivibrator are shown in Fig. 7-5.

Capacitor C2 is included in the positive feedback path to introduce a slight delay in the circuit's response. If this capacitor was not included, I could have the following situation:

> The circuit is in positive saturation. The leading edge of a trigger pulse changes the output to negative saturation. The trailing edge of *the same input trigger pulse* returns the circuit to the original positive saturation state.

The 0.1 μF capacitor will be sufficient for trigger pulses up to about 50 milliseconds. This corresponds to an input frequency of 20 Hz. For longer trigger pulses, you will need a larger capacitance.

A variation on this basic circuit is illustrated in Fig. 7-6. Again

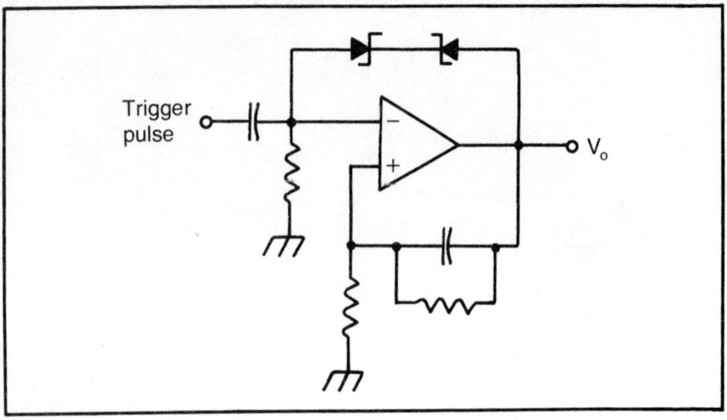

Fig. 7-6. This variation of the basic circuit of Fig. 7-4 offers improved performance.

the zener diodes in the negative feedback path limit the output voltage swings.

One popular application for an ac-coupled bistable multivibrator is for frequency division. If a constant frequency square wave is fed into the input, the output will be a square wave with exactly one half of the original input signal. Other typical applications include electronic switching and simple data storage (memory).

MONOSTABLE MULTIVIBRATORS

A monostable multivibrator has only one stable output state. The output is normally in one of the saturation voltage states. When a trigger pulse is fed into the circuit input, the output goes to the opposite saturation voltage state for a specific period of time, then it returns to the original state.

This type of circuit is also known as a *one-shot* and as a pulse stretcher. Typical input and output signals are illustrated in Fig.

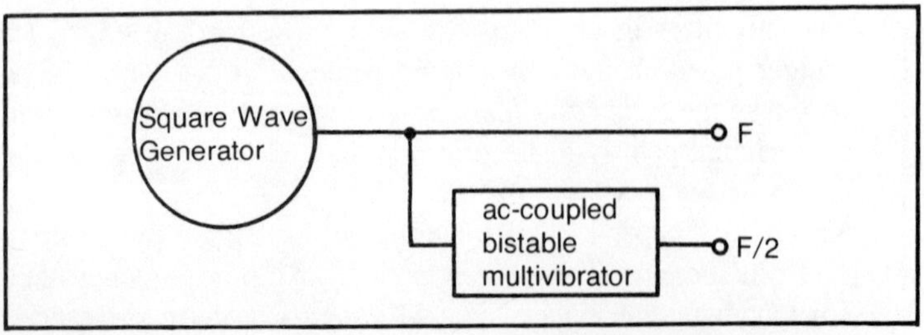

Fig. 7-7. An ac-coupled bistable multivibrator can be used for frequency division.

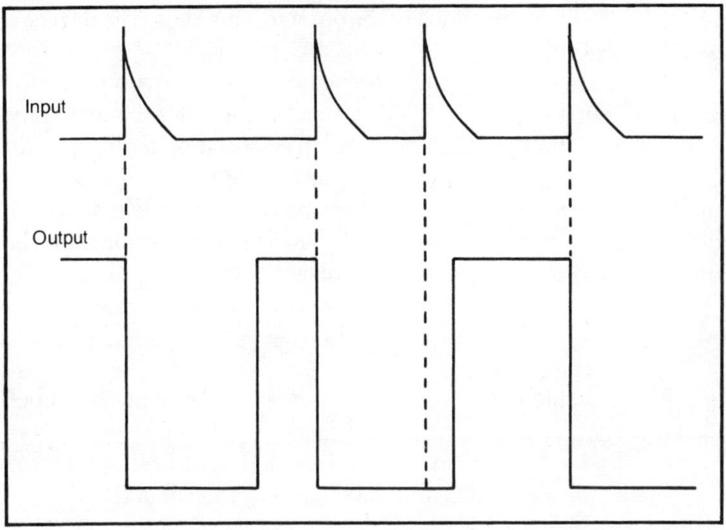

Fig. 7-8. A monostable multivibrator produces a single, fixed-length output pulse for each input trigger pulse.

7-8. The output pulse is the same length, regardless of the length of the trigger pulse. If a second trigger pulse is applied before the output pulse is finished, it will be ignored in most monostable multivibrator circuits. In a few circuits, a second trigger pulse will extend the length of the output pulse.

A simple monostable multivibrator circuit is shown in Fig. 7-9. Resistors R2 and R3 form a voltage divider that applies a negative

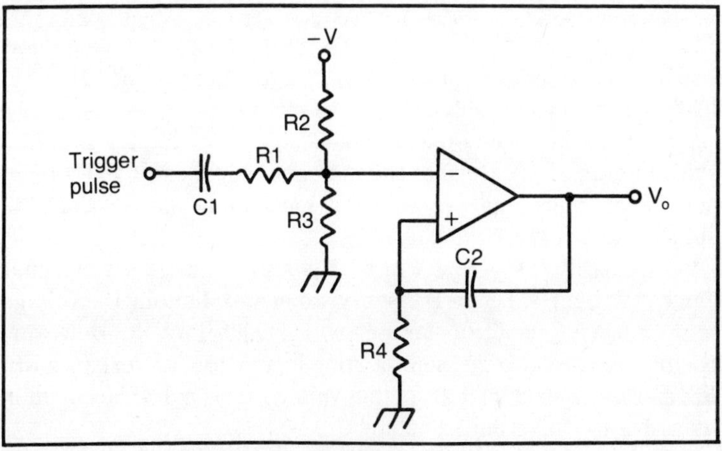

Fig. 7-9. A simple monostable multivibrator.

voltage to the inverting input of the op amp. This keeps the device in positive saturation.

A positive trigger pulse fed into the input overcomes the negative voltage at the inverting input. Since this input is now positive, the output swings into negative saturation until capacitor C2 charges enough to make the noninverting input more positive than the inverting input, returning the op amp to its positive saturation state. The time it takes for this to happen is determined by the time constant of capacitor C2 and resistor R4.

$$t = C2R4$$

For example, a 0.1 μF capacitor and a 1 megohm (1,000,000 ohms) resistor will give an output pulse of 1 second.

$$t = 0.0000001 \times 1{,}000{,}000 = 1 \text{ second}$$

For longer output pulses, increase the value of either or both of these components. For shorter output pulses, decrease either or both component values. By using a potentiometer for R4, you can manually adjust the length of the output pulse.

Typical values for the other components in this circuit are as follows:

R1 = 10 K
R2 = 22 K
R3 = 1 K
C1 = 0.1 μF

For these component values the trigger pulse will need to be at least +2 volts to reliably trigger the circuit.

A more advanced op amp monostable multivibrator circuit is shown in Fig. 7-10. Despite the increased complexity and component count, this circuit works pretty much in the same way as the simpler design shown in Fig. 7-9.

Ordinarily the output will be a positive voltage set by zener diodes D3 and D4. Diode D1 is forward biased, limiting the voltage at the inverting input of the op amp to about 0.6 volt. However, positive feedback to the noninverting input (through resistors R5, R3, and R4) is about half the output voltage. This holds the circuit in its stable (positive output) state.

A negative trigger pulse at the input will temporarily exceed

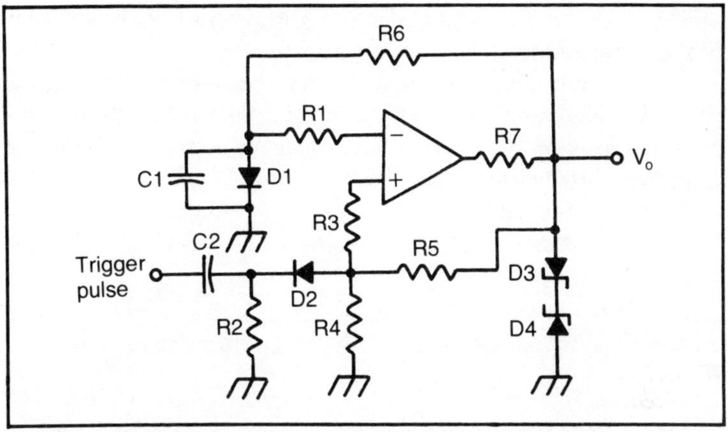

Fig. 7-10. A more advanced monostable multivibrator circuit.

the positive feedback voltage, and the op amp will switch to a negative output. The negative output voltage is also controlled by the zener diodes (D3 and D4).

The negative output condition is held for awhile because the negative voltage is fed back to the noninverting input. Meanwhile, capacitor C1 is being charged with a negative voltage. At some point the negative voltage at the inverting input will be greater than the negative voltage at the noninverting input. At this point the circuit will revert to its positive output (stable) state.

Positive trigger pulse inputs are ignored by this circuit.

The length of the output pulse is determined primarily by resistor R6 and capacitor C1, but the amount of positive feedback (set by resistors R4 and R5) also has an influence on the output pulse time. For most designs, matters are greatly simplified by making R4 equal to R5. This will make the positive feedback signal to the noninverting input equal to half the output voltage. In this case, the time constant equation looks like this:

$$t = 0.7R6C1$$

For example, let's say I have a 0.22 μF capacitor for C1 and a 1 megohm resistor for R6. In this instance, the output pulse will have a length of .154 seconds.

$$t = 0.7 \times 1{,}000{,}000 \times 0.00000022 = 0.154 \text{ second}$$

Once again, for longer output pulses increase the value of

either or both of these components. For shorter times, one or both of the component values should be decreased.

For a practical design, I will know the desired output pulse time, and will need to select appropriate component values. It's generally easiest to select a reasonable value for C1, and then rearrange the formula to solve for resistor R6.

$$R6 = \frac{t}{0.7C1}$$

Let's say I want an output pulse that is 0.25 second long. I will try a 0.47 μF capacitor for C1. This means R6 will need a value of 750 K.

$$R6 = \frac{0.25}{0.7 \times 0.00000047} = \frac{0.25}{0.000000329}$$

$$R6 = 759{,}878 \text{ ohms} \cong 750 \text{ K}$$

Figure 7-11 shows a more advanced monostable multivibrator circuit using two op amps.

PULSE MODULATORS

The monostable multivibrator is sometimes called a pulse

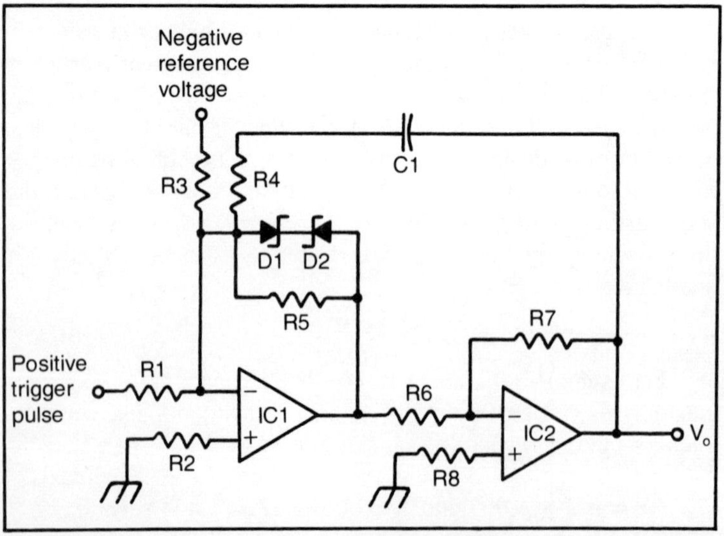

Fig. 7-11. An improved monostable multivibrator circuit can be built around two op amps.

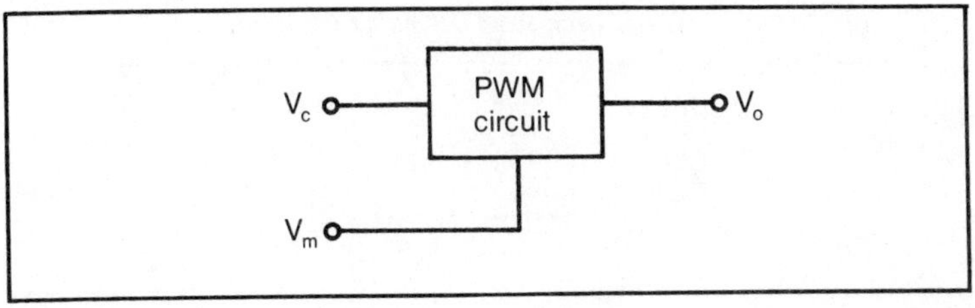

Fig. 7-12. A PWM circuit requires two input signals.

stretcher because a short input pulse is turned into a long output pulse. Other circuits that manipulate, or modulate rectangle waves, or trigger pulses, can also be built around op amps.

Pulse width modulation (PWM) is used in a number of data transmission and recording applications. Two input signals are required. V_c is the carrier. This is the string of square waves to be modulated. The other input signal is the modulating signal or V_m. It is sometimes called the program. A block diagram is shown in Fig. 7-12.

The instantaneous voltage of the V_m signal at a given instant will determine the width of the pulse at the output. The output is a string of rectangle waves with varying duty cycles, as shown in Fig. 7-13. The duty cycle varies in step with the V_m signal.

A typical pulse width modulator circuit is illustrated in Fig. 7-14. Typical component values are given in Table 7-1.

Once a PWM signal has been transmitted or stored, the receiver will need to demodulate the signal to recover the original program signal (V_m).

Since the amplitude of the pulses remains constant, this isn't very difficult. Simple low-pass filtering will produce an output voltage that is proportional to the pulse width. Since the pulse width corresponds to the program signal, the output of the low-pass filter will resemble the signal fed into the V_m input of the pulse width modulator.

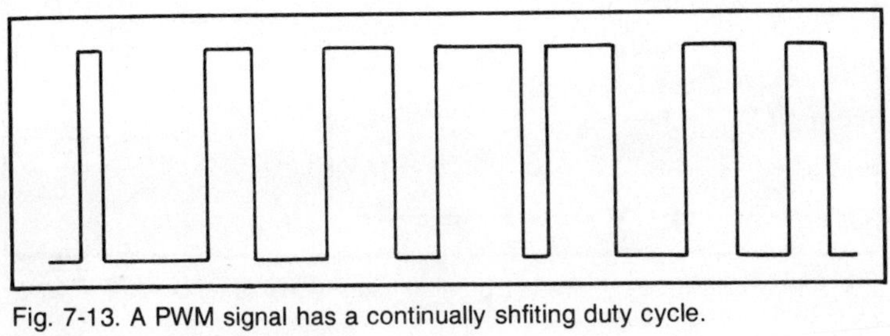

Fig. 7-13. A PWM signal has a continually shfiting duty cycle.

Table 7-1. Component Values for Fig. 7-14.

R1	27 K
R2	1 megohm
R3	10 K
R4	10 K
C1	0.01 μF
D1, D2	6.8 volt zener diode

Another form of pulse modulation is illustrated in Fig. 7-15. In this case, the program signal (V_m) controls the amplitude of the output pulses.

A typical pulse amplitude modulator (PAM) circuit is shown in Fig. 7-16. In this circuit, the program signal (V_m) must be kept negative, since it acts as the collector voltage supply for the transistor gate (Q1). A positive signal here would turn the transistor off.

The square wave signal (V_c) switches the transistor on and off. The program signal is alternately fed to the inverting input of the op amp through R3 and R4, or through R3 and the collector-to-emitter path of Q1 to ground. The program signal reaches the op amp only when Q1 is cut off by the square wave (V_c) input signal. The op amp acts as a unity-gain amplifier. The output will be a string of pulses with instantaneous amplitude levels corresponding to the program (V_m) signal.

The resistor values should conform to the following relationships:

$$R3 = R4$$

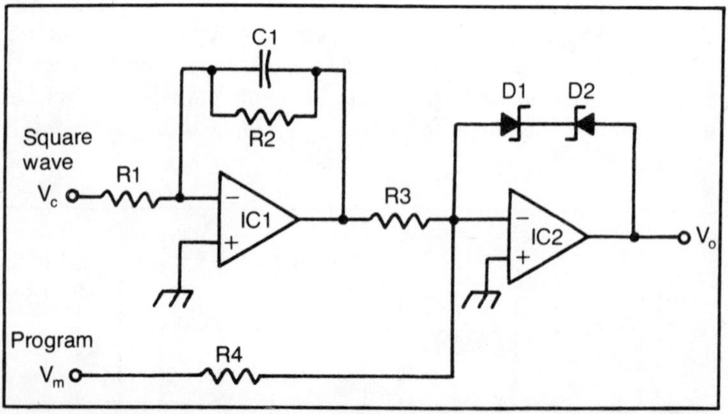

Fig. 7-14. This circuit will use the program input signal to modulate the pulse width of the square wave input signal.

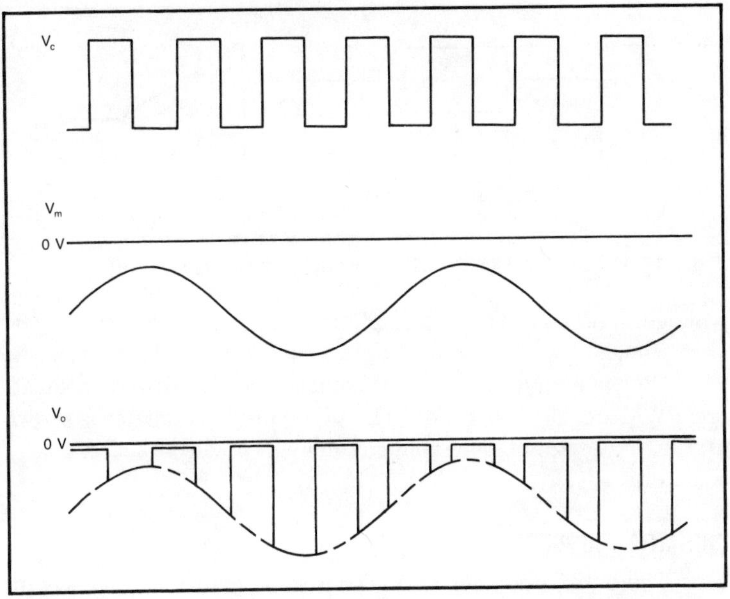

Fig. 7-15. In PAM, the amplitude of the output signal is modulated by the program signal.

$$R6 = 2\,R3$$
$$R6 = R7$$
$$R5 = \frac{2R3}{3}$$

Potentiometer R9 is adjusted to provide the desired reference

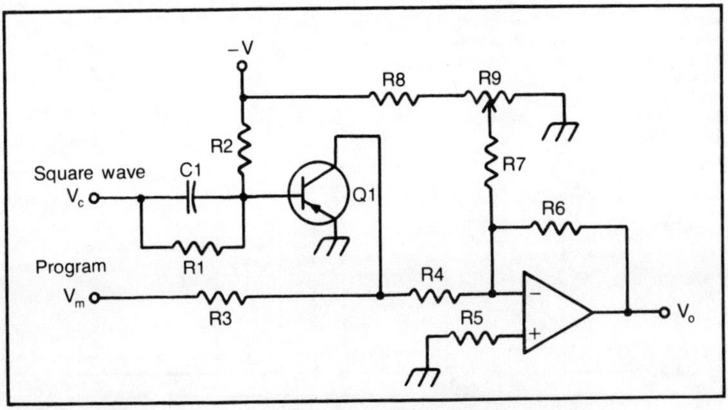

Fig. 7-16. This circuit will perform pulse amplitude modulation.

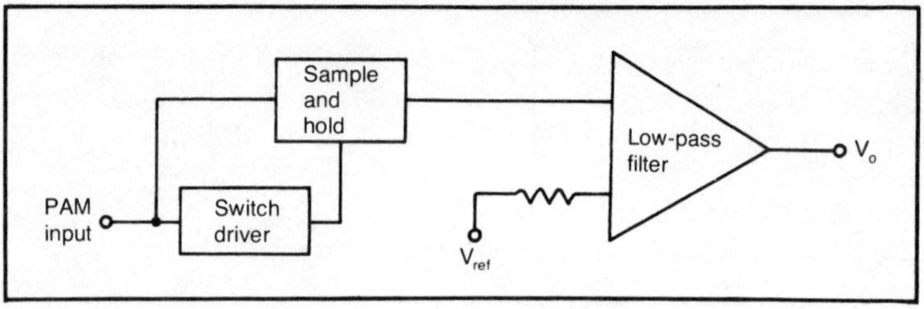

Fig. 7-17. The program signal of a PAM signal may be recovered in this manner.

voltage. It acts as an overall amplitude control for the output signal as a whole.

Demodulating PAM signals is more complex than demodulating PWM signals. A block diagram of a PAM demodulator is shown in Fig. 7-17. Sample-and-hold circuits will be discussed in a later chapter.

SCHMITT TRIGGERS

There are a number of other op amp circuits that work with pulse type signals. These are related to, but not equivalent to the multivibrator circuits described earlier.

Schmitt triggers are used to create clean rectangle wave (pulse) signals from any input signal. This is especially useful in data transmission applications where noise can create confusion and/or false triggering.

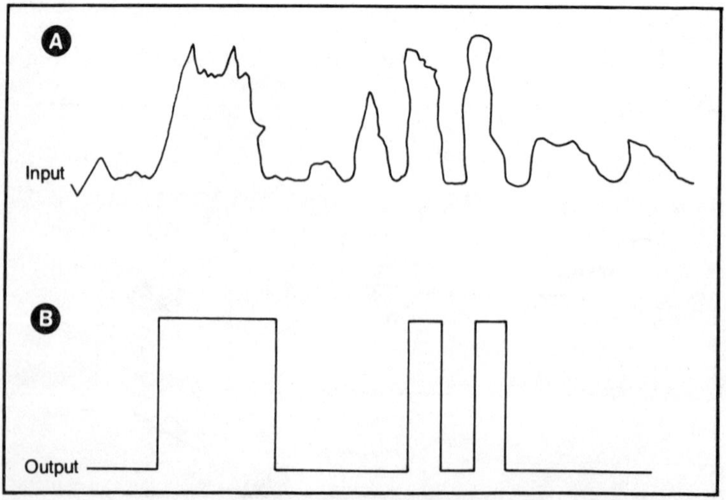

Fig. 7-18. Often the desired pulse-wave signals are buried in noise.

Look at the noisy input signal in Fig. 7-18A. It is not easy to determine just where the desired pulses are located. A circuit being triggered by these pulses might also be confused by the noisy signal. It may not trigger when it is supposed to, or it may be falsely triggered by a noise spike where no trigger pulse was intended.

A Schmitt trigger circuit can clean up most noisy signal problems and recover the desired trigger pulses, as shown in Fig. 7-18B.

The simple circuit shown in Fig. 7-19 is not a true Schmitt trigger, but it will help to illustrate the basic principle involved. Actually this circuit is a form of comparator. Comparators will be discussed in a later chapter.

There is no feedback in this circuit. The op amp operates at its full open-loop gain. For an ideal op amp, the open-loop gain is infinite. Since the output voltage is limited by the saturation voltage levels of the device, I can assume near infinite gain with clipping.

The inverting input is grounded. The input voltage here is zero. Remember that the output of an op amp is equal to the difference between the noninverting and inverting input signals multiplied by the gain.

$$V_o = (V_1 - V_2)G$$

V_o is the output voltage, V_1 is the noninverting input voltage, V_2 is the inverting input voltage, and G is the gain.

In this circuit V_2 is zero, and G is essentially infinity. Therefore, the equation can be simplified.

$$V_o = V_1 \times \infty$$

This means there are three possible outputs. If V_1 is zero, the

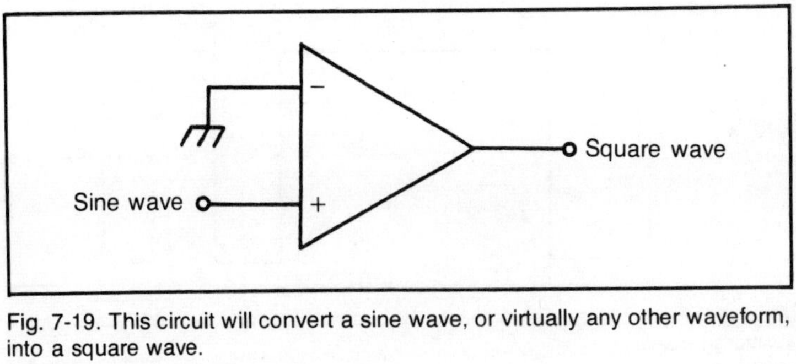

Fig. 7-19. This circuit will convert a sine wave, or virtually any other waveform, into a square wave.

output voltage (V_o) will be zero. If V_1 is even slightly greater than zero (that is, positive), the output will theoretically be an infinitely large positive voltage. With practical devices, the actual output will be the positive saturation voltage of the op amp.

Finally, if V_1 is even slightly less than zero (that is, negative), the theoretical output will be an infinitely large negative voltage. The practical output will be the op amp's negative saturation voltage. These are the only possible input and output conditions.

Figure 7-20 shows how this circuit will work with random noise as the input.

This circuit is most useful in converting other waveforms into rectangle waves. For instance, if the input is a sine wave, as illustrated in Fig. 7-21, the output will be a square wave. The circuit of Fig. 7-19 is often referred to as a sine wave to square wave converter.

A true Schmitt trigger circuit is designed to apply a trigger pulse to another circuit when the input signal reaches a specific value. A basic Schmitt trigger circuit is illustrated in Fig. 7-22.

Most practical Schmitt triggers feature controlled *hysteresis*. This is basically a lag, or grey area between the cuton and cutoff points. Signals within the hysteresis region are ignored by the

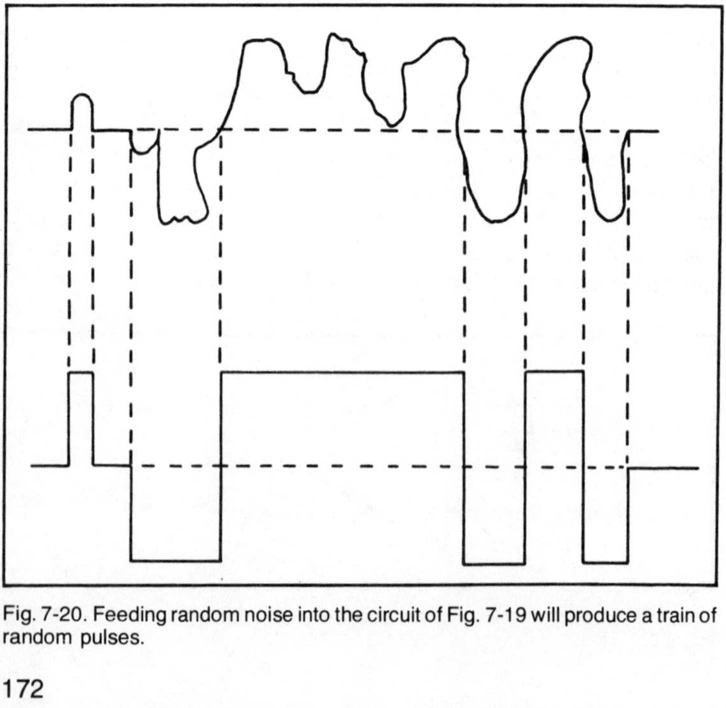

Fig. 7-20. Feeding random noise into the circuit of Fig. 7-19 will produce a train of random pulses.

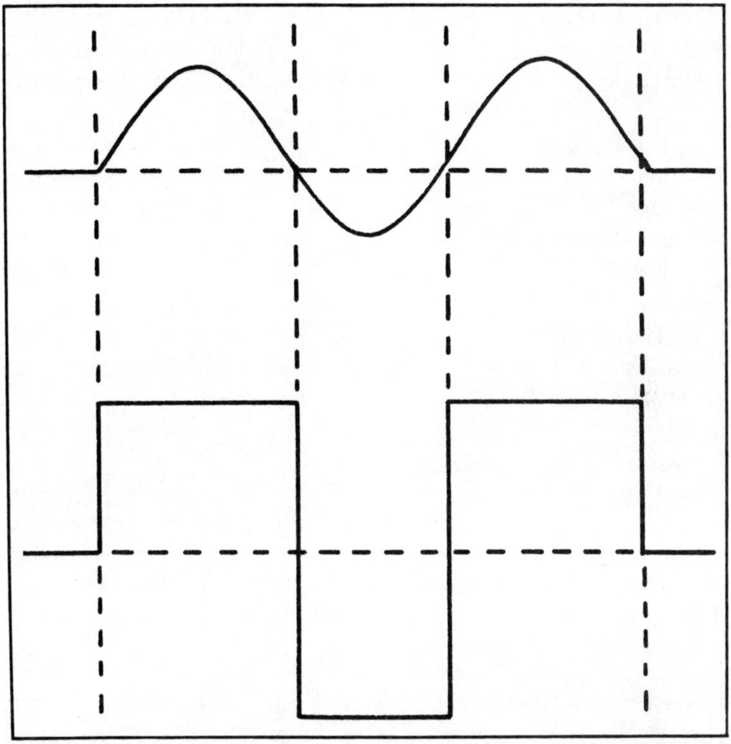

Fig. 7-21. The circuit of Fig. 7-19 converts a sine wave into a square wave with the same frequency.

Schmitt trigger circuit. This makes the circuit more reliable, and less prone to false triggering due to noise spikes. This is illustrated in Fig. 7-23. The output is switched on by a higher voltage than is used to switch it back off. Minor fluctuations in the input signal do not affect the output signal.

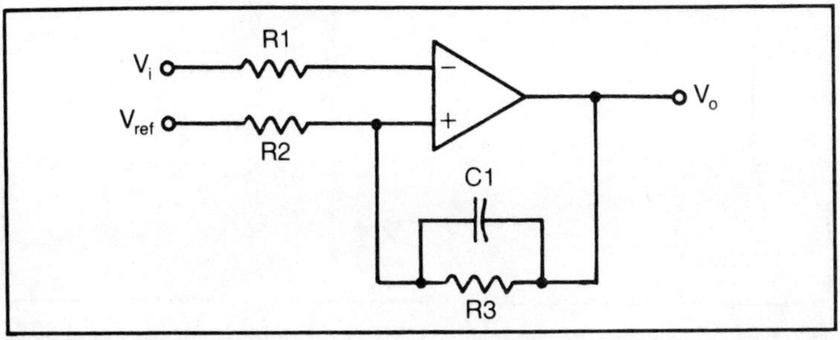

Fig. 7-22. This is a simple Schmitt trigger circuit.

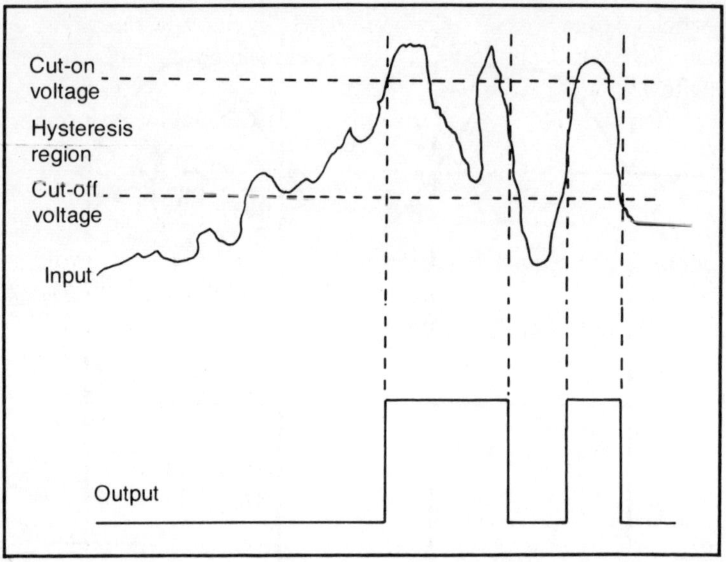

Fig. 7-23. The action of a Schmitt trigger on a random noise input is illustrated here.

Returning to the circuit of Fig. 7-22, I can see that positive feedback is applied to the noninverting input of the op amp through resistor R3. This is done to speed up the switching rate. This circuit can respond very rapidly to changes in the input signal. The

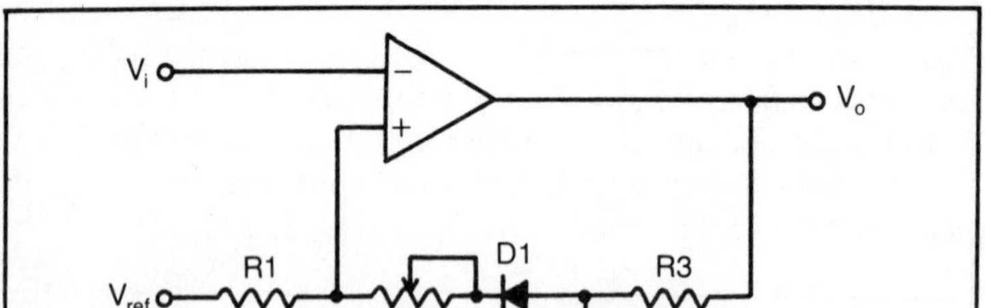

Fig. 7-24. In this Schmitt trigger circuit, the amount of hysteresis is manually controllable.

switching time is essentially limited only by the slew rate of the op amp IC itself. Capacitor C1 may be eliminated if extremely fast switching is not essential.

The switch-on voltage can be found with the following formula:

$$V_{on} = V_r + (\overline{V}_{ps} - V_r) \times \left(\frac{R2}{R3 + R2} \right)$$

where V_{on} is the switch-on voltage level, V_r is the reference voltage fed to the noninverting input of the op amp through R2, and V_{ps} is the positive saturation voltage of the op amp.

For determining the switch-off voltage level, the equation becomes the following:

$$V_{off} = V_r - (V_{ns} - V_r) \times \left(\frac{R2}{R3 + R2} \right)$$

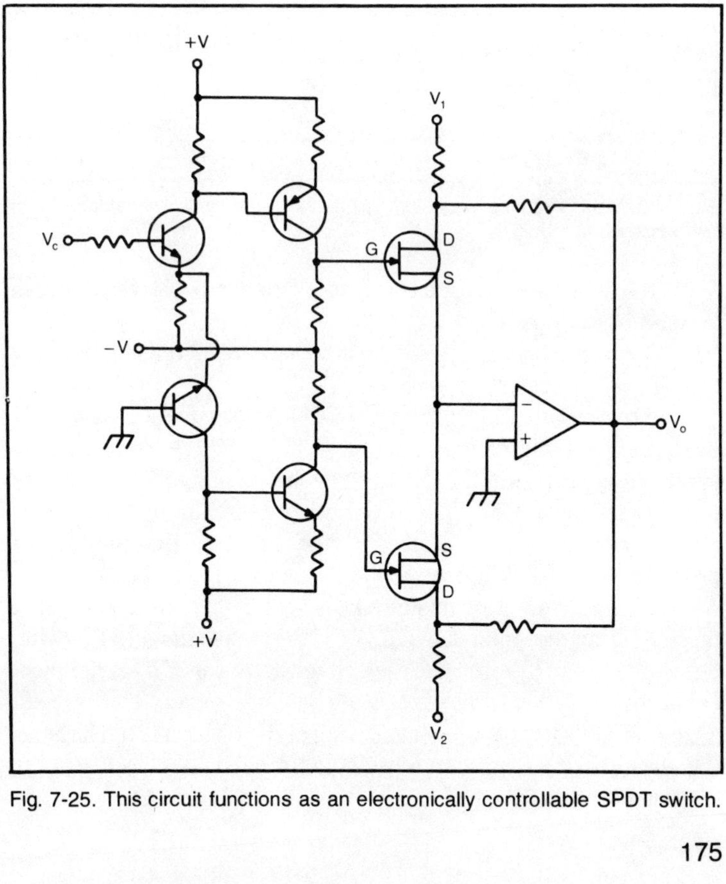

Fig. 7-25. This circuit functions as an electronically controllable SPDT switch.

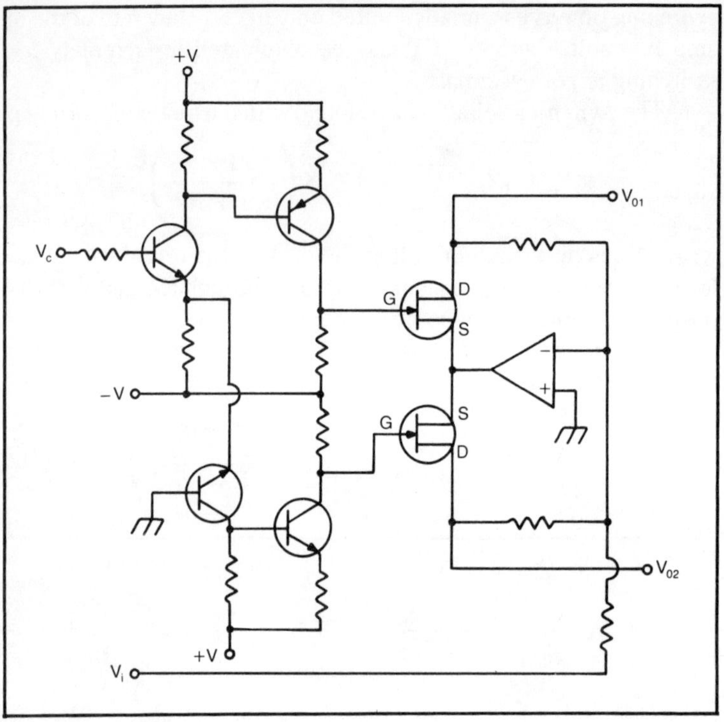

Fig. 7-26. In this circuit, a single input signal is electronically switched between two separate outputs.

V_{off} is the switch-off voltage level, and V_{ns} is the negative saturation voltage of the op amp.

The hysteresis region can be increased by increasing the value of R3.

In the circuit shown in Fig. 7-24, it is possible to change the hysteresis without changing the switching voltage level. In some applications, the switch-on or the switch-off voltage level will need to be precisely set by V_r, and the other switching level is determined by the amount of desired hysteresis. Adjusting the value of potentiometer R2 will change the hysteresis of the circuit.

A fringe benefit of this circuit is that a second output can be tapped off at the point labeled A. This output signal will switch between ground (0 volts) and the voltage determined by zener diode D3 in step with the action of the main Schmitt trigger. This is useful when the full saturation voltage swing of the op amp is too large for the circuit being triggered by the output pulses.

ELECTRONIC SWITCHES

An op amp can also be used as an electronic switch. In a sense, all of the circuits in this chapter are essentially true of the bistable multivibrators and Schmitt triggers.

The circuit shown in Fig. 7-25 can switch either of two input signals (V_1 or V_2) to the output. It can be used to replace a SPDT (single-pole double-throw) switch in any application where automatic control of such a switch is needed. Which signal will appear at the output depends on whether the control input signal (V_c) is positive or negative.

Figure 7-26 shows a similar circuit for switching a single input between two separate outputs (V_{o1} and V_{o2}).

8

Experiments II

I have covered quite a bit of material in the last three chapters. It's time to get some practical hands-on experience with another set of experiments.

The components and equipment you will need to perform these experiments are listed in Table 8-1. Since I will be working with various ac waveforms, a vom will not be sufficient. You will need an oscilloscope for many of the experiments. A frequency counter is optional. It would be helpful, but not absolutely essential.

EXPERIMENT 1—SINE WAVE OSCILLATOR

In my first experiment I will design and breadboard an op amp sine wave oscillator. The components and equipment called for in this experiment are summarized in Table 8-2.

I will be working with the circuit shown in Fig. 8-1. This is the twin-T sine wave oscillator circuit from Chapter 6.

The design equations for this circuit are summarized in Table 8-3. Use these equations to design a 1000 Hz oscillator. Assign a value of 0.047 μF to capacitor C2. You may substitute a capacitor marked 0.05 μF. The difference between these values is negligible.

Round off your component values to the nearest standard values (refer to Tables 8-4 and Table 8-5). Fill in your calculated values in Table 8-6, then compare your results with mine, which are listed in Table 8-7.

Table 8-1. Components and Equipment List.

Item	Quantity
Breadboard system	
positive power supply (+12 to 18 volts) solderless socket	
negative power supply (−12 to −18 volts)	
Components	
741 op amp IC	(need 2)
100 ohm resistor	
330 ohm resistor	(need 2)
620 ohm resistor	
820 ohm resistor	
1 K resistor	
1.2 K resistor	
2.2 K resistor	
2.7 K resistor	
3.3 K resistor	(need 2)
3.9 K resistor	
4.7 K resistor	
6.8 K resistor	
8.2 K resistor	(need 2)
10 K resistor	(need 4)
12 K or 15 K resistor	
18 K resistor	
22 K resistor	
27 K resistor	
33 K resistor	(need 2)
39 K resistor	
47 K resistor	
100 K resistor	(need 2)
1 MΩ resistor	
2.5 K potentiometer	
10 K potentiometer	
0.0047 μF capacitor	(need 2)
0.01 μF capacitor	(need 3)
0.022 μF capacitor	
0.047 μF capacitor	(need 2)
0.1 μF capacitor	(need 2)
0.5 μF capacitor (or larger)	
5 μF capacitor	
LEDs (light emitting diodes)	(need 2)
SPDT switch	
SPST momentary action pushbutton switch (normally open)	
Other	
hook-up wire	
small speaker	
oscilloscope	
voltmeter or VOM	
frequency counter (optional)	

Parts
Breadboard system
741 op amp IC
100 ohms resistor
820 ohms resistor
2.7 K resistor
3.3 K resistor (2 needed)
10 K resistor (2 needed)
2.5 K potentiometer
10 K potentiometer
0.01 μF capacitor (2 needed)
0.022 μF capacitor
0.047 μF capacitor (2 needed)
0.1 μF capacitor
0.5 μF capacitor (or larger)
hook-up wire
small speaker
oscilloscope
frequency counter (optional)

Table 8-2. Parts List for Experiment 1.

Carefully breadboard this circuit, as explained in Chapter 4. Double-check all connections before applying power to the circuit.

Connect a small speaker across the output and ground, as shown in Fig. 8-2. The 100 ohm resistor attenuates the output signal, and the capacitor blocks any dc component of the output signal which could damage the speaker. Set potentiometer R4 to its

Table 8-3. Equations for Experiment 1.

$$F = \frac{1}{2\pi R1\,C2} \qquad R4 \cong \frac{R1}{2}$$

$$R1 = \frac{1}{2\pi FC1} \qquad C1 = 2C2$$

$$R2 = R1 \qquad C3 = C2$$

$$R3 = \frac{R1}{4}$$

Table 8-4. Standard Capacitor Values.

0.001 μF	0.01 μF	0.1 μF
0.0022 μF	0.022 μF	0.22 μF
0.0033 μF	0.033 μF	0.33 μF
0.0047 μF	0.047 μF	0.47 μF
0.0056 μF	0.056 μF	0.56 μF
0.0068 μF	0.068 μF	0.68 μF
0.0082 μF	0.082 μF	0.82 μF

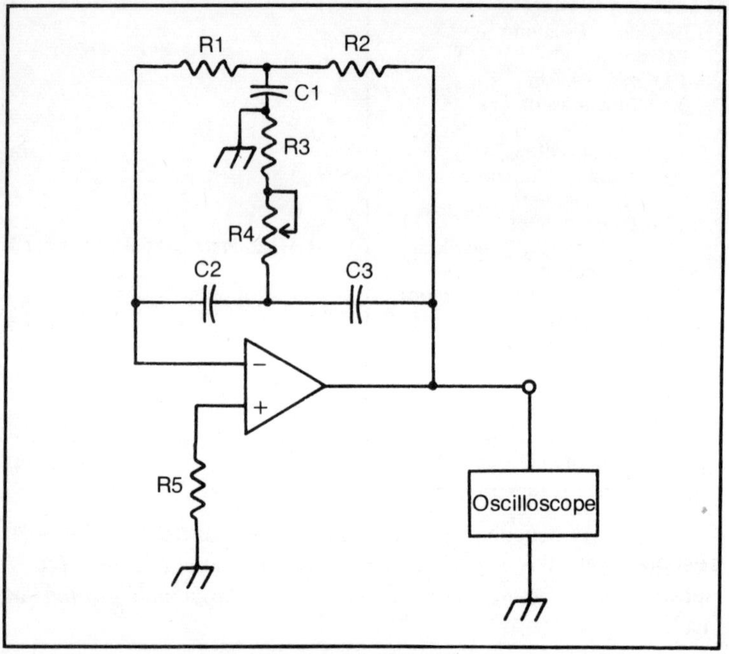

Fig. 8-1. This circuit is used for Experiment 1.

Table 8-5. Standard Resistor Values.

100 Ω	1 K (1000)	10 K	100 K	1 M Ω
120 Ω	1.2 K	12 K	120 K	1.2 M Ω
150 Ω	1.5 K	15 K	150 K	1.5 M Ω
180 Ω	1.8 K	18 K	180 K	1.8 M Ω
220 Ω	2.2 K	22 K	220 K	2.2 M Ω
270 Ω	2.7 K	27 K	270 K	2.7 M Ω
330 Ω	3.3 K	33 K	330 K	3.3 M Ω
390 Ω	3.9 K	39 K	390 K	3.9 M Ω
470 Ω	4.7 K	47 K	470 K	4.7 M Ω
560 Ω	5.6 K	56 K	560 K	5.6 M Ω
680 Ω	6.8 K	68 K	680 K	6.8 M Ω
820 Ω	8.2 K	82 K	820 K	8.2 M Ω
910 Ω	9.1 K	91 K	910 K	9.1 M Ω
				10 M Ω

Table 8-6. Worksheet for Experiment 1.

F	1000 Hz	C1	________
R1	________	C2	0.047 μF
R2	________	C3	________
R3	________		
R4	________		

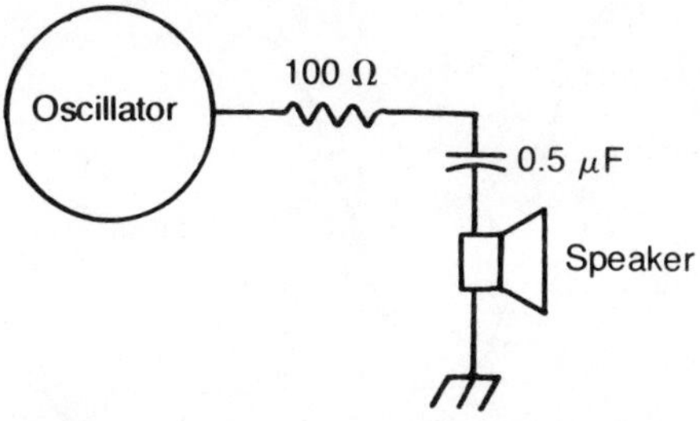

Fig. 8-2. To hear the sine wave signal, add these components to the circuit of Fig. 8-1.

maximum setting. Slowly reduce this resistance until a pure whistling tone is heard from the speaker. If you cannot hear a tone at any setting of R4 something is wrong. Disconnect the power supply and recheck all circuit connections.

Once the correct setting of R4 is found, remove the output resistor, capacitor, and speaker, being careful not to disturb the setting of the potentiometer. Any changes in the circuitry should be made with the power supply disconnected!

Examine the displayed waveform. It should be very similar to the basic sine wave illustrated in Fig. 8-3. Adjust R4 and note the distortion that appears in the displayed waveform.

Readjust R4 for the purest sine wave display. Calculate the frequency from the oscilloscope display, or by using a frequency counter. Is it close to the design goal of 1000 Hz? There may be some slight variation due to component tolerances and rounding off in the calculations. For the component values listed in Table 8-7, the nominal output frequency (ignoring the effects of component tolerances) should be approximately equal to:

$$F = \frac{1}{2\pi \, R1C2} = \frac{1}{6.28 \times 3300 \times 0.000000047}$$

F	1000 Hz
R1	3.3 K
R2	3.3 K
R3	820 ohms
R4	2.5 K potentiometer (1693 ohms)
C1	0.1 μF
C2	0.047 μF
C3	0.047 μF

Table 8-7. Calculated Values for Experiment 1.

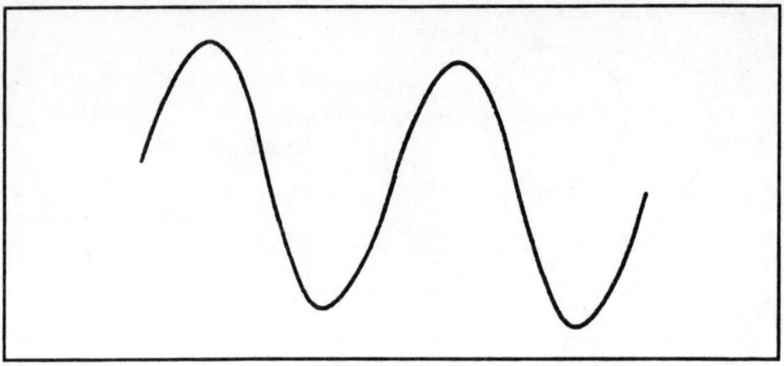

Fig. 8-3. The oscilloscope display in the circuit of Fig. 8-1 should be a fairly clean sine wave.

$$F = \frac{1}{0.0009745} = 1026 \text{ Hz}$$

Remove the power supply and replace the capacitors with new units with the following values:

$$C1 = 0.022\ \mu F$$
$$C2 = 0.01\ \mu F$$
$$C3 = 0.01\ \mu F$$

Turn on circuit power and readjust potentiometer R4 to the point where the circuit just begins to break into oscillation. With an oscilloscope, adjust this control for minimum distortion in the displayed waveform.

What is the output frequency now?

$$F = \frac{1}{2\pi R1C2} = \frac{1}{6.28 \times 3300 \times 0.00000001}$$

$$F = \frac{1}{0.0002073} \cong 4823 \text{ Hz}$$

Decreasing the capacitor values (or the resistance values) increases the output frequency. All like components (all capacitors or all resistors) must be changed proportionately.

Logically increasing one set of component values should decrease the output frequency. Let's test this idea by replacing the resistors with the following values:

R1 = 10 K
R2 = 10 K
R3 = 2.7 K
R4 = 10 K potentiometer (5 K)

The capacitor values should remain the same as in the last part of this experiment.

Adjust potentiometer R4 for the purest possible sine wave display on the oscilloscope. What is the output frequency with these component values?

$$F = \frac{1}{2\pi\ R1C2} = \frac{1}{6.28 \times 10{,}000 \times 0.00000001}$$

$$F = \frac{1}{0.0006283} \cong 1592\ \text{Hz}$$

Increasing the resistor values (or the capacitor values) decreases the output frequency.

Using the components from Fig. 8-2, can you design a twin-T sine wave oscillator with a nominal output frequency of about 340 Hz?

EXPERIMENT 2—SQUARE WAVE GENERATOR

Gather together the parts and equipment listed in Table 8-8. In this experiment, you will work with the square wave generator circuit shown in Fig. 8-4. A square wave, as you should recall from Chapter 6 is a much more complicated waveform than the sine

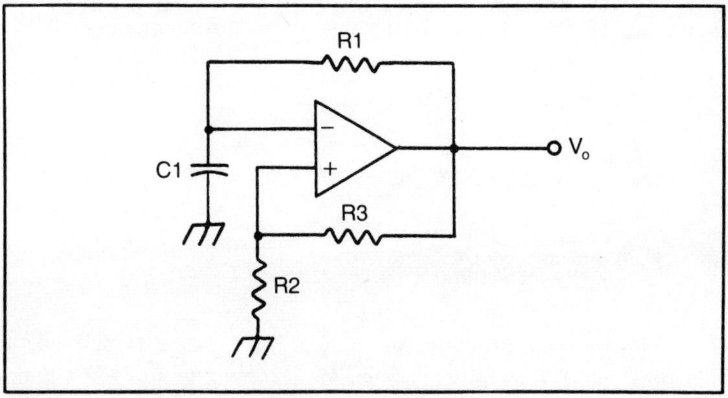

Fig. 8-4. The circuit for Experiment 2.

Breadboard system
741 op amp
100 ohms resistor
10 K (two needed) resistors
47 K resistor
100 K (two needed) resistors
0.01 μF capacitor
0.047 μF capacitor
0.1 μF capacitor
0.5 μF capacitor
hook-up wire
speaker
oscilloscope
frequency counter (optional)

Table 8-8. Parts List for Experiment 2.

wave. A square wave includes a fundamental and many harmonics that are odd integer multiples of the fundamental. No even harmonics (multiples of two) are included in a square wave.

The output frequency of the circuit of Fig. 8-4 is determined primarily by resistor R1 and capacitor C1, but the positive feedback network (resistors R2 and R3) also has an influence. To simplify matters in this experiment, I will assign a permanent ratio of 10:1 for these resistors. Resistor R2 will be given a fixed value of 10 K, and R3 will be kept at 100 K. In this case, the frequency determining equation is simply:

$$F = \frac{5}{R1C1}$$

For the first part of this experiment I will design a square wave oscillator with an output frequency of 1000 Hz. Use a 0.1 μF capacitor for C1. What value resistor should be used for R1? This value can be found simply by rearranging the frequency equation.

$$R1 = \frac{5}{FC1}$$

Plugging in the values for this example I find the following:

$$R1 = \frac{5}{1000 \times 0.0000001} = \frac{5}{0.0001} = 50{,}000 \text{ ohms}$$

This is quite close to the standard resistance value of 47 K (47,000 ohms). Carefully breadboard the circuit using these component values.

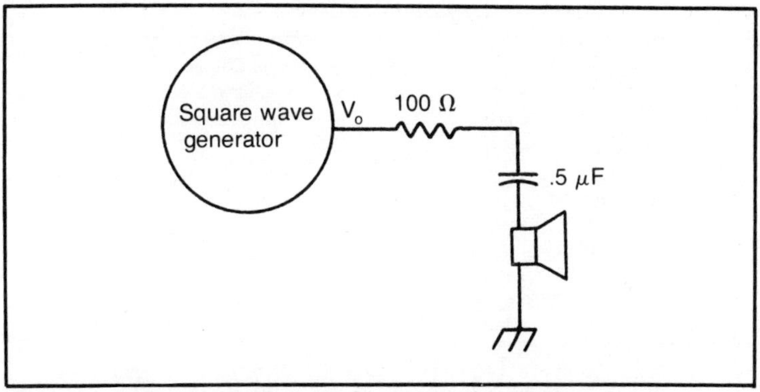

Fig. 8-5. Adding a speaker to the square wave generator of Fig. 8-4.

R1 = 47 K
R2 = 10 K
R3 = 100 K
C1 = 0.1 μF

First connect a speaker to the output, as shown in Fig. 8-5. The adder resistor and capacitor are used to limit volume, and protect the speaker from any dc component of the output signal. Their values are not particularly critical. A 100 ohm resistor and a 0.5 μF capacitor should do fine.

Double-check all connections before applying power to the circuit. There are no adjustments to make in this circuit, and it should begin oscillating as soon as power is applied. The tone you hear should be raspier and less shrill than the sine wave in Experiment 1. The sound of a square wave is somewhat suggestive of an oboe.

Disconnect the power and remove the speaker along with the output resistor and capacitor. Connect an oscilloscope between the

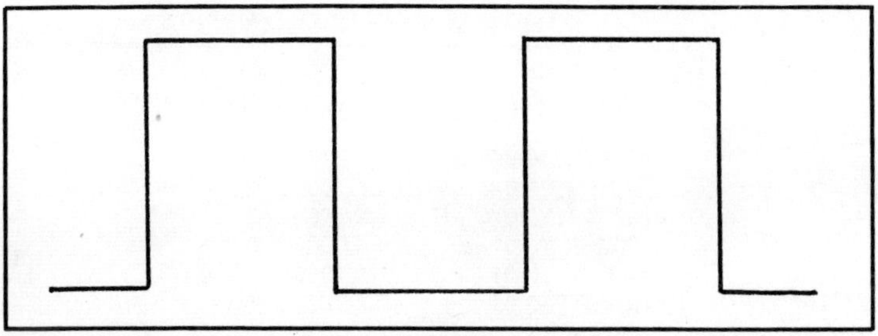

Fig. 8-6. The circuit of Fig. 8-4 generates square waves like the signal illustrated here.

Table 8-9. Worksheet for Experiment 2.

R1	C1	Measured Output Frequency
47 K	0.047 μF	______
10 K	0.047 μF	______
100 K	0.047 μF	______
100 K	0.01 μF	______
47 K	0.01 μF	______
10 K	0.01 μF	______
100 K	0.1 μF	______
10 K	0.1 μF	______

circuit output and ground. Study the waveform displayed. It should resemble the ideal square wave illustrated in Fig. 8-6. The high level time should be equal to the low level time.

Measure the output frequency with the oscilloscope, or a frequency counter. It should be close to our specified value of 1000 Hz, although there will probably be some error due to component tolerances, and the rounding off of the value of R1. Ignoring the effects of component tolerances, the output frequency should be approximately:

$$F = \frac{5}{R1C1} = \frac{5}{47{,}000 \times 0.0000001} = \frac{5}{0.0047} = 1064 \text{ Hz}$$

The actual measured value of the output frequency should be close to this calculated value.

Turn off the power to the circuit and replace capacitor C1 with a 0.047 μF capacitor. Turn on the power supply and measure the output frequency. Write your result in the appropriate space in Table 8-9.

Repeat these steps for each of the combinations of values for

Table 8-10. Calculated Values for Experiment 2.

R1	C1	Calculated Output Frequency
47 K	0.047 μF	2263 Hz
10 K	0.047 μF	10638 Hz
100 K	0.047 μF	1064 Hz
100 K	0.01 μF	5000 Hz
47 K	0.01 μF	10638 Hz
10 K	0.01 μF	50000 Hz
100 K	0.1 μF	500 Hz
10 K	0.1 μF	5000 Hz

R1 and C1 listed in Table 8-9. Compare your measured results with the nominal values listed in Table 8-10. They should be very close in each case.

When you have completed this experiment, do not disassemble the breadboarded circuit. It will be needed in the next few experiments.

EXPERIMENT 3—RECTANGLE WAVE GENERATORS

By adding two components to the square wave generator circuit of Fig. 8-4, you can convert it into a rectangle wave generator. The revised circuit is shown in Fig. 8-7. The parts and equipment required for this experiment are listed in Table 8-11.

Breadboard the square wave generator circuit from Fig. 8-4. This circuit should still be set up from the last experiment. Use the following component values:

R1	10 K
R2	10 K
R3	100 K
C1	0.1 μF

Measure the output frequency with an oscilloscope or frequency counter. You should get a value close to 5000 Hz.

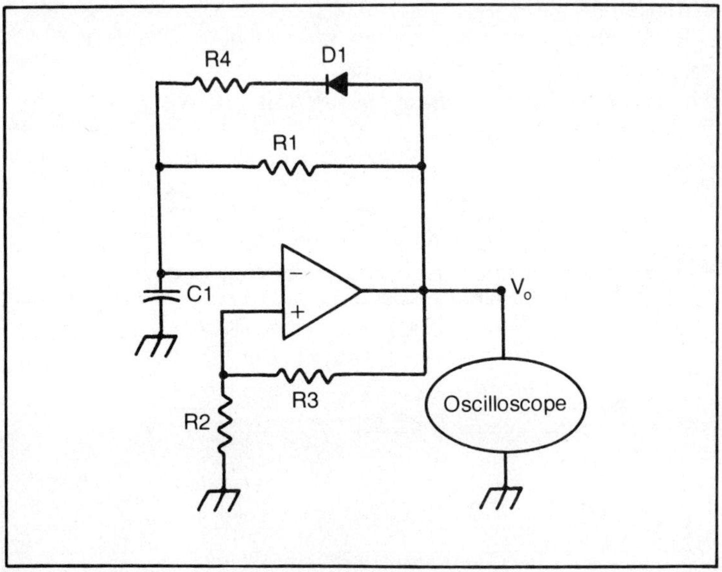

Fig. 8-7. Experiment 3 converts the square wave generator circuit of Fig. 8-4.

Breadboard system 741 op amp 1 K resistor 2.2 K resistor 3.3 K resistor 4.7 K resistor 6.8 K resistor 10 K resistor (need 3) 22 K resistor 47 K resistor 100 K resistor (need 2) 0.1 μF capacitor hook-up wire oscilloscope frequency counter

Table 8-11. Parts List for Experiment 3.

$$F = \frac{5}{R1C1} = \frac{5}{10{,}000 \times 0.0000001}$$

$$F = \frac{5}{0.001} = 5000 \text{ Hz}$$

Notice the symmetry of the waveform as displayed on the oscilloscope.

Now temporarily disconnect the power supply and add a 4.7 K resistor and a 1N914 (or similar) diode in parallel with resistor R1, as shown in Fig. 8-7. Reconnect the power supply and examine the new waveform on the oscilloscope. It is no longer symmetrical. The duty cycle is something other than 1:2.

Measure the output frequency. It will no longer be at 5000 Hz of the first portion of the experiment. Fill in the measured duty cycle and frequency of the new rectangle waveform in the appropriate space in Table 8-12.

Table 8-12. Worksheet for Experiment 3.

	MEASURED VALUES		CALCULATED VALUES	
R4	**Frequency**	**Duty cycle**	**Frequency**	**Duty Cycle**
4.7 K	______	1:____	7185	1:4
1 K	______	1:____	______	1:____
2.2 K	______	1:____	______	1:____
3.3 K	______	1:____	______	1:____
6.8 K	______	1:____	______	1:____
10 K	______	1:____	______	1:____
22 K	______	1:____	______	1:____
47 K	______	1:____	______	1:____
100 K	______	1:____	______	1:____

On positive half-cycles R1 and R4 are effectively in parallel, presenting a combined value of 3.2 K.

$$R_t = \frac{R1 \times R4}{R1 + R4} = \frac{10{,}000 \times 4700}{10{,}000 + 4700} = \frac{47{,}000{,}000}{14{,}700}$$

$$R_t = 3197 \text{ ohms} \cong 3.2 \text{ K}$$

For negative half-cycles, the diode blocks the feedback signal from flowing through R4. Only R1 has an effect on this portion of the waveform. Its value is 10 K.

Since the feedback resistance changes for each half cycle, the frequency is effectively changed for each portion of the cycle. For the positive portion of the cycle, calculate the frequency with this equation:

$$F = \frac{5}{R_t C1}$$

For the negative portion of the cycle, only R1 is used in the frequency equation.

$$F = \frac{5}{R1C1}$$

This is easier to see and work with by converting to the time of a single cycle. This can be done simply by inverting the frequency.

$$t_T = \frac{1}{F} = \frac{1}{\frac{5}{R1C1}} = \frac{R1C1}{5}$$

Since you are concerned with the length of half-cycles (not full-cycles) here, divide by two. The positive half-cycle time can be found with this equation:

$$t_1 = \frac{R_t C1}{5} \times \frac{1}{2} = \frac{R_t C1}{10}$$

Similarly, for the negative half-cycle, the equation is as follows:

$$t_2 = \frac{R1C1}{10}$$

The time for the total cycle is simply the sum of the two half-cycles.

$$t_T = t_1 + t_2$$

You can take the reciprocal of t_T to find the composite frequency.

$$F = \frac{1}{t_T}$$

Plug in the values for the first experimental example.

$$t_1 = \frac{R_tC1}{10} = \frac{3197 \times 0.0000001}{10} = \frac{0.0003197}{10}$$

$$t_1 = 0.00003197 \text{ seconds} = 0.03197 \text{ milliseconds}$$

$$t_2 = \frac{R1C1}{10} = \frac{10{,}000 \times 0.0000001}{10} = \frac{0.001}{10}$$

$$t_2 = 0.0001 \text{ second} = 0.1 \text{ millisecond}$$

$$t_T = t_1 + t_2 = 0.00003197 + 0.0001 = 0.00013197 \text{ second}$$

$$t_T = 0.13197 \text{ millisecond}$$

$$F = \frac{1}{t_T} = \frac{1}{0.00013197} \cong 7185 \text{ Hz}$$

You can combine these equations into a single equation to find the output frequency directly.

$$F = \frac{1}{t_T} = \frac{1}{t_1 + t_2} = \frac{1}{\left(\frac{R_tC1}{10}\right) + \left(\frac{R1C1}{10}\right)}$$

$$F = \frac{1}{\frac{(R_t + R1)C1}{10}} = \frac{10}{(R_t + R1)C1}$$

This formula can be tested by plugging in the same component values and seeing if you get the same result.

$$F = \frac{10}{(R_t + R1)C1} = \frac{1}{(3197 + 10{,}000) \times 0.0000001}$$

$$F = \frac{10}{(13{,}197 \times 0.0000001)} = \frac{10}{0.0013917}$$

$$F \cong 7185 \text{ Hz}$$

Yes, you can get the same results using either method.

What about the duty cycle of the new rectangle wave generated by this circuit? The duty cycle of a rectangle wave, as you should recall, is the ratio of the positive half-cycle time (t_1) to the total cycle time (t_T).

$$\text{duty cycle} = \frac{t_T}{t_1}$$

$$\text{duty cycle} = \frac{0.00013197}{0.00003197} = 4.13$$

For all intents and purposes, this is a duty cycle of 1:4 for your circuit.

Do these calculated results come close to the measured values you wrote down in Table 8-12? While there may be some variation due to component tolerances, the results should be close to the figures worked out here.

Repeat the experiment for each of the values of R4 listed in

Parts
Breadboard system
741 op amp (need 2)
100 ohm resistor
1 K resisto:
2.2 K resistor
2.7 K resistor
4.7 K resistor
6.8 K resistor
10 K resistor (need 4)
12 K or 15 K resistor (see text)
22 K resistor
33 K resistor
47 K resistor
100 K resistor
0.01 μF capacitor (need 2)
hook-up wire
small speaker
oscilloscope

Table 8-13. Parts List for Experiment 4.

Table 8-12. Compare the measured and calculated values in each instance.

You may also want to hook a separate speaker to the output of the circuit, as you did in Fig. 8-5. As the duty cycle figure gets larger (that is, as the positive half-cycle makes up a smaller percentage of the total cycle time), the sound will get raspier as more harmonics are added to the signal.

EXPERIMENT 4—LOW-PASS FILTER

Remove resistor R4 and diode D1 from the circuit used in the last experiment. Do this with the power off, of course. This should return you to the original square wave generator circuit shown back in Fig. 8-4. You will be reusing this circuit in the next few experiments on filters.

First you will examine the effects of a low-pass filter on a square wave signal. The complete circuit for this experiment is shown in Fig. 8-8. IC1 with its associated components (R1, R2, R3, and C1) is the square wave generator. The values for these components are as follows:

R1 = 47 K
R2 = 10 K
R3 = 100 K
C1 = 0.01 μF

The second section of the circuit (IC2, R4, R5, R6, and C2) forms a low-pass active filter. This circuit was originally introduced as Fig. 5-19.

Resistor R7 (100 ohms) attenuates the output signal before feeding it through the speaker. Capacitor C3 (0.5 μF or larger) blocks any dc component of the output signal. The speaker will allow you to hear the unfiltered square wave signal when switch S1 is in position A. Moving the switch to position S2 will feed the filtered signal to the speaker. This setup allows you to listen and compare the sound of the signals as harmonics are filtered out.

Connecting the oscilloscope probe to point X will let you display the straight square wave signal. The filtered signal can be seen by moving the oscilloscope probe to point Y. The frequency can be measured at either point. You should get the same reading either way.

Since this circuit is more complex than the ones used in the earlier experiments, be extra careful to double-check all connec-

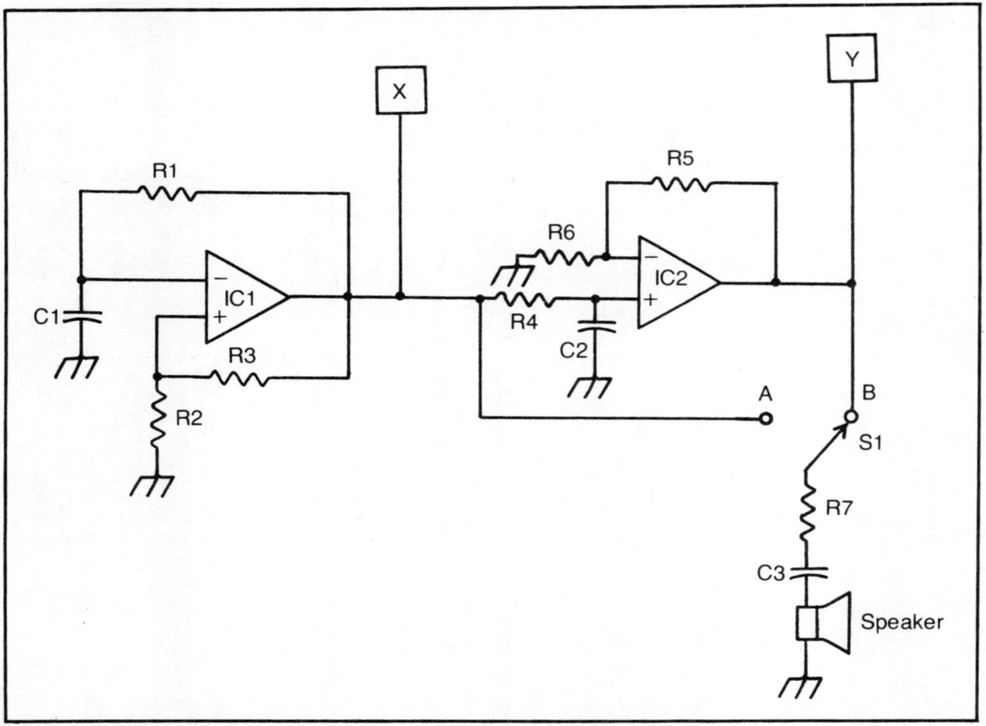

Fig. 8-8. In Experiment 4, the output from a square wave generator is passed through a low-pass filter.

tions before applying power. Be sure to make the appropriate power supply connections to both of the ICs.

Resistors R5 and R6 determine the gain of the low-pass filter. For this experiment, you will want unity gain, so equal value resistors should be used. I used a pair of 10 K resistors, but this isn't particularly critical as long as R5 = R6.

The cutoff frequency for the filter can be found with the following formula:

$$F = \frac{1}{2\pi R4C2}$$

To achieve a specific cutoff frequency, select a reasonable value for C2, and rearrange the equation to solve for the appropriate value of resistor R4.

$$R4 = \frac{1}{2\pi FC2}$$

Throughout this experiment, you will assign a constant value of 0.01 μF to capacitor C2. This allows us to simplify the equation by precalculating the constants:

$$R4 = \frac{1}{6.28 \times 0.00000001 \times F} = \frac{1}{0.0000000628\ F}$$

From the previous experiments you know that the square wave generator (IC1) is putting out a 1000 Hz signal. The harmonic content of this signal is summarized in Table 8-14.

Let's say you want to pass the fundamental (1000 Hz) at full strength, but filter all of the harmonics. You won't be able to

Table 8-14. 1000 Hz Square Wave Harmonics.

Fundamental	1000 Hz
3rd harmonic	3000 Hz
5th harmonic	5000 Hz
7th harmonic	7000 Hz
9th harmonic	9000 Hz
11th harmonic	11000 Hz
13th harmonic	13000 Hz
15th harmonic	15000 Hz
17th harmonic	17000 Hz

Higher harmonics are low enough in amplitude that they may be ignored for our purposes.

eliminate them completely with this simple filter circuit, but you can reduce their amplitudes significantly.

Maximum reduction in the harmonics can be achieved by setting the filter's cutoff frequency slightly higher than the fundamental frequency. You'll use a nominal value of 1200 Hz. According to the equation, resistor R4 should have a value of about 13,263 ohms.

$$R4 = \frac{1}{0.0000000628F} = \frac{1}{0.0000000628 \times 1200}$$

$$\frac{1}{0.0000754} = 13{,}263 \text{ ohms}$$

Either a 12 K or a 15 K resistor may be used with reasonably good results.

Set up the circuit with the appropriate components and apply power to both ICs. Flip switch S1 back and forth to compare the sound of the unfiltered signal (position A) and the filtered signal (position B). Because of the weakened harmonics, the filtered signal (B) should have a much purer tone than that of the unfiltered square wave (A).

Compare the oscilloscope display at points X and Y. At point X, of course, you should simply see a square wave, as in Experiment 2. At point Y you will probably see something more like a distorted triangle wave. A typical triangle wave is shown in Fig. 8-9. Remember from Chapter 6 that a triangle wave has the same harmonic structure as a square wave, but the harmonics are considerably weaker. Passing a square wave signal through a low-pass filter weakens the harmonic amplitudes of the signal, effectively syn-

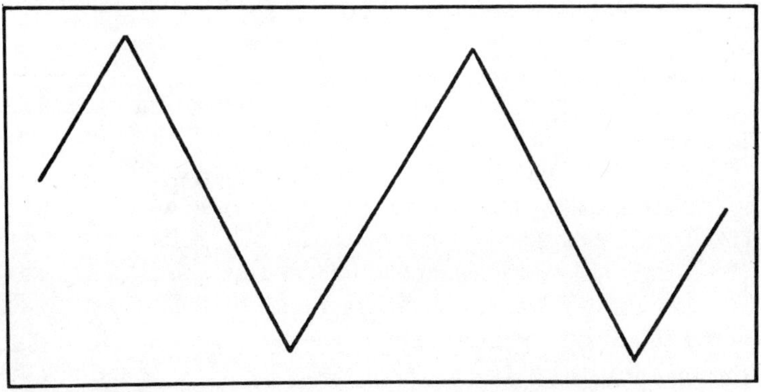

Fig. 8-9. Passing a square wave through a low-pass filter creates a triangle wave.

thesizing a triangle wave. The output signal here will not quite be a true triangle wave because the harmonics will not be reduced by the proper amount.

Now, let's see what happens when you raise the cutoff frequency to 6000 Hz. This will require changing the value of R4.

$$R4 = \frac{1}{0.0000000628 \times 6000} = \frac{1}{0.000377}$$

$$R4 = 2653 \text{ ohms}$$

A 2.7 K resistor will be close enough.

Be sure to turn the power off before changing any of the components in the circuit. Don't risk damaging your ICs or power supplies.

Compare the sound of the filtered and unfiltered signals by moving switch S1 back and forth. Look at the output waveform displayed on your oscilloscope at point Y. With a cutoff frequency of 6000 Hz the fundamental, the third harmonic, and the fifth harmonic are passed to the output without attenuation. Higher harmonics (seventh and above) are progressively reduced in amplitude.

Repeat this experiment for each of the following values:

R4	cuttoff frequency (approx)
10 K	1600 Hz
6.8 K	2340 Hz
4.7 K	3390 Hz
2.2 K	7230 Hz
1 K	15,900 Hz
22 K	723 Hz
33 K	482 Hz

Compare the filtered and unfiltered sounds for each case. Examine the displayed output waveforms (point Y) and notice how they differ from the original square wave signal (point X).

Notice that the last two cases (R4 = 22 K and R4 = 33 K) have cutoff frequencies less than the fundamental of the input signal (1000 Hz). The entire signal is attenuated.

For the last section of this experiment, replace resistor R1 with a 100 K potentiometer. This will allow you to manually change the input frequency. Use a 4.7 K resistor for R4. Examine how the output waveform (point Y) changes as the input frequency changes and the cutoff frequency remains constant.

EXPERIMENT 5—HIGH-PASS FILTER

This experiment will be similar to the last one except that you will be passing your square wave through a high-pass filter instead of a low-pass filter. The circuit is illustrated in Fig. 8-10. The components and equipment you will need for this experiment are listed in Table 8-15.

In this experiment, you will use a second order VCVS circuit. This circuit was first shown in Fig. 5-26. In Fig. 8-10 the filter is made up of IC2, R4, R5, R6, R7, C2, and C3.

The following components will be assigned fixed values throughout this experiment:

R1 = 47 K	R7 = 100 ohms
R2 = 10 K	C1 = 0.01 μF

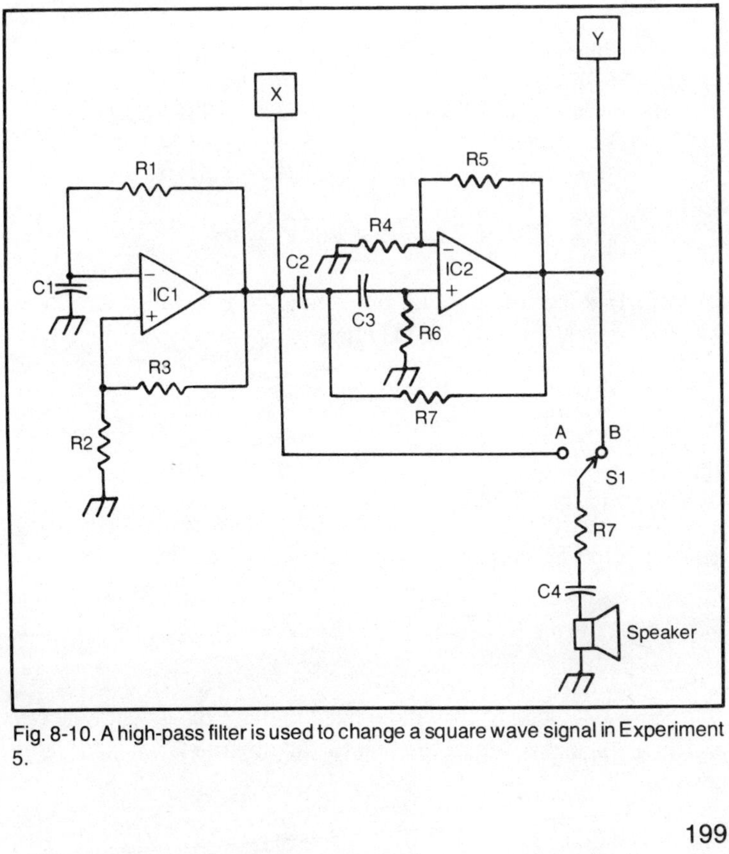

Fig. 8-10. A high-pass filter is used to change a square wave signal in Experiment 5.

Breadboard system
741 op amp (need two)
100 ohm resistor
3.3 K resistor (need two)
8.2 K resistor (need two)
10 K resistor (need two)
18 K resistor
33 K resistor (need two)
47 K resistor
100 K resistor
0.0047 μF capacitor (need two)
0.01 μF capacitor
hook-up wire
small speaker
oscilloscope
frequency meter (optional)

Table 8-15. Parts List for Experiment 5.

R3 = 100 K　　C2 = 0.0047 μF
R4 = 18 K　　C3 = 0.0047 μF
R5 = 10 K

This leaves resistors R6 and R7 as variable components to determine the cutoff frequency of the high-pass filter.

As explained in Chapter 5, the cutoff frequency of a second order VCVS active filter can be found using this formula:

$$F_c = \frac{1}{2\pi \text{ R6C2}}$$

For this formula to work out right, the following component value relationships must be maintained:

$$\text{R5} = 0.586\text{R4}$$
$$\text{R6} = \text{R7}$$
$$\text{C1} = \text{C2}$$

This equation can be rewritten to solve for R6 when the desired cutoff frequency and capacitance are known.

$$\text{R6} = \frac{1}{2\pi \text{ FC2}}$$

Since the value of C2 will be fixed in this experiment, you can precalculate some of this equation to simplify the formula for later designs.

$$R6 = \frac{1}{2 \times 3.14 \times F \times 0.0000000047}$$

$$R6 = \frac{1}{0.0000000295F}$$

As in the previous experiment, the square wave generator (IC1, R1, R2, R3, and C1) is putting out a signal with a frequency of 1000 Hz.

$$F = \frac{5}{R1C1} = \frac{5}{4700 \times 0.00000001}$$

$$F = \frac{5}{0.0047} = 1064 \text{ Hz} \cong 1000 \text{ Hz}$$

Since it is a square wave, this signal includes the odd harmonics (3000 Hz, 5000 Hz, 7000 Hz, 9000 Hz, 1000 Hz, and so forth).

Begin your experiment by setting the cutoff frequency for the high-pass filter at 4000 Hz. This will filter out the fundamental, and the third harmonic, but leave the fifth and all higher harmonics. Actually, because of the filter's rolloff slope, there may be some negligible vestige of the fundamental and third harmonic frequencies in the output signal as weak undertones, or subharmonics.

You now have to determine the appropriate resistance value for R6 and R7. Simply plug the desired cutoff frequency value into the simplified equation given earlier.

$$R6 = \frac{1}{0.0000000295 \times 4000} = \frac{1}{0.0001181}$$

$$R6 = 8466 \text{ ohms}$$

Round off to the nearest standard resistor value of 8.2 K resistors for R6 and R7.

Carefully breadboard the circuit shown in Fig. 8-10, using the component values outlined above. Double-check all connections before applying power.

Compare the tone produced by the speaker as you move switch S1 back and forth from position A (unfiltered) to position B (filtered) and back again. The filtered signal will be much softer than the unfiltered signal because the strongest components of the original

Table 8-16. Overtone Structure for Part 1 of Experiment 5.

Frequency Component	New Position	Position in Original Signal
(1000 Hz)	very weak undertone	fundamental
(3000 Hz)	very weak undertone	third harmonic
5000 Hz	fundamental	fifth harmonic
7000 Hz	nonharmonic	seventh harmonic
9000 Hz	nonharmonic	ninth harmonic
11000 Hz	nonharmonic	eleventh harmonic
13000 Hz	nonharmonic	thirteenth harmonic
15000 Hz	third harmonic	fifteenth harmonic
17000 Hz	nonharmonic	seventeenth harmonic
19000 Hz	nonharmonic	nineteenth harmonic

signal (the fundamental and the third harmonic) have been deleted without amplifying the remaining harmonics.

You should also notice that the filtered tone has a higher apparent pitch than the original unfiltered square wave.

Measure the frequency at point X using either an oscilloscope or a frequency counter. This is the original unfiltered square wave signal. You should get a reading of about 1000 Hz (1064 Hz). Now, if you measure the output frequency (at point Y), you should get a reading of approximately 5000 Hz (5320 Hz). Since the original fundamental, and the third harmonic, have been deleted from the signal, the fifth harmonic is now dominant, and acts as the new fundamental.

The tonal quality of the signal will be changed quite a bit, because the overtones do not bear a harmonic relationship to the new fundamental frequency. Because of the strong nonharmonic content of this new signal, the sound will be rather harsh, and the sense of pitch will be somewhat indefinite.

Compare the waveforms displayed on your oscilloscope at points X (unfiltered square wave) and Y (filtered output signal) in the test circuit. Notice how drastically the waveshape has been altered.

Now, let's move the cutoff frequency up to 10,000 Hz. All harmonics from the ninth (9000 Hz) down will be filtered. The original fundamentals will be almost completely absent from the output signal. You would need some sophisticated equipment to detect what is left of this frequency component.

For this new cutoff frequency you will need to change the values of R6 and R7 to 3.3 K.

$$R6 = \frac{1}{0.0000000295F} = \frac{1}{0.0000000295 \times 10{,}000}$$

$$R6 = \frac{1}{0.000295} = 3386 \text{ ohms} \cong 3.3 \text{ K}$$

Disconnect power from the breadboarded circuit and replace R6 and R7 with 3.3 K resistors. Now apply power and repeat the measurements described for the first part of this experiment. Before you measure the output frequency, can you determine what the new fundamental of the output signal will be?

Let's try another example. This time set the cutoff frequency at 1100 Hz. This means R6 and R7 must take on new values.

$$R6 = \frac{1}{0.0000000295 \times 1100} = \frac{1}{0.0000325}$$

$$R6 = 30784 \text{ ohms}$$

The nearest standard resistor value is 33 K.

Repeat the measurement steps described above. Notice that the output frequency is still 1000 Hz (1064). The fundamental frequency is too close to the cutoff frequency of the filter to be significantly attenuated. The square wave may or may not be visibly distorted when measured with an oscilloscope at point Y, depending on the component tolerances throughout the circuit, but it will still be recognizably a square wave.

This emphasizes the fact pointed out in Chapter 5 that any filter has a gradual rolloff slope. The better the filter, the steeper this rolloff will be, but no practical filter circuit will ever have an

Table 8-17. Parts List for Experiment 6.

Parts
Breadboard system
741 op amp IC (need 2)
100 ohm resistor
620 ohm resistor
1.2 K resistor
2.2 K resistor
2.7 K resistor
3.9 K resistor
6.8 K resistor
8.2 K resistor (need 2)
10 K resistor (need 2)
27 K resistor
47 K resistor
100 K resistor
0.01 μF capacitor (need 3)
0.5 μF (or larger) capacitor
hook-up wire
small speaker
oscilloscope

infinitely steep rolloff. In a high-pass filter, frequencies that are below the cutoff frequency but are close to it will only be slightly attenuated rather than removed.

You may want to experiment further with this circuit. Try additional values for R6 and R7. Remember that these two resistors should always have identical values.

You might also want to see what happens when you alter the input frequency by changing the value of R1. Using a potentiometer for R1 as in the preceding experiment will allow you to manually change the input frequency while watching the changing waveform being displayed on your oscilloscope.

EXPERIMENT 6—BANDPASS FILTER

The circuit for the next experiment is shown in Fig. 8-11. IC2, R4, R5, R6, R7, C2, and C3 make up a bandpass filter. This is the multiple-feedback active-filter circuit which was originally shown in Fig. 5-31.

There are three major equations for designing this type of circuit. If C3 = C2, the gain is determined by this formula:

$$G = \frac{R6}{2R4}$$

The center frequency can be determined by using this equation:

$$F_c = 0.159 \times \sqrt{\frac{1}{R6 \times 2C2} \times \left(\frac{1}{R4} + \frac{1}{R5}\right)}$$

Finally, the equation for determining the Q of the bandpass filter is as follows:

$$Q = \frac{F_c}{BW}$$

BW is the desired bandwidth to be passed by the filter.

You can algebraically combine these equations to find the three basic resistance values (R4, R5, and R6):

$$R4 = \frac{Q}{2\pi F_c GC}$$

$$R5 = 2GR4$$

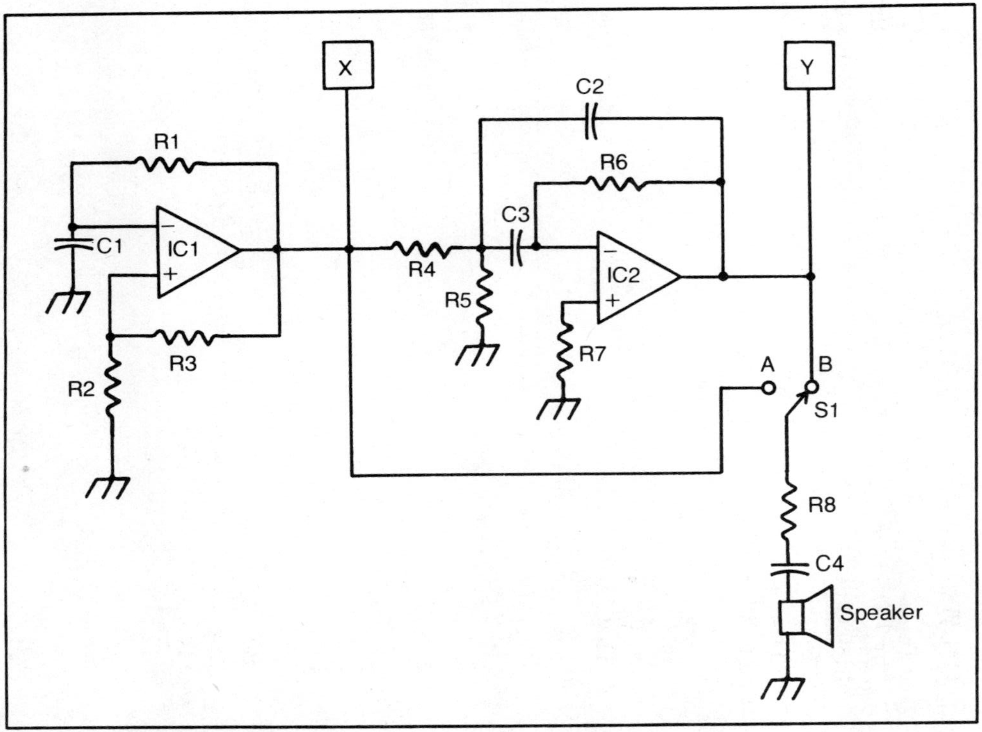

Fig. 8-11. In Experiment 6, a square wave is fed through a bandpass filter.

$$R6 = \frac{Q}{2\pi F_c C(2Q^2 - G)}$$

In this experiment, set the gain at 2, which simplifies the equations to the following:

$$R4 = \frac{Q}{12.56F_c C}$$

$$R5 = 4R4$$

$$R6 = \frac{Q}{6.28F_c C(2Q^2 - 2)}$$

Select 0.01 μF as the value for the two filter capacitors (C2 and C3). The equation for R5 remains unchanged, but the other two equations can be simplified further.

$$R4 = \frac{Q}{0.0000001256F_c}$$

$$R6 = \frac{Q}{0.0000000628F_c(2Q^2 - 2)}$$

In the first section of this experiment, you will set up a center frequency of 7000 Hz, and a bandwidth of 3000 Hz. This makes the Q equal to 2.33.

$$Q = \frac{7000}{3000} = 2.33$$

Solving for resistor R4, you find that the closest standard value is 2.7 K.

$$R4 = \frac{2.33}{0.0000001256 \times 7000} = \frac{2.33}{0.0008792}$$

$$R4 = 2650 \text{ ohms}$$

Now, R5 takes on a value of four times R4.

$$R5 = 4 \times 2700 = 10{,}800 \text{ ohms}$$

You can use a 10 K resistor here. Finally, solve for R6.

$$R6 = \frac{2.33}{0.0000000628 \times 7000 \times (2 \times 2.33^2 - 2)}$$

$$R6 = \frac{2.33}{0.0004396 \times (10.88 - 2)} = \frac{2.33}{0.0004396 \times 8.88}$$

$$R6 = \frac{2.33}{0.0039} = 597 \text{ ohms}$$

A standard 620 ohm resistor should be close enough for our purposes.

Carefully breadboard the circuit shown in Fig. 8-11 using the component values listed in Table 8-18.

Since the center frequency is 7000 Hz, and the bandwidth is 3000 Hz, frequency components from about 5500 Hz to approximately 8500 Hz will be passed. For our 1000 Hz square wave input signal, this only affects the seventh harmonic (7000 Hz).

Compare the sound produced by switching S1 back and forth from A (unfiltered square wave) to B (filtered output signal). You will probably have to listen very carefully to hear the difference. It will be subtle.

Check the waveform at the points marked X and Y, using your oscilloscope. Point X, of course, is the straight, unfiltered square wave. The waveform at point Y should be a slightly distorted square wave due to the deletion of the seventh harmonic.

More dramatic effects can be achieved with a wider bandwidth. Let's change the bandwidth of our circuit to 500 Hz instead of 3000 Hz.

Table 8-18. Component Values for Part 1.

Center Frequency = 7000 Hz
Bandwidth = 3000 Hz
Q = 2.33
G = 2
Input frequency = 1000 Hz square wave

R1	47 K	C1	0.01 μF
R2	10 K	C2	0.01 μF
R3	100 K	C3	0.01 μF
R4	2.7 K	C4	0.5 μF or larger
R5	10 K		
R6	620 ohms		
R7	100 ohms		

$$Q = \frac{7000}{6500} = 1.08$$

This change requires you to change the resistor values also.

$$R4 = \frac{1.08}{0.0000001256 \times 7000} = \frac{1.08}{0.0008792}$$

$$R4 = 1228 \text{ ohms} \cong 1.2 \text{ K}$$

$$R5 = 4 \times 1200 = 3600 = 3.9 \text{ K}$$

$$R6 = \frac{1.08}{0.0000000628 \times 7000 \times (2 \times 1.08^2 - 2)}$$

$$R6 = \frac{1.08}{0.0004396 \times (2 \times 1.1664 - 2)}$$

$$R6 = \frac{1.08}{0.0004396 \times (2.3328 - 2)} = \frac{1.08}{0.0004396 \times 0.3328}$$

$$R6 = \frac{1.08}{0.0001463} = 7382 \cong 6.8 \text{ K}$$

Now, with the expanded bandwidth, frequencies from 3750 Hz to 10,250 Hz are affected. The original harmonic content of the input signal was as follows:

1000 Hz	11,000 Hz
3000 Hz	13,000 Hz
5000 Hz	15,000 Hz
7000 Hz	17,000 Hz
9000 Hz	19,000 Hz

When three of the harmonics (5000 Hz, 7000 Hz, and 9000 Hz) are filtered out, the harmonic content of the output signal is reduced seven frequencies:

1000 Hz	15,000 Hz
3000 Hz	17,000 Hz
11,000 Hz	19,000 Hz
13,000 Hz	

This should have a much more dramatic effect on both the sound and the displayed waveform of the output signal.

Now change the center frequency to 4000 Hz, and set the bandwidth to 3500 Hz. Solve for the new values.

$$Q = \frac{4000}{3500} = 1.14$$

$$R4 = \frac{1.14}{0.0000001256 \times 4000} = \frac{1.14}{0.0005024}$$

$$R4 = 2269 \text{ ohms} \cong 2.2 \text{ K}$$

$$R5 = 4 \times 2200 = 8800 \cong 8.2 \text{ K}$$

$$R6 = \frac{1.14}{0.0000000628 \times 4000 \times (2 \times 1.14^2 - 2)}$$

$$R6 = \frac{1.14}{0.0002512 \times (2 \times 1.2996 - 2)}$$

$$R6 = \frac{1.14}{0.0002512 \times (2.5992 - 2)} = \frac{1.14}{0.0002512 \times 0.5992}$$

$$R6 = \frac{1.14}{0.0001505} = 7574 \text{ ohms} = 7.5 \text{ K} \cong 8.2 \text{ K}$$

The filtered band ranges from approximately 2250 to about 5750 Hz. The third (3000 Hz) and the fifth (5000 Hz) harmonics are deleted from the filtered output signal.

Let's lower the center frequency to 2000 Hz, and use a 1200 Hz bandwidth. The Q and the resistor values must again be changed accordingly.

$$Q = \frac{2000}{1200} = 1.67$$

$$R4 = \frac{1.67}{0.0000001256 \times 2000} = \frac{1.67}{0.0002512}$$

$$R4 = 6635 \text{ ohms} \cong 6.8 \text{ K}$$
$$R5 = 4 \times 6800 = 27{,}200 \text{ ohms} \cong 27 \text{ K}$$

$$R6 = \frac{1.67}{0.0000000628 \times 2000 \times (2 \times 1.67^2 - 2)}$$

$$R6 = \frac{1.67}{0.0001257 \times (2 \times 2.7778 - 2)}$$

$$R6 = \frac{1.67}{0.0001257 \times (5.5556 - 2)} = \frac{1.67}{0.0001257 \times 3.5556}$$

$$R6 = \frac{1.67}{0.0004469} = 3729 \text{ ohms} \cong 3.9 \text{ K}$$

With these values frequencies from 800 Hz to 3200 Hz are filtered out of the signal. The fundamental (1000 Hz) and the third harmonic (3000 Hz) are filtered out of your input signal. Since the lower cutoff frequency is lower than the lowest frequency component of the input signal, the bandpass filter behaves as if it was simply a high-pass filter with a cutoff frequency equal to the upper cutoff frequency (3200 Hz). The third harmonic (3000 Hz) will probably be only partially attenuated.

Experiment with other component values, center frequencies and bandwidths. You might also want to experiment with the value of resistor R1 (perhaps replacing it with a potentiometer) to see the effects of the bandpass filter on input signals of different frequencies.

EXPERIMENT 7—BISTABLE MULTIVIBRATOR

You will now leave oscillators and filters and experiment with some of the pulse circuits from Chapter 7. The first experiment in this section will deal with the bistable multivibrator circuit shown in Fig. 8-12. The list of parts and equipment needed for this experiment is given in Table 8-19.

Resistors R4 and R5 form a voltage divider to place about 2 to 3 volts on the input when the switch (S1) is moved to position A.

Carefully breadboard this circuit, double-checking for any errors before applying power. Set switch S1 to position B (grounded input) and turn on the power to the circuit.

One of the LEDs will light up, and the other will remain dark. It is impossible to predict which one will light up. For purposes of discussion, let's assume LED D1 lit up when power was applied, and D2 stayed off.

Briefly flick switch S1 to position A. D1 should now be extin-

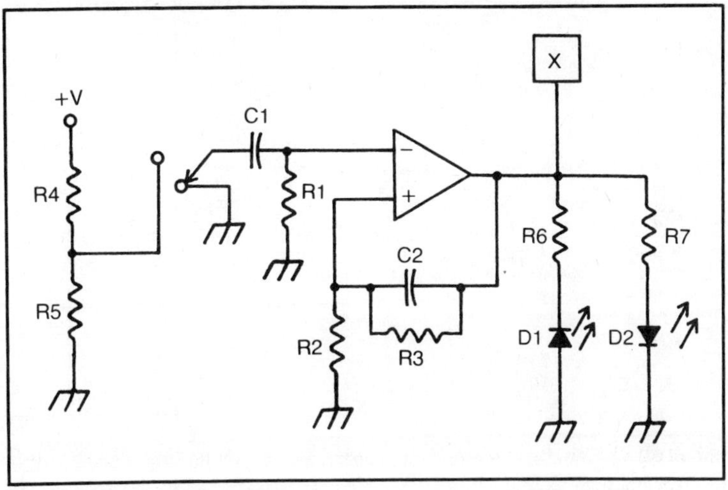

Fig. 8-12. This bistable multivibrator circuit is used in Experiment 7.

guished, and D2 should light. Immediately return the switch to position B. The state of the LEDs should not change.

The next time you move the switch to position A and back to B, the LEDs should return to their original state (D1 on, D2 off). Repeat this several times. Each time you put an input pulse through the circuit (by flicking switch S1) the state of the LEDs should be reversed.

Measure the output voltage of the circuit at point X with a vom. When one LED is lit, the output voltage should be about a volt less than the positive power supply voltage. When the other LED is lit, the output voltage will be about a volt less than the negative power supply voltage. The output voltage should always be in one or the other of these states, with no inbetween voltages ever occurring.

Breadboard system	
741 op amp IC	
R1	1 K resistor
R2	100 K resistor
R3	1 MΩ resistor
R4	39 K resistor
R5	10 K resistor
R6, R7	330 ohm resistors (need 2)
C1, C2	0.1 μF capacitors (need 2)
D1, D2	LEDs (light emitting diodes)
S1	SPDT switch
hook-up wire	
voltmeter	

Table 8-19. Parts List for Experiment 7.

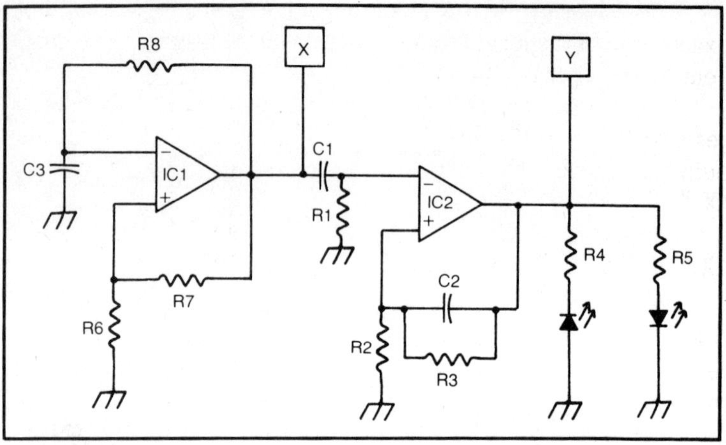

Fig. 8-13. Experiment 8 demonstrates how a bistable multivibrator can be used for frequency division.

EXPERIMENT 8—FREQUENCY DIVIDER

A bistable multivibrator can also be used for frequency division. The circuit for this experiment is shown in Fig. 8-13. The components and equipment you'll need are summarized in Table 8-20.

This is basically the same circuit you used in the preceding experiment (Fig. 8-12), except the input signal comes from a square wave generator composed of IC1, R8, R9, R10, and C3.

Once you have breadboarded the circuit and double-checked it, apply power and measure the input frequency (point X) with a frequency counter or an oscilloscope. For the component values listed this should be approximately 1000 Hz.

Table 8-20. Parts List for Experiment 8.

Breadboard system	
741 op amp ICs (need 2)	
R1	1 K resistor
R2, R7	100 K resistor (need 2)
R3	1 MΩ resistor
R4, R5	330 ohm resistor (need 2)
R6	10 K resistor
R8	47 K resistor (see text)
C1, C2	0.1 μF capacitor (need 2)
C3	0.01 μF capacitor
D1, D2	LEDs (light emitting diodes)
hook-up wire	
frequency counter, or oscilloscope	

Now measure the output frequency at point Y. It should be exactly half of the input frequency, or about 500 Hz for the component values shown.

You can change the input frequency by altering the value of resistor R8. Experiment with different values. In each case, the output frequency (Y) should be exactly half of the input frequency (X).

EXPERIMENT 9—MONOSTABLE MULTIVIBRATOR

In this chapter, you have experimented with astable multivibrators (square wave generators) and bistable multivibrators. Now, let's work with the third type of multivibrator—the monostable multivibrator, or one shot.

The components and equipment you will need for this experiment are listed in Table 8-21. Carefully breadboard the circuit shown in Fig. 8-14, using the following component values:

R1 = 39 K	R6 = 1 mΩ
R2 = 10 K	R7 = 330 ohms
R3 = 10 K	C1 = 0.1 μF
R4 = 22 K	C2 = 0.1 μF
R5 = 1 K	

Momentarily depress switch S1, and immediately release it.

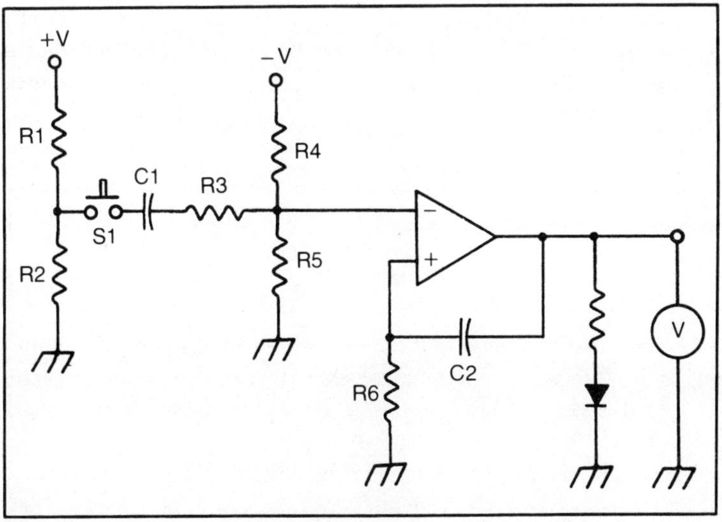

Fig. 8-14. This monostable multivibrator circuit is used in Experiment 9.

Table 8-21. Parts List for Experiment 9.

Breadboard system
741 op amp IC
330 ohm resistor
1 K resistor
10 K resistor (need 3)
22 K resistor
39 K resistor
470 K resistor
1 MΩ resistor
0.022 μF capacitor
0.1 μF capacitor (need 2)
5 μF capacitor
LED (light emitting diode)
SPST normally open momentary pushbutton switch
hook-up wire
voltmeter or VOM

The voltage should go negative and the LED should be extinguished for about a tenth of a second.

The time constant is determined by resistor R6 and capacitor C2.

$$t = C2R6 = 0.0000001 \times 1{,}000{,}000 = 0.1 \text{ second}$$

Raise the value of capacitor C2 to 5 μF. This lengthens the time constant to 5 seconds.

$$t = 0.000005 \times 1{,}000{,}000 = 5 \text{ seconds}$$

Press S1 several times. Notice that the output goes negative for the same length of time regardless of how long the switch is held closed.

Now replace resistor R6 with a 470 K resistor. What is the time constant now?

$$t = 0.000005 \times 470{,}000 = 2.35 \text{ seconds}$$

Experiment with other values of R6 and C2. Increasing either or both of these component values will make the time constant longer. Decreasing either or both of these component values will reduce the output time constant.

Very short time constants may be difficult to see. For instance, if R6 equals 10 K and C2 equals 0.022 μF, then the time constant will be very short.

$$t = 0.000000022 \times 10{,}000 = 0.00022 \text{ second}$$
$$t = 0.22 \text{ millisecond}$$

The circuit will work using these small values, but the resulting time constant is far too brief for human perception. The LED will blink too fast to be seen, and the voltmeter pointer will not have time to respond to the change in the output voltage before the circuit times out and returns to its normal rest state.

9

Power Supplies

An op amp IC can be damaged by a defective or improperly selected power supply. Most op amp circuits have low enough current drain to make the use of batteries feasible, at least for the short-term. This is often undesirable. Replacing batteries can become expensive very rapidly. In addition, there is also the possibility of installing a battery with the polarity reversed, which can destroy the op amp IC.

It is generally desirable to use an ac to dc power supply whenever possible. Such power supplies must meet certain requirements to avoid damaging the IC and to prevent erratic operation. In most cases, a tightly regulated supply will be necessary.

In this chapter, I will discuss power supplies and protection for op amp circuits. I will also explain how op amps can be used within power supply circuits as control circuitry.

DUAL POWER SUPPLIES

Most op amp ICs require a power supply that provides a positive voltage and a negative voltage with respect to ground. This requirement is illustrated in Fig. 9-1. For convenience, the power supplies are shown as batteries.

In most cases, the dual-polarity power supply should be symmetrical. That is, +V should be equal to −V. If these voltages are unequal, a dc voltage offset proportional to the difference will appear at the output. This will introduce errors in dc outputs and

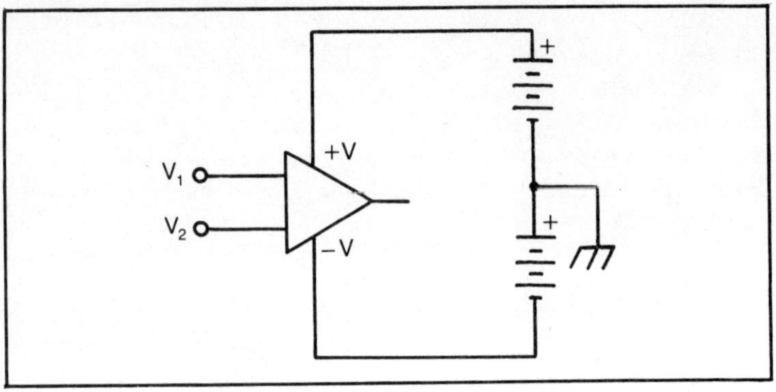

Fig. 9-1. Most op amps require dual-polarity power supplies.

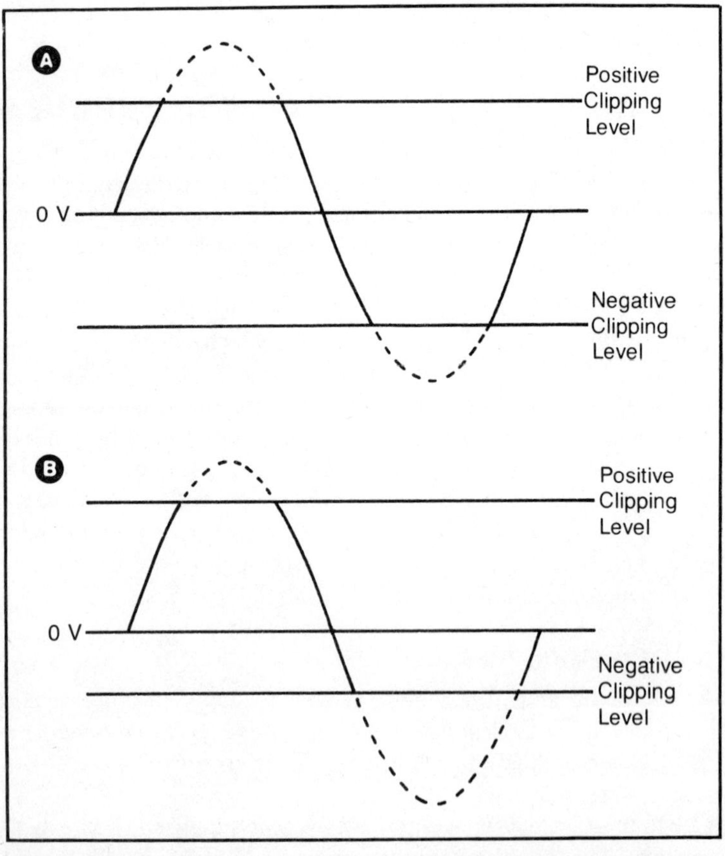

Fig. 9-2. With the correct dual-polarity supply, the effects of clipping will be symmetrical (A). If the supply voltages are unequal, signals will be unevenly clipped (B).

will cause ac signals to ride on a dc level. Often, output capacitors will be necessary to remove this offset dc voltage.

Nonsymmetrical supply voltages can also result in unequal clipping, as shown in Fig. 9-2.

Voltage regulation of the power supply is generally desirable because fluctuating supply voltages can cause momentary nonsymmetrical conditions and varying clipping (saturation) levels that cause erratic operation.

A basic dual-polarity power supply circuit is illustrated in Fig. 9-3. Notice that the power transformer is centertapped. This is mandatory for this type of circuit.

The bridge rectifier (D1 through D4) simultaneously charges capacitor C1 positively and capacitor C2 negatively. Since the charging of these two capacitors comes from the same source, they are equally charged with opposite polarities with respect to ground.

Of course C1 equals C2 and R1 equals R2 to maintain symmetry. The capacitors filter the pulsating dc from the diodes in the bridge rectifier, but additional filtering or voltage regulation is generally needed for most op amp circuits. Voltage regulation will be discussed a little later in this chapter.

It is possible to build a dual-polarity power supply using a transformer without centertap, although the results won't be quite as good. Such a circuit is shown in Fig. 9-4. The transformer in this circuit should have about half the secondary voltage as you'd use in the centertapped circuit of Fig. 9-3.

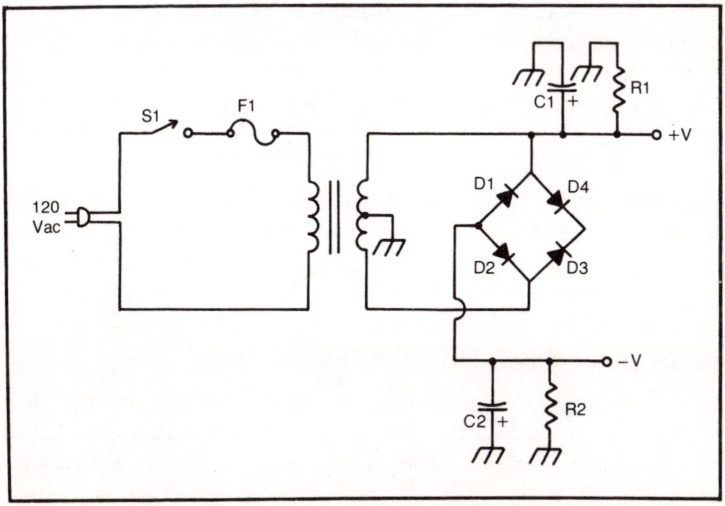

Fig. 9-3. This is a typical dual-polarity power supply suitable for use with op amps.

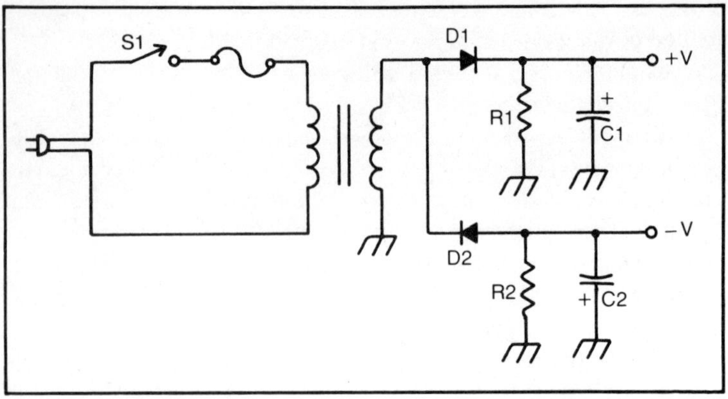

Fig. 9-4. This circuit demonstrates how a dual-polarity power supply can be built around a transformer without a centertap.

On one half-cycle capacitor, C1 is charged positively through diode D1. On the other half-cycle, C2 is charged negatively through diode D2. Since both half-cycles are equal (the sine wave is symmetrical waveform), the two output voltages will be symmetrical.

Since each output is only half-wave rectified, the pulsations of the output voltage is greater, requiring larger filter capacitors and tighter regulation.

Most op amp ICs will accept power supply voltages from about ±9 volts to ±18 volts. In some applications, slightly lower supply voltages may be used. It is generally a good idea to keep the supply voltage as low as possible to permit the full output voltage swings required. A power supply voltage will place less strain on the op amp reduce the effects of internal noise. Lowering the supply voltages will not affect circuit gain, but the positive and negative output saturation levels will be lowered, increasing the possibility of clipping problems.

The specific requirements of the circuit you are working with will dictate whether a lower power supply may be used. Many manufacturer's spec sheets and many circuit diagrams list +V as V_{cc} and −V as V_{ee}, or V_{dd}.

USING SINGLE-ENDED POWER SUPPLIES

In some applications, you may be able to get away with a single-ended (one polarity) power supply, such as those used with most transistor circuits. This is true of some ac coupled circuits where the dc errors in offset are not too critical.

A few IC op amps are designed to permit free use of a single-ended power supply. For example, the 324 (a quad 741) needs one power supply connection and a ground connection.

It is also possible to adapt a single-ended power supply so that it will behave like a dual-polarity power supply. The simplest technique is to use a voltage divider resistor network, as illustrated in Fig. 9-5. Notice that the output includes an artificial, *floating* ground. This point should not be connected to a true earth ground.

The two resistors should have equal values to provide the required symmetrical outputs. Let's work through a typical case to explain why.

I'll assume each of the resistors in the voltage divider has a value of 10 K (10,000 ohms). Therefore, the series combination equals 10 K + 10 K = 20 K (20,000 ohms). I will place 24 volts across the voltage divider. (For simplicity, I will ignore the load resistance.) According to Ohm's Law the current flowing through the voltage divider resistors is 1.2 mA.

$$I = \frac{V}{R} = \frac{24}{20000} = 0.0012 \text{ amp} = 1.2 \text{ mA}$$

This current flows through each of the resistors individually. Therefore the voltage drop across each of the resistors is 12 volts.

$$V = IR = 0.0012 \times 10{,}000 = 12 \text{ volts}$$

The junction between the two resistors is taken as the common reference point. This means +V is 12 volts above the common point, and −V is 12 volts below.

I have converted a 24 volt single-ended power supply into a ±12 volt dual-polarity power supply. For the best results, precision, low tolerance resistors should be used in the voltage divider to

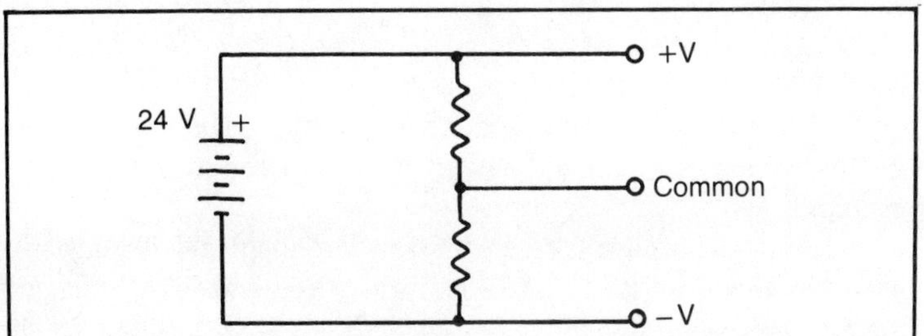

Fig. 9-5. A voltage divider resistance network can create an artificial ground to form a pseudo dual-polarity power supply.

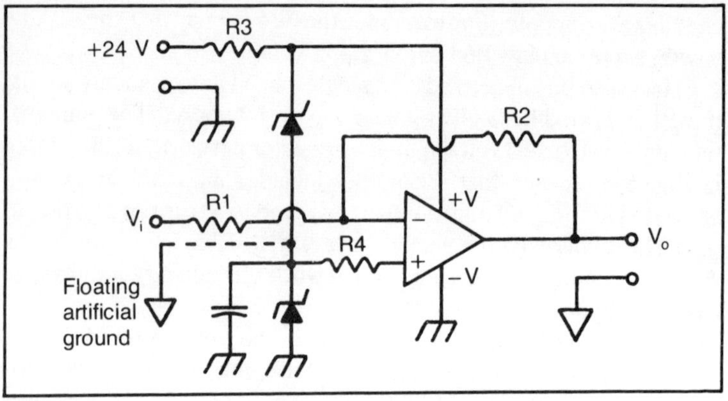

Fig. 9-6. Zener diodes can be used to convert a single-ended power supply into dual-polarity supply voltages with an artificial ground.

ensure symmetry. If high tolerance components must be used, they should be measured with an ohmmeter and matched as closely as possible.

Another approach to converting a single-ended power supply into a dual-polarity power supply is shown in Fig. 9-6.

Here zener diodes are used to place the noninverting input in an inverting amplifier circuit at virtual ground. Of course, the two zener diodes should be identical types to divide the supply voltage in half.

Notice that the applied single-polarity voltage must be twice the desired split-polarity voltages. That is, a +30 volt power supply must be employed to run an op amp circuit at ±15 volts.

Resistor R4 is used for offset isolation. The resistance of this component should equal that of the feedback resistor (R2). Since the bias currents flowing through the noninverting input are typically quite small, no appreciable voltage drop will appear across R4. Resistor R3 carries a much larger current, so it can be used to drop the supplied voltage (2X+V)down to whatever specific voltage is needed at the +V terminal on the IC.

The apparent (floating) ground is the junction between the two diodes. This point should never be connected to (true) earth ground.

The two capacitors are used for power supply decoupling, and are very important for reliable operation. Remember that op amps are high gain, high frequency devices. Any stray signals that may be coupled back through the power supply could lead to severe instability problems. Decoupling capacitors can filter out most stray signals that might sneak into the power supply lines.

The decoupling capacitors should always be placed physically as close as possible to the IC itself, or there may be resonance problems due to the inductance of the connecting wires.

These single-ended power supplies with an aritifical ground will be perfectly functional for many op amp circuits, however there are some practical limitations. The most important is that all signals must be referenced to the artificial floating ground. True earth grounds must not be used. This isn't a problem in many applications, but when it is, you will have to use a true dual-polarity power supply.

VOLTAGE REGULATION

Power supplies for op amp circuits should be voltage regulated. The output of a typical power supply circuit, such as those discussed earlier in this chapter, contain a certain amount of ripple, as illustrated in Fig. 9-7. A remnant of the original ac signal rides on top of the desired voltage. You can reduce the amount of ripple somewhat, by increasing the size of the filter capacitors. Unfortunately, filter capacitors can not completely eliminate the ripple.

Why does this ripple matter? The actual supply voltages vary from instant to instant. This causes the output saturation levels of the op amp to fluctuate, which could result in erratic clipping and distortion of high-level output signals. The ripple effect can also result in noisy operation in many circuits. Moreover, if the ripple's phase gets shifted in one half of the supply (either V+ or V−), the dual voltages will not be continuously symmetrical.

Another problem is voltage transients, or power spikes. These are momentary surges of power that could damage the IC if its maximum supply voltage is exceeded. Voltage regulation holds the

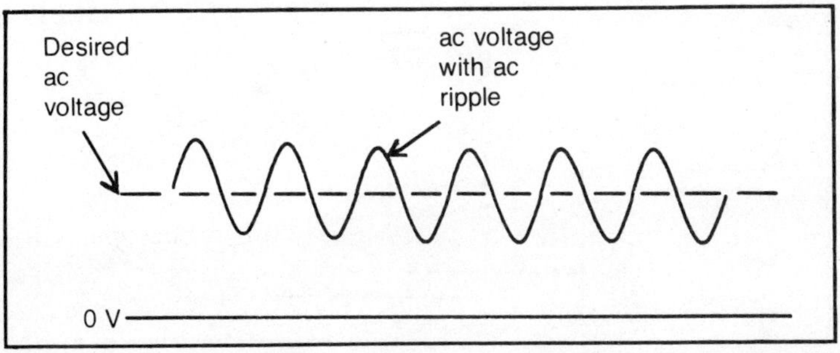

Fig. 9-7. Most practical power supplies exhibit some degree of ac ripple in their output voltages.

supply voltage(s) to a constant dc level, even if the current drawn by the load changes.

Figure 9-8 shows a typical power supply/load circuit. The load (circuit being driven by the power supply) is basically a resistance across the power supply outputs. The voltage dropped across this resistance (the output voltage) is determined by the current drawn by the load circuit and internal resistances (R_x) within the power supply itself.

Assume I have a power supply circuit that generates 10 volts when the load current is 5 mA(0.005 amp). The internal resistances total up to 500 ohms. According to Ohm's Law the same current passes through each of the resistors in the circuit. Therefore, the voltage drop across R_x is I × R.

$$V = IR = 0.005 \times 500 = 2.5 \text{ volts}$$

The apparent value of the load resistance R1 is determined by the current drawn through it and the voltage drop across it. For my example, normal operation results in a load resistance of 2000 ohms.

$$R = \frac{V}{I} = \frac{10}{0.005} = 2000 \text{ ohms}$$

If the current drawn by the load circuit increases to 10 mA (0.01 amp), the voltage drop across the internal resistances (R_x) increases.

$$V = IR = 0.01 \times 500 = 5 \text{ volts}$$

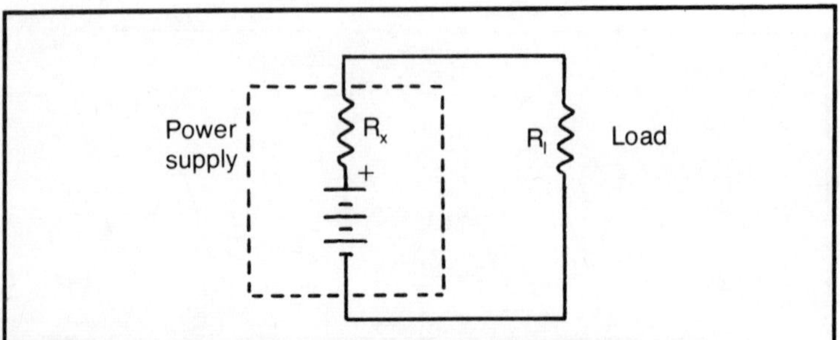

Fig. 9-8. The voltage to a load depends on the load resistance and the internal resistances of the power supply itself.

In the normal case described previously, the power supply is actually generating 12.5 volts—2.5 volts is dropped across the internal resistances (R_x), and 10 volts is dropped across the apparent load resistance (R1). Now, the power supply continues to generate the same voltage. Since 5 volts is being dropped across R_x, that leaves 12.5 – 5 = 7.5 volts to be dropped across the load. The apparent resistance has changed to 750 ohms.

$$R = \frac{V}{I} = \frac{7.5}{0.01} = 750 \text{ ohms}$$

As the current drawn by the load increases, the apparent load resistance drops, causing the supplied voltage to fall off.

Obviously this is undesirable in any kind of precision application where the current draw might vary more than a few percent. The fluctuating supply voltage can cause erratic operation and multiple obnoxious problems.

Adding a voltage regulation stage to the power supply makes the output voltage virtually independent of the load. The simplest form of voltage regulation is simply a zener diode. The output voltage will not be permitted to exceed the zener voltage, however, it may drop below this point.

An improved voltage regulator is shown in Fig. 9-9. When a positive unregulated voltage is applied, capacitor C1 begins to charge up to a level limited by zener diode D1. If the input voltage increases for any reason, D1 will shunt the excess voltage to ground. If the input voltage drops momentarily, the time constant of the capacitor will slow the reaction. Therefore, the voltage being applied to the base of the npn transistor is kept at a fixed level.

The transistor effectively multiples the capacitance of C1, causing it to be even slower to any drops in input voltage. The capacitance of C1 is multiplied by the dc beta of the transistor.

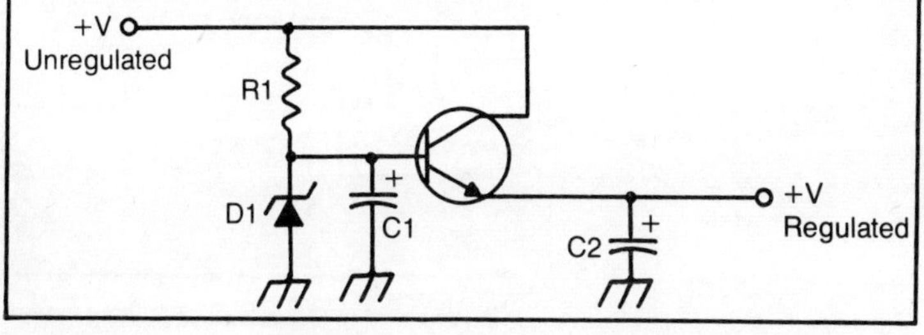

Fig. 9-9. This simple circuit can be used to regulate a single-ended power supply.

The transistor also serves as an emitter follower. The regulated voltage at the emitter of the transistor is equal to the zener voltage less the forward drop across the base/emitter junction of the transistor.

Capacitor C2 is another filter capacitor to help absorb the effects of any transients that manage to get through the transistor. Both capacitors (C1 and C2) should be quite large so they are slow to discharge if the input voltage should drop off for any reason.

The regulated output voltage of this circuit is determined entirely by the zener diode and the transistor, regardless of the current drawn by the load. Even fairly large changes in the current drawn will not have a significant effect on the output voltage.

This circuit can only handle a single-polarity power supply. The resistor of zener diode voltage divider network may be added after the regulator if a single-sided supply is being used.

For true dual-polarity (true earth ground) operation, dual-polarity voltage regulation is obviously needed. A typical circuit is illustrated in Fig. 9-10. Notice that transistor Q1 and its associated components form the same circuit that I just discussed (Fig. 9-9). Transistor Q2 and its associated components make up a mirror image of the circuit for the opposite polarity (–V).

The output voltages are referenced to true earth ground.

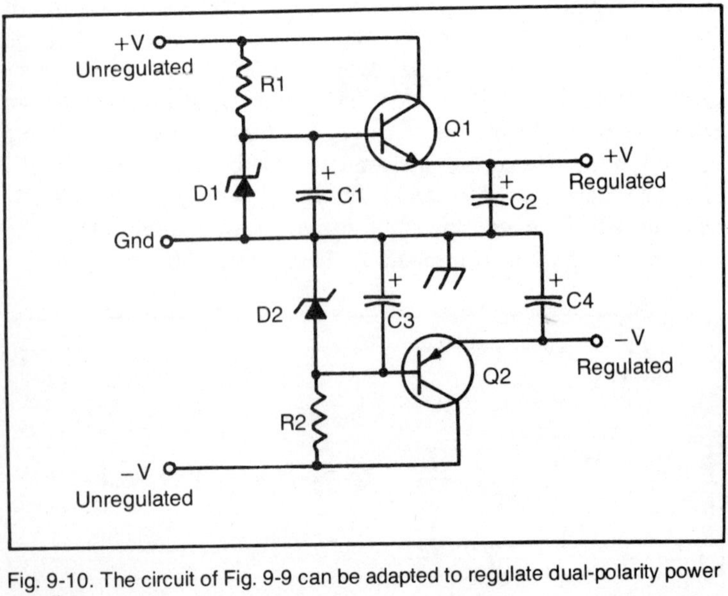

Fig. 9-10. The circuit of Fig. 9-9 can be adapted to regulate dual-polarity power supplies.

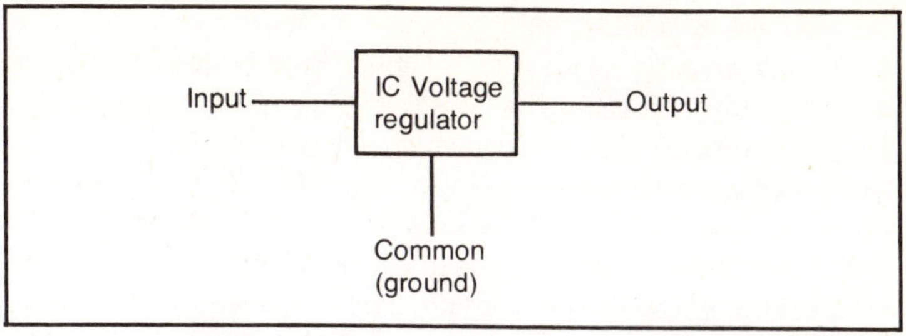

Fig. 9-11. Voltage regulator ICs have three leads: input, output, and common.

Notice that a pnp transistor is used in the −V portion of the circuit.

A number of IC voltage regulators are available. They cover a wide variety of supply voltages and offer very good regulation over a wide range. These devices have three leads, as shown in Fig. 9-11. They are input (unregulated voltage), common (ground), and output (regulated voltage). Generally, the input voltage should be somewhat greater than the desired output voltage to cover internal losses. IC voltage regulators are available in versions for positive voltages and negative voltages.

Input and output filter capacitors should be used with IC voltage regulators to ensure the best regulation. This is shown in Fig. 9-12. These filter capacitors should have moderate to large values. Large capacitances are difficult to place on a semiconductor chip, so external components are required.

Additional capacitors should be placed near the voltage supply leads of the IC(s) in the circuit being driven for decoupling and to aid in protection against transients. These decoupling capacitors may be considerably smaller than the main filter capacitors which are usually placed near the power supply itself. Typical values for decoupling capacitors range from about 0.01 μF to 0.1 μF.

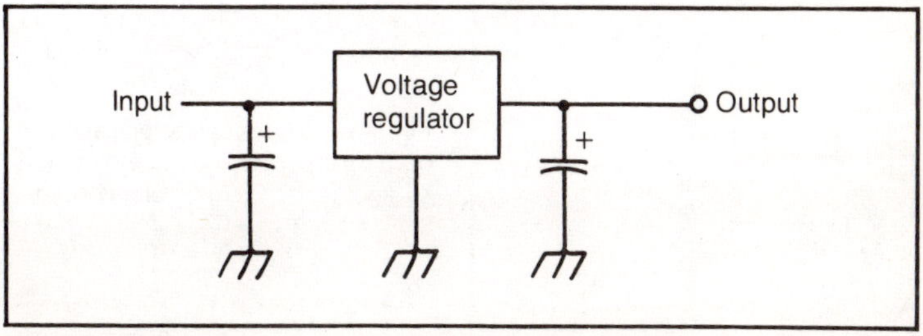

Fig. 9-12. Filter capacitors should be used at the input and output of a voltage regulator IC.

PROTECTING OP AMPS

There are several ways an op amp IC can be damaged by the power supply. Probably the most obvious is overvoltage. The manufacturer's spec sheet for each device will specify a maximum supply voltage which should never be exceeded. If the maximum supply voltage is ±18 volts, **do not** use a ±20 volt power supply. In fact, it would be a good idea not to exceed about 80 percent to 90 percent of the specified maximum supply voltage. This leaves a little headroom for errors due to component tolerances, momentary transients, and ripple.

Two other potential problems due to power supplies are polarity reversal and transient spikes. Either of these can destroy an op amp IC in a fraction of a second.

The supply voltage polarities must not under any circumstances be reversed. That is a negative voltage should not be placed on the +V terminal, and a positive voltage should not be fed to the −V terminal. This is usually not much of a problem in equipment using electronic power supplies. Once the power supply leads are correctly connected they are left alone, and the polarity remains constant.

Circuits using batteries require some care, however. It is all too easy to install a battery backwards, with the polarity reversed. This can almost instantly zap an op amp IC.

One way to protect against inadvertant polarity reversal would be to add a pair of diodes in the power supply lines, as shown in Fig.9-13. Each diode permits current flow only in one direction, insuring correct polarity. If the polarity is reversed, the diodes will block the supply voltages from reaching the IC.

Transient spikes can be a trickier problem. Essentially, these are bursts of noise that appear on the power supply lines. They can

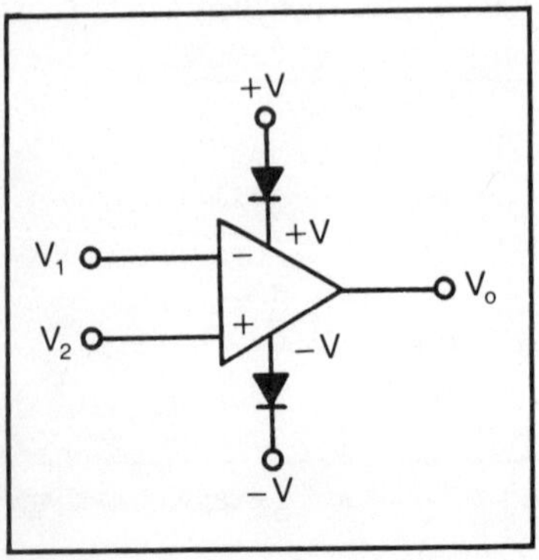

Fig. 9-13. Diodes can be placed in an op amp's supply lines to protect the IC against accidental polarity reversal.

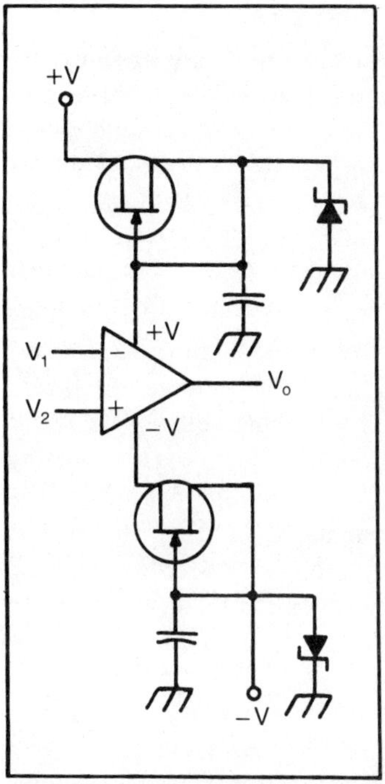

Fig. 9-14. This circuitry can be used to protect an IC from transient spikes.

be due to many different causes. Rf pick-up or coupling back to the power supply are two possibilities. Some transient spikes can be many times higher than the nominal supply voltage. They can easily be large enough to exceed the maximum momentary rating of the IC chip. Transients are extremely brief, perhaps even less than a millionth of a second, but if they are large enough, they have plenty of time to blow out an IC.

In an application where this is likely to be a problem, it may be a good idea to add a transient suppression circuit to the power supply lines, as illustrated in Fig. 9-14. The FETs act as voltage absorbing current sources, and the zener diodes clamp the voltage reaching the ICs terminals to a safe value. These zeners should be selected for clamping values that are somewhat greater than the nominal supply voltage, but less than the manufacturer's specified maximum supply voltage. For instance, in a circuit running off of a ±12 volt supply, using an op amp IC with a maximum permissible supply voltage of ±18 volts, 15 volt zener diodes would be a good choice. This will leave the zener diodes in a normally off condition except when they are needed during a supply voltage transient spike.

USING OP AMPS IN POWER SUPPLY CIRCUITS

Op amp devices have many applications. They can even be used within power supply circuits themselves. Figure 9-15 shows a voltage regulator circuit built around an op amp. The op amp serves as a reference voltage source to control the base of the npn series-pass transistor. The reference voltage is set by the zener diode and resistor R1. This voltage is fed into the noninverting input of the op amp.

At the same time, a voltage proportional to the output voltage is tapped off of the voltage divider, made up of resistors R2, R3, and R4. R2 can be adjusted to calibrate the output voltage. A 10 turn potentiometer is preferred for the finest adjustment.

The sample voltage tapped off from R3 is fed into the inverting input of the op amp. If the inverting input voltage is equal to the noninverting input voltage, the base signal of the transistor will be held constant. If the output voltage is increased or decreased, the op amp will be thrown out of balance, causing its output signal to change, altering the voltage driving the base of the transistor, correcting the output voltage.

For the resistor values listed in Table 9-1, this circuit will accept a 18 to 20 volt positive unregulated input and convert it to a regulated +9 volt signal. Obviously, this circuit is not particularly efficient. Half of the input voltage is wasted. This is true of many simple voltage regulator circuits.

This circuit will supply output currents of up to 100 mA (0.1 amps).

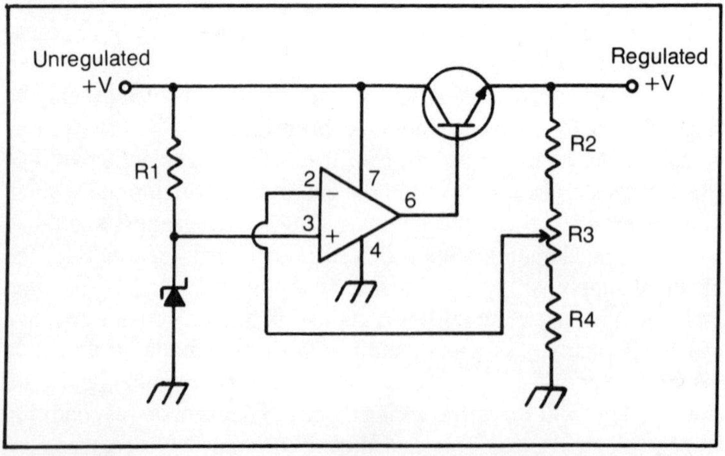

Fig. 9-15. Op amps can be used at the heart of voltage regulator circuits.

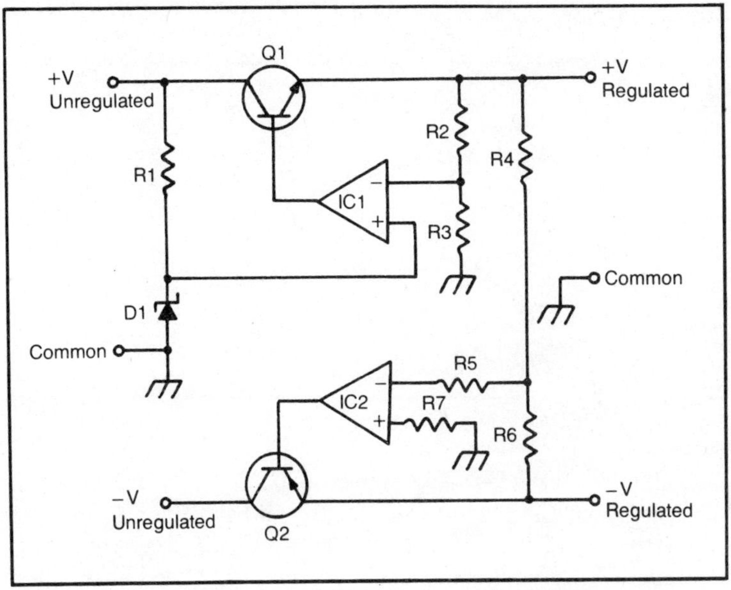

Fig. 9-16. This circuit will regulate dual-polarity supply voltages.

A dual-polarity op amp voltage regulator circuit is shown in Fig. 9-16. A dual-polarity unregulated supply serves as the input. The output is a regulated dual-polarity voltage source referenced to true earth ground.

Let's examine the positive portion of the regulator first. This section of the circuit is comprised of IC1, Q1, D1, R1, R2, and R3. For simplicity, this part of the circuit is shown separately in Fig. 9-17.

Transistor Q1 is an npn series pass element. The base current of this transistor is controlled directly by the output of the op amp (IC1).

The noninverting input signal to this op amp is a reference voltage determined by resistor R1 and zener diode D1. A portion of

Table 9-1. Typical Resistance Values for Fig. 9-15.

Input Voltage		+18 to +20 volts, unregulated
Output Voltage		+9 volts, regulated
R1	12 K	
R2	1.2 K	
R3	2.5 K 10 turn potentiometer	
R4	1.2 K	

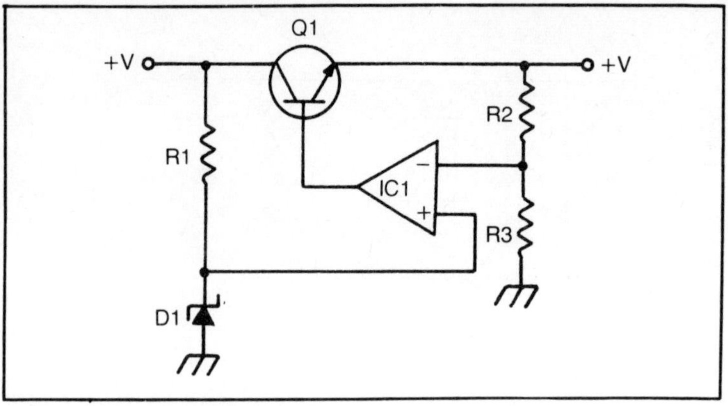

Fig. 9-17. The positive voltage regulator of Fig. 9-16 is shown separately here.

the output voltage is fed back to the op amp's inverting input through the voltage divider made up of resistors R2 and R3.

If the inverting input is fed the same voltage as the noninverting input, the amplifier will be in a stable condition, holding the output constant. If the output voltage should increase or decrease for any reason, the voltage at the inverting input will be changed proportionately. It will no longer be equal to the reference voltage set by R1 and D1. This causes the op amp's output signal to change, altering the base current of the transistor. The transistor's emitter voltage (the output) will be brought back into line with the nominal output voltage.

Even a very small change in the output voltage will cause the circuit to respond extremely quickly to correct the error, making output ripple negligible and virtually nonexistent.

You should be able to see that this portion of the circuit is very similar to the single-ended op amp voltage regulator circuit that was shown in Fig. 9-15. Now, what about the negative half of the circuit? I could just construct a mirror image of the positive voltage regulator with the polarities reversed. However, to reduce the parts count and to insure that the dual-polarity outputs will be symmetrical around earth ground, a slightly different approach is employed in this circuit.

Figure 9-18 shows just the negative portion of the dual-polarity op amp voltage regulator circuit. This section of the circuit includes IC2, Q2, R4, R5, R6, and R7. No zener diode is needed to set a reference voltage in this part of the circuit. The positive regulated output voltage is used as the reference voltage source. Notice that Q2 is a pnp unit to account for the change of polarity.

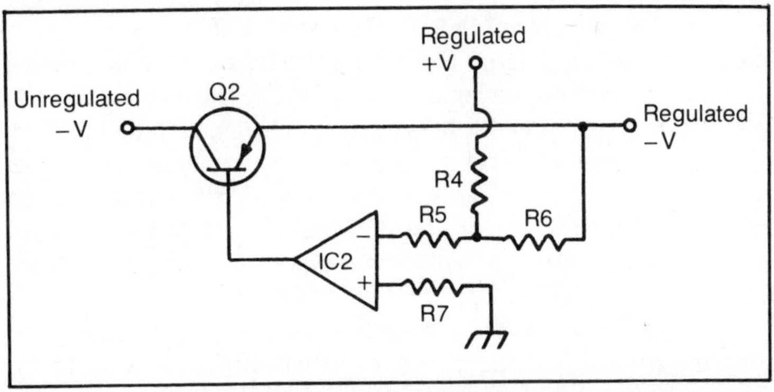

Fig. 9-18. The negative voltage regulator of Fig. 9-16 is shown separately here.

Resistor R7 references the noninverting input of the op amp (IC2) to ground. Resistors R4 and R6 have identical values. This means the voltage at their junction will be 0 if the regulated +V output voltage is equal to the regulated −V output voltage. This signal is fed to the op amp's inverting input through R5.

If the two regulated output voltages are not equal, a nonzero signal will be fed into the inverting input, causing the op amp's output voltage to be shifting, controlling the base current of transistor Q2.

Since the +V voltage is regulated by the other half of the circuit, changes in Q2's base current cause it to adjust the −V output voltage.

One major advantage of this circuit is that even if the output voltage change momentarily, the outputs will maintain symmetry, since the −V regulated voltage is referenced to the +V regulated voltage. This is of major importance in precision op amp circuits.

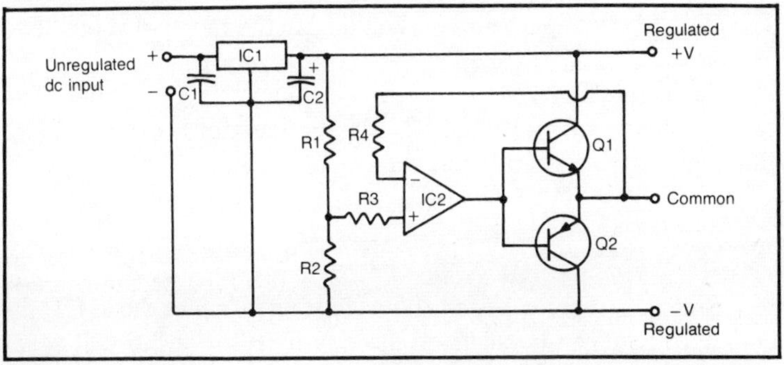

Fig. 9-19. This circuit can convert a single-ended power supply into a dual-polarity power supply.

The circuit shown in Fig 9-19 converts a single-ended power supply into a dual-polarity power supply with an aritifical (floating ground). As with other single-to-dual-polarity converters, the output common should never be shorted to true earth ground.

This circuit requires a regulated input voltage. A voltage regulator (IC1) and two capacitors (C1 and C2) allow you to use an unregulated signal. Since the input voltage is regulated, both halves of the dual polarity output will also be regulated. An advantage here is that only a single voltage regulated is required.

Q1 and Q2 are a pair of complimentary (opposite polarity but with equal electrical characteristics) shunt regulator transistors. They create artificial ground to be used by the load circuit. This artificial ground is occassionally referred to as a *counterpoise* ground.

Resistor R4 ties the inverting input of the op amp to the artifical ground point. The noninverting input is fed a reference voltage set by the voltage divider network made up of resistors R1 and R2. If the artificial ground level shifts for any reason, the op amp's output will change proportionately. If the op amp's output becomes more negative, then Q2 (pnp) will start to conduct more heavily. If the op amp's output becomes more positive Q1 (npn) will start to conduct more heavily.

When the output is correctly balanced, the two transistors should conduct equally. The circuit will always try to remain in this condition. The transistors should be selected to dissipate a fairly large amount of power.

For the dual polarity output voltages to be symmetrical, resistors R1 and R2 should have equal values and R3 should equal R4.

The circuit shown in Fig. 9-20 is another single-ended voltage regulator, but the output voltage is variable from about 5 to 20 volts. The unregulated input voltage should be in the range of about 25 to 30 volts.

Basically this circuit is a modified noninverting amplifier. The gain is determined by R1 and R3. A potentiometer is used for R3 to allow for variable gain. The gain formula is as follows:

$$G = \frac{R1 + R3}{R1}$$

Since the input voltage is held constant by zener diode D1, the output voltage can be adjusted directly by R3. If a 3.6 volt zener diode is used, and R1 equals 470 ohms, the output voltage can be easily solved, using this equation:

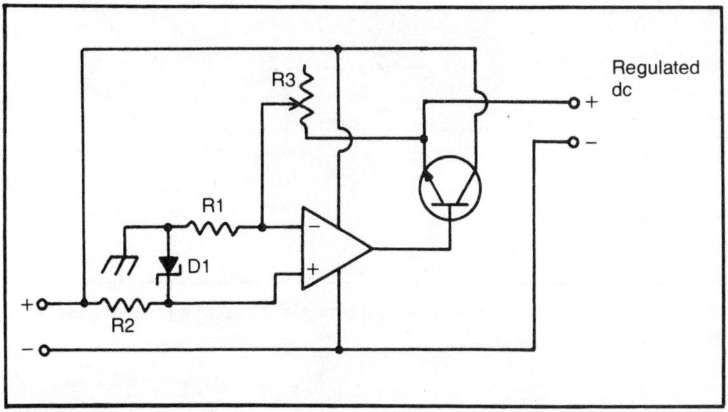

Fig. 9-20. The regulated output voltage of this circuit can be manually adjusted by potentiometer R3.

$$V_o = 3.6 + (7.7 \times R3)$$

The value of R3 should be given in kilohms in this equation to get the result directly in volts.

As an example, let's assume R3 has a value of 0.5 K (500 ohms). This would give an output voltage of 7.45 volts.

$$V_o = 3.6 + (7.7 \times .5) = 3.6 + 3.85 = 7.45 \text{ volts}$$

Raising the value of R3 to 2 K (2000 ohms) changes the output voltage to 19 volts.

$$V_o = 3.6 + (7.7 \times 2) = 3.6 + 15.4 = 19 \text{ volts}$$

A 2.5 K potentiometer should be used for resistor R3. Resistor R2 should be about 4.7 K (4700 ohms). The output voltage of this circuit will be very stable even with large changes in the load resistance.

Since the output of a 741 op amp IC is current limited, power transistor Q1 is added as a current amplifier. The output current of this circuit can range up to 0.5 amp (500 mA).

VOLTAGE/CURRENT CONVERTERS

It is sometimes necessary to convert a current into a voltage, or vice versa. Op amps can perform this function just fine.

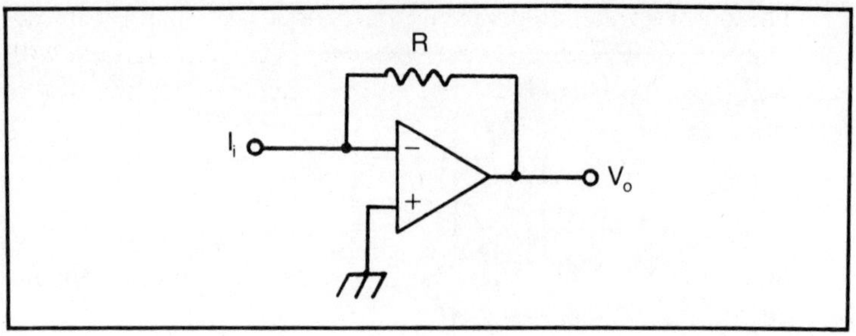

Fig. 9-21. A current-to-voltage converter circuit is similar to an inverting amplifier without the input resistor.

A simple current-to-voltage converter circuit is illustrated in Fig. 9-21. Does it look familiar? It certainly should. It is a basic inverting amplifier, but without an input resistor. The output voltage is determined by the input current, and the value of the feedback resistor.

$$V_o = I_i \times R$$

This is because of Ohm's law and the fact that the input current also flows through the feedback resistor. Let's say R has a value of 330 Ω and the input current is 10 mA (0.01 amp). In this case, the output voltage would be equal to 3.3 volts.

$$V_o = 0.01 \times 330 = 3.3 \text{ volts}$$

Raising the input current to 25 mA (0.025 amp), increases the output voltage.

$$V_o = 0.025 \times 330 = 8.25 \text{ volts}$$

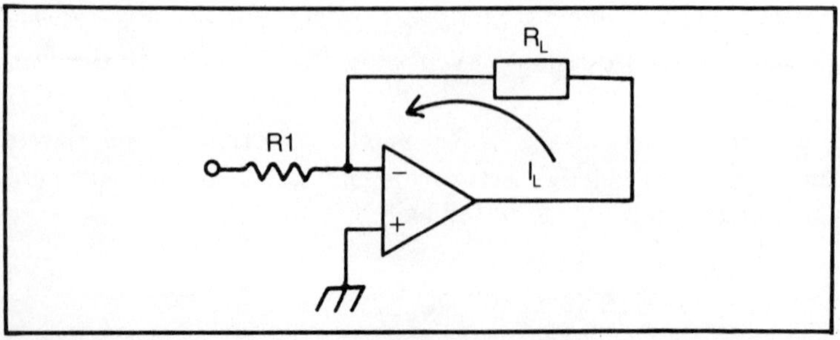

Fig. 9-22. The load current in a voltage-to-current converter is independent of the load resistance.

The output voltages changes in proportion to changes in the input current. With practical components, input bias currents will have an effect. The actual voltage will be equal to the sum of the currents time the resistance.

$$V_o = (I_i + I_b) \times R$$

An op amp with the lowest possible input bias currents should be selected for obvious reasons.

Figure 9-22 shows a circuit for going in the other direction. This is a voltage-to-current converter. It is used to drive relays and analog meters. The feedback resistance, labeled R1 is the coil of the relay or ammeter. The current flowing through the load is determined by the input voltage and the value of resistor R1, according to this formula:

$$I_l = -\frac{V_i}{R1}$$

The load current (I_l) is entirely independent of the value or the load resistance (R_L), which is usually quite small. This voltage-to-current converter circuit is also known as a transmittance amplifier.

The circuit of Fig. 9-22 uses the inverting input of the op amp, so the signal polarity is inverted. If this is undesirable, the noninverting voltage-to-current converter circuit shown in Fig. 9-23 may be used. For the noninverting version, the equation for the load current (I_l) remains unchanged except for the polarity.

$$I_l = \frac{V_i}{R1}$$

CONSTANT-CURRENT SOURCE

Some circuit applications require a reference current which

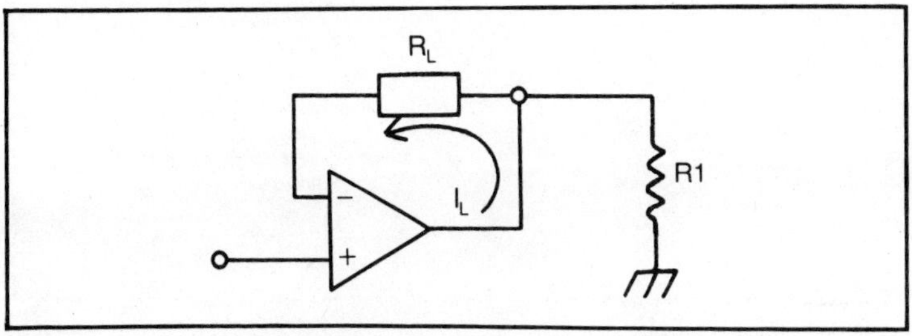

Fig. 9-23. This current is a noninverting voltage-to-current converter.

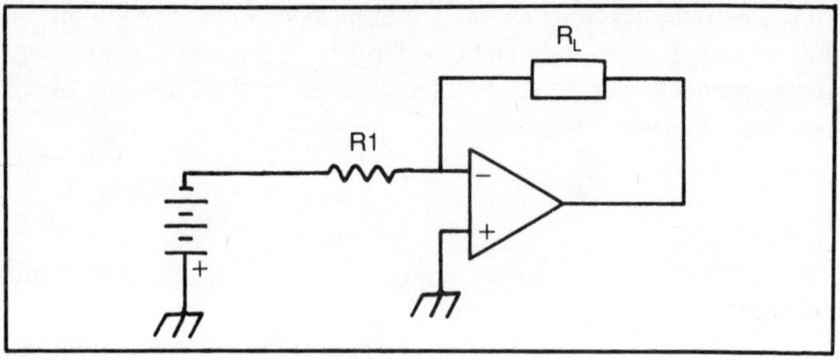

Fig. 9-24. A constant-current source is quite similar to the voltage-to-current converter circuit of Fig. 9-22.

should not change with the load. The op amp can be used here too. Basically, I use a slight variation of the voltage-to-current converter to create a constant-current source. The circuit is shown in Fig. 9-24.

The input voltage is a stable reference voltage (shown as a battery in the diagram). As with the voltage-to-current converters discussed above, the load current (I_l) is determined solely by the input voltage (V_i) and resistor R1.

$$I_l = \frac{V_i}{R1}$$

Since both the input voltage and R1 are constant values, the load current (I_l) has a constant, unvarying value. It is not effected by the load resistance (R_L). Even if the load resistance changes drastically, the current flowing through the load will remain unchanged as long as V_i and R1 are held constant.

10

Detector Circuits

Yet another common application for the versatile operational amplifier is to detect certain conditions in the input signal. Most op amp detector circuits are variations on a circuit known as a comparator. A few of the circuits in earlier chapters used comparator stages.

COMPARATORS

A comparator is a circuit that compares two signals and indicates which is larger. Generally op amps in comparator applications are operated either open loop (no feedback path), or with some positive feedback (feedback loop from output to noninverting input).

Typically, one of the signals being compared is held constant and serves as a reference level (V_{ref}). The other input varies and is compared to the reference signal.

A basic comparator circuit is illustrated in Fig. 10-1. The battery serves as a reference voltage that is fed to the inverting input of the op amp. The fluctuating input signal is fed to the noninverting input.

Notice that there is no feedback loop. This means the op amp operates at its full open-loop gain. In an ideal op amp, the open-loop gain will be infinity. For practical op amps the open-loop gain is finite, but very large, so I can reasonably assume infinite gain in my discussion. In either case, a small differential input signal will drive the output of the op amp into saturation.

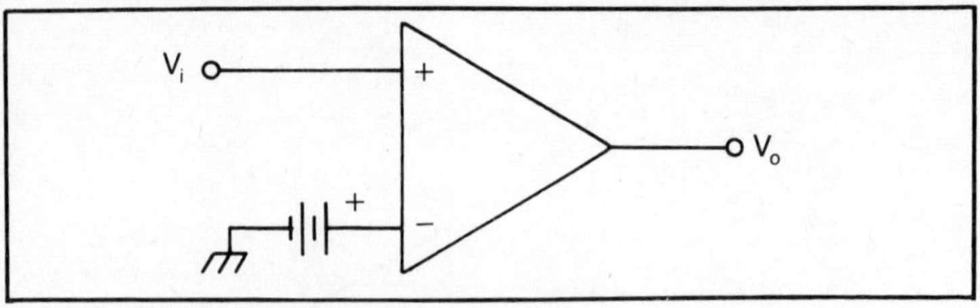

Fig. 10-1. A basic comparator circuit compares an input signal to a reference signal and determines which is larger, or if they are equal.

Assume the battery (V_{ref}) is +3.0 volts. Let's examine what happens with various values for V_i. First, assume V_i also equals +3.0 volts. The differential input is the difference between the noninverting (V_i) and the inverting (V_{ref}) voltages.

$$V_d = V_i - V_{ref}$$

When these two input voltages are equal, the differential input is zero.

$$V_d = 3.0 - 3.0 = 0$$

A differential input of 0 results in an output of 0, regardless of the gain (ignoring the effects of offsets and bias currents).

Now, raise V_i to 3.1 volts. The differential input voltage is found in the same manner.

$$V_d = 3.1 - 3.0 = 0.1$$

The output voltage (V_o) is the product of the differential input voltage and circuit gain.

$$V_o = V_d \times G$$

Since the open-loop gain here is effectively infinite, however, the output is driven into saturation when V_d deviates from zero.

$$V_o = V_d \times \infty = \infty = V_{sat}$$

Since in this example the differential input voltage is positive, the op amp is driven into positive saturation. The positive polarity of the output voltage indicates that V_i is greater than the reference voltage.

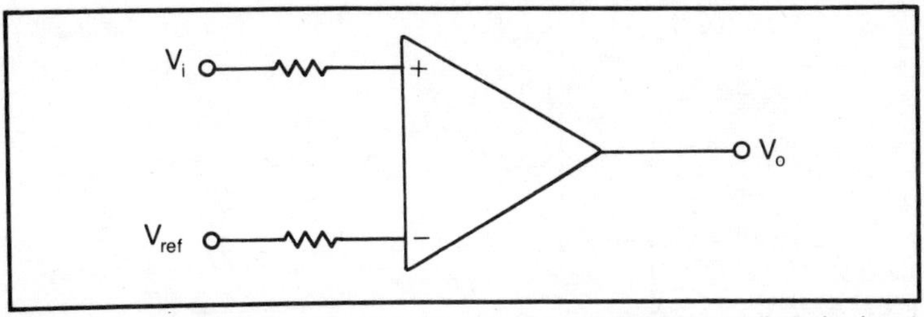

Fig. 10-2. A practical comparator circuit uses input resistors to limit the input current.

If V_i is reduced to 2.8 volts, the differential input voltage becomes negative.

$$V_d = 2.8 - 3.0 = -0.2$$

Once again the small differential input signal drives the op amp into saturation because of the high open-loop gain. Since the differential input voltage happens to be negative, the output will be forced into negative saturation. Of course, the negative polarity of the output signal indicates that V_i is smaller than V_{ref}. Even a small difference between the input signals will cause the output to switch between the positive and negative saturation conditions.

A practical op amp comparator circuit is illustrated in Fig. 10-2. The input resistors are added for current limiting. This circuit functions in the same way as the basic circuit of Fig. 10-1.

The reference voltage should not exceed the maximum common-mode voltage rating of the IC being used. Let's say the op amp can accept differential input voltages of up to 3.5 volts. If the reference voltage is 4 volts and the input signal (V_i) drops to zero, this limit will be exceeded, and the device could be damaged or destroyed.

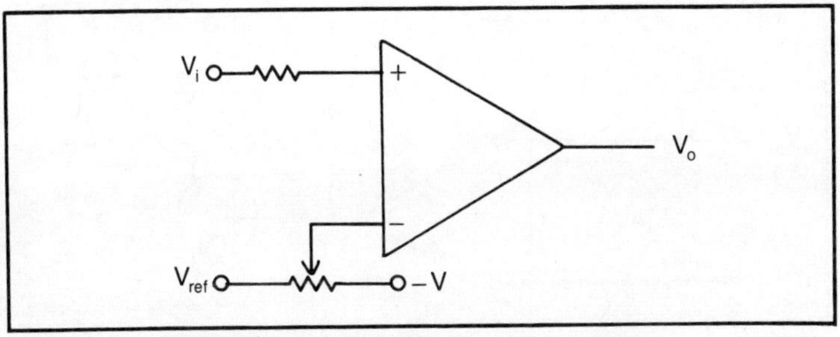

Fig. 10-3. A comparator circuit can be made more versatile if the reference voltage is manually variable.

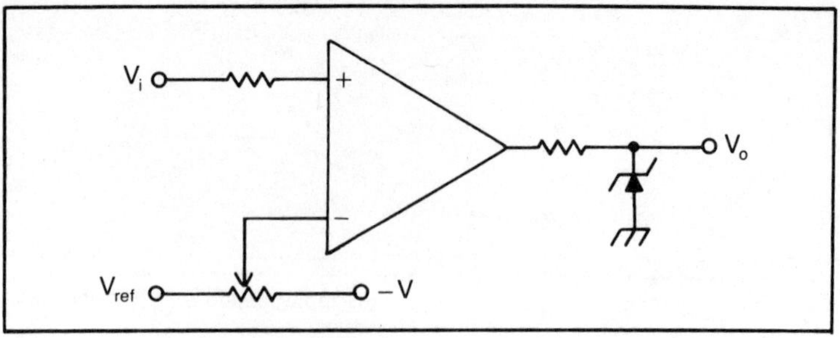

Fig. 10-4. It is sometimes necessary to limit the output voltage swings of a comparator.

Replacing the fixed resistor in the V_{ref} input resistor with a potentiometer permits a manually adjustable reference voltage. This is shown in Fig. 10-3. Obviously, this makes the comparator circuit more versatile.

In some applications the saturation voltages of the op amp may be too large for the circuit being driven. Adding the output resistor and zener diode, as shown in Fig. 10-4, will limit the output swings to a level that is acceptable to the load circuit.

Take a look at one final variation of the basic comparator. This circuit is shown in Fig. 10-5. Transistor Q1 amplifies the output current. When the op amp's output switches from negative to positive, the transistor is turned on, and lamp L1 glows.

Resistor R2 sets the base current driving the transistor, and R3 limits the current surge that can result when the cold lamp is first turned on. Diode D1 prevents the op amp from putting excessive bias on the base/emitter junction of the transistor when it is in the off condition.

The lamp will light whenever the input voltage (V_i) is greater

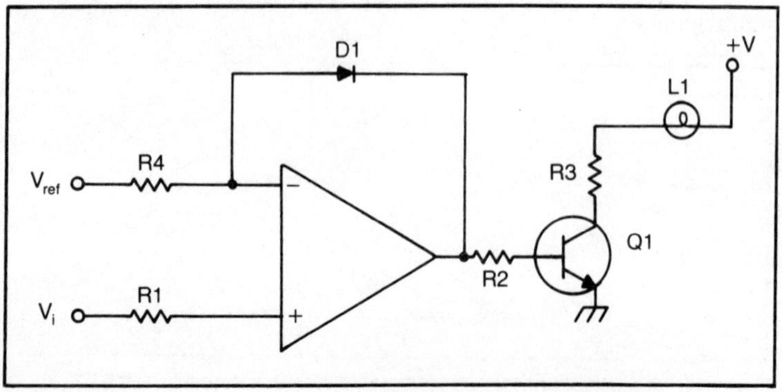

Fig. 10-5. In this circuit, the lamp will light when V_i is greater than V_{ref}.

Table 10-1. Typical Values for Fig. 10-5.

R1	1 K resistor
R2	8.2 K resistor
R3	100 ohm resistor
R4	1 K resistor
Q1	2N2222 or similar npn transistor
L1	25 volt, 40 mA flashlight bulb

than the reference voltage (V_{ref}). Typical component values for this circuit are given in Table 10-1.

ZERO-CROSSING DETECTORS

Closely related to the basic comparator is the zero-crossing detector. This is essentially a comparator that uses the circuit ground potential as the reference voltage. The basic zero-crossing detector circuit is shown in Fig. 10-6. The two resistors should have equal values. 1 K (1000 ohms) is a typical value.

Since the op amp is being operated in an open-loop configuration, the gain is very high. A small difference in the input voltage causes a very large change in the output voltage. There are three possible output conditions—zero output (or very close to it), positive saturation, and negative saturation. As the input voltage is moved away from saturation, the output will quickly pass through the intermediate voltages between zero and the saturation point.

If the input is zero, the output will be zero because $V_i = V_{ref}$. As the input passes through zero in a positive direction, the output soon jumps to the positive saturation output voltage. Similarly, if the input passed through zero towards negative voltages, the op amp's output leaps to the negative saturation condition.

This circuit's output tells if the input equals, is greater than (positive) or is less than (negative) zero. It can be called a polarity detector.

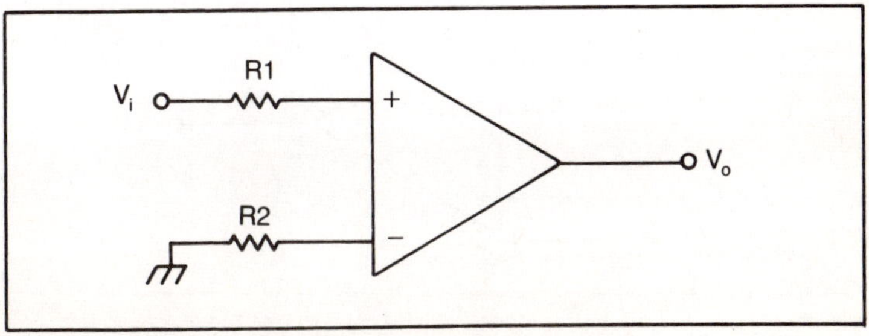

Fig. 10-6. This circuit is a basic zero-crossing detector.

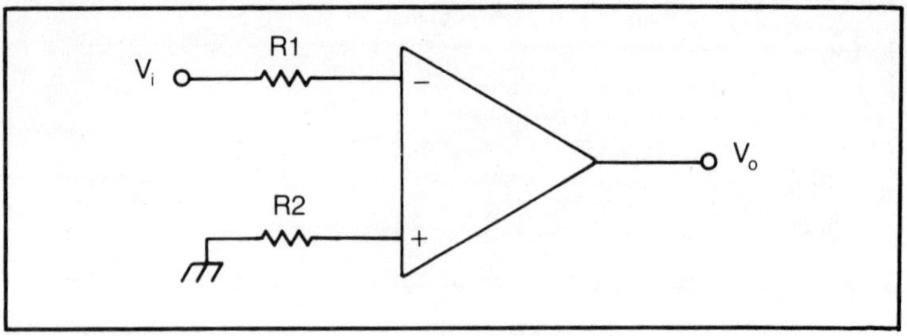

Fig. 10-7. Here is an inverting zero-crossing detector.

An inverting zero-crossing detector is easy to achieve by simply reversing the inputs. This is shown in Fig. 10-7. In this circuit, as the input goes above zero (positive), the output goes to negative saturation. When the input drops below zero (negative), the op amp is placed in positive saturation.

In many applications, the saturation voltages for the op amp being used are too large for the circuit or device being driven. I can limit the output voltage swings within a specified range by adding some feedback limiters.

The simplest approach is negative feedback through a pair of back-to-back zener diodes as illustrated in Fig. 10-8. While simple to implement and quite inexpensive, this approach leaves a lot to be desired. It is not particularly effective for low-level signals.

When the zeners are operating below the knee of their curve, the gain of the circuit drops off. This makes the circuit perform rather sluggishly. There will also be a symmetrical leading and trailing edge on each output pulse, as shown in Fig. 10-9.

A more efficient method of output limiting is to use one of the bridge circuits illustrated in Figs. 10-10 through Fig. 10-12. These

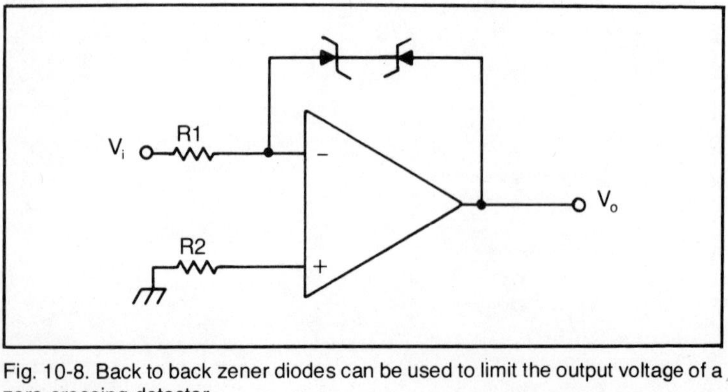

Fig. 10-8. Back to back zener diodes can be used to limit the output voltage of a zero-crossing detector.

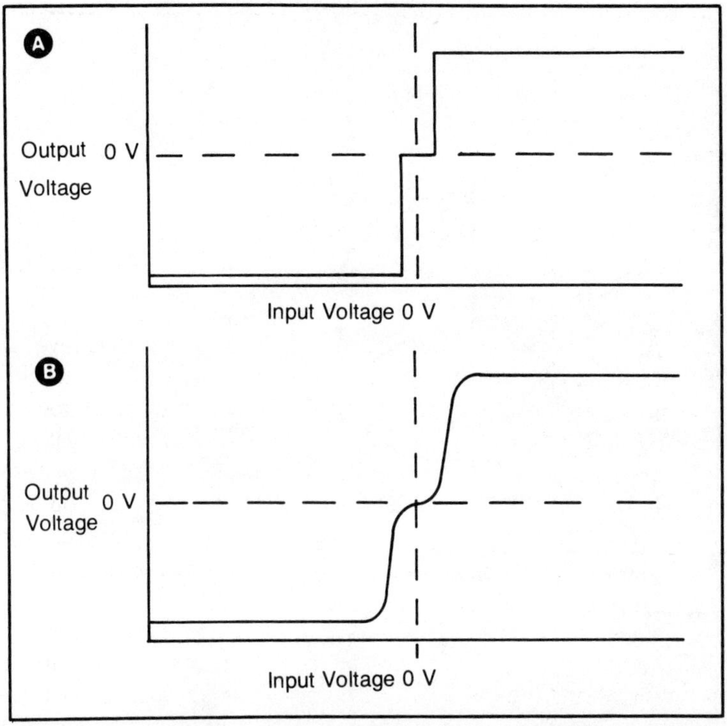

Fig. 10-9. (A) is the output graph for an ideal zero-crossing detector, while (B) represents the output of the practical circuit shown in Fig. 10-8.

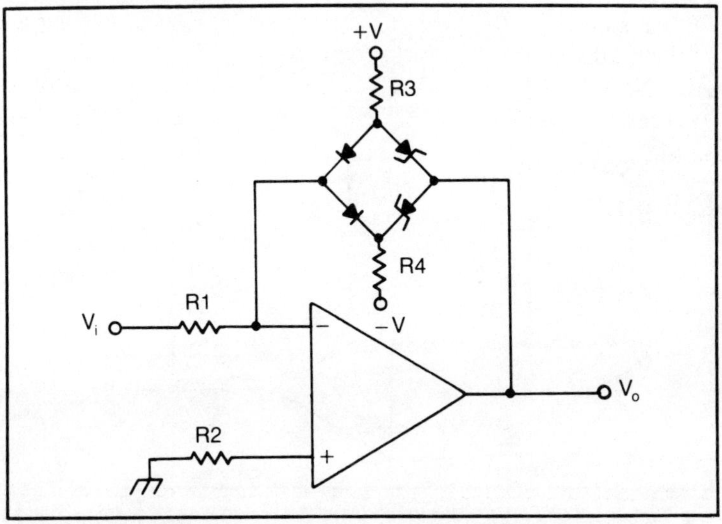

Fig. 10-10. Bridge limiting can also be used in zero-crossing detector circuits.

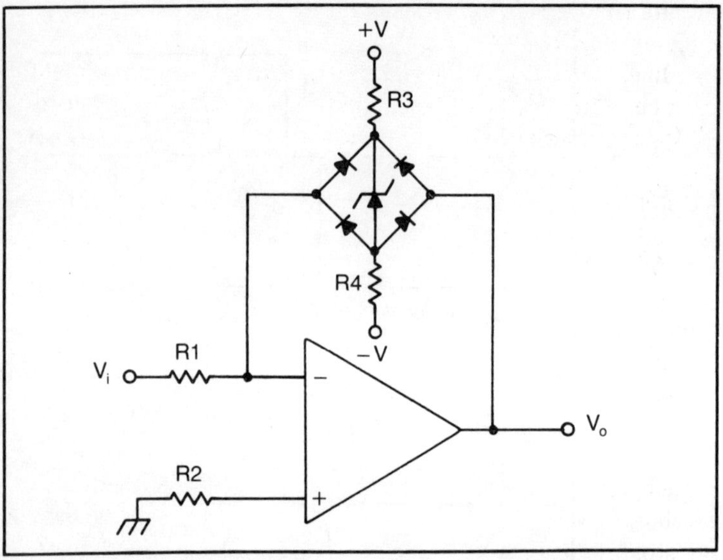

Fig. 10-11. Here is another bridge limited zero-crossing detector circuit.

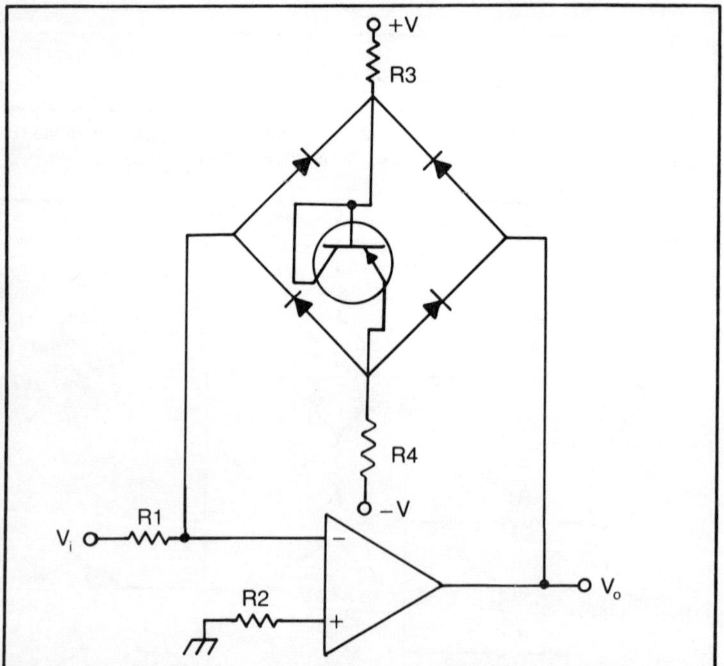

Fig. 10-12. This is yet another example of bridge limiting in a zero-crossing detector.

circuits feature hard amplitude limiting with a sharp crossover at zero, even at low-level input signals.

In each of these circuits, the input resistor is used to limit the input current. The input resistance of the circuit is determined by the value of this resistor. The reference voltage (ground connection) is made through a resistor with the same value as the input resistor to compensate for the input bias current errors within the IC itself.

All of the zero-crossing detector circuits presented so far (indeed, most zero-crossing detector circuits that have been designed) can suffer from output *chatter,* or instability when the input signal crosses zero. This is especially true if any noise gets into the signal. The problem most frequently shows up with slowly varying low-level input signals.

I can obtain some immunity from the effects of noise by adding some hysteresis to the circuit. I can do this with positive feedback, as shown in Fig. 10-13. The positive feedback path is through resistors R3 and R2. A constant voltage of the opposite polarity is fed to the noninverting input of the op amp.

Let's assume the input is negative. The output will be the positive voltage limited by zener diode D1. The positive feedback voltage at the noninverting input is determined by the following equation:

$$V_{f+} = V_{D1} \times \frac{R2}{(R2 + R3)}$$

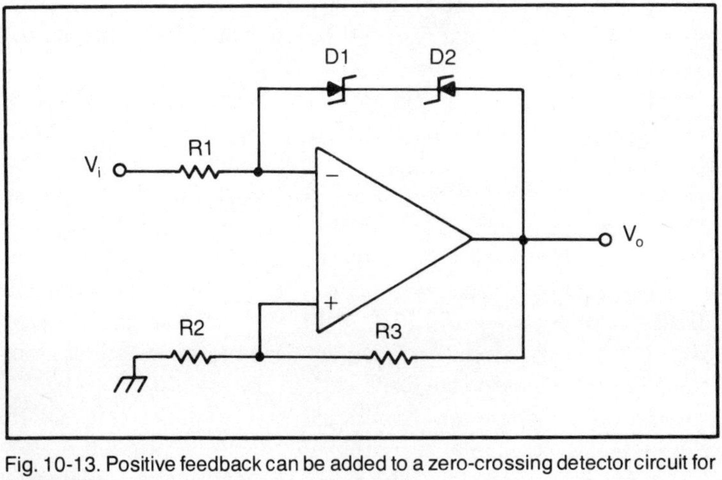

Fig. 10-13. Positive feedback can be added to a zero-crossing detector circuit for hysteresis.

Now the input can become less negative, and pass through zero, then increase positively until it equals the voltage at the noninverting input. If the input signal rises even slightly above this point, the differential input voltage being fed to the op amp becomes negative, and the high gain pulls the output of the IC towards its negative saturation voltage, which is limited by zener diode D2. Because of the positive feedback this happens very rapidly.

Now the positive feedback signal to the noninverting input becomes negative.

$$V_{f+} = -V_{D2} \times \frac{R2}{(R2 + R3)}$$

The output will be held constant until the input signal passes back through zero and negatively exceeds the voltage being fed to the noninverting input. When this happens, the output goes positive once more, bringing me back to where I started.

The positive feedback voltage is kept relatively small. The delayed response (or hysteresis) created by this positive feedback gives the circuit considerable noise immunity.

WINDOW COMPARATORS

Another important type of comparator circuit is the window comparator. The basic comparators already discussed compare the input signal to a single reference voltage, and the output indicates whether the input voltage is greater than or less than the reference voltage. Window comparators compare the input voltage to a band of reference voltages and indicate whether the input is within this band or outside of it.

There are three basic types of window comparators. One type simply indicates when the signal is within the reference band. Another type gives an indication when the input signal is outside of (either above or below) the reference band. The third type has different output conditions to indicate if the input signal is above or below the reference band).

The simplest type of window comparator is the threshold detector circuit, shown in Fig. 10-14. As with some of the zero-crossing detector circuits presented earlier, a diode bridge is included in the negative feedback path of the circuit.

An output graph for this circuit is illustrated in Fig. 10-15. Notice that it is symmetrical around the point of origin (O).

A window comparator with a nonsymmetrical band of interest,

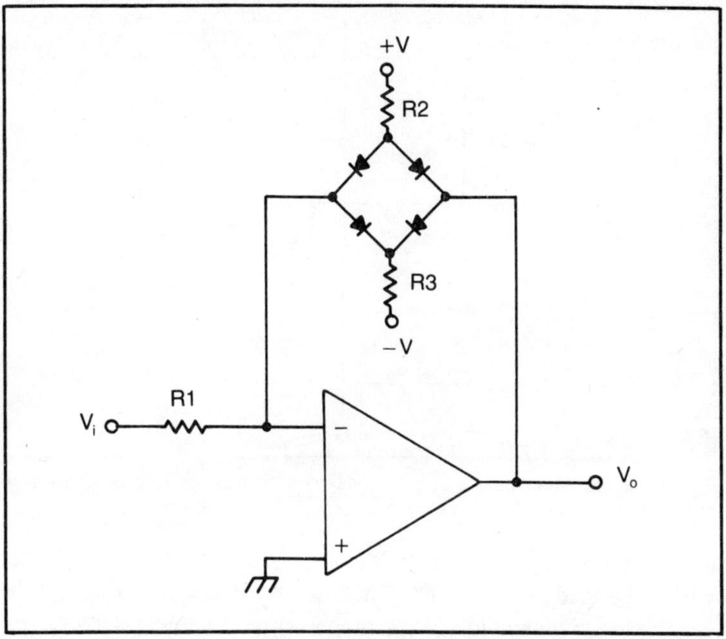

Fig. 10-14. This circuit is a threshold detector.

as shown in Fig. 10-16, can be created simply by adding an external reference voltage through a second input resistor, as illustrated in Fig. 10-17.

For small input signals near zero, the bridge supply voltages

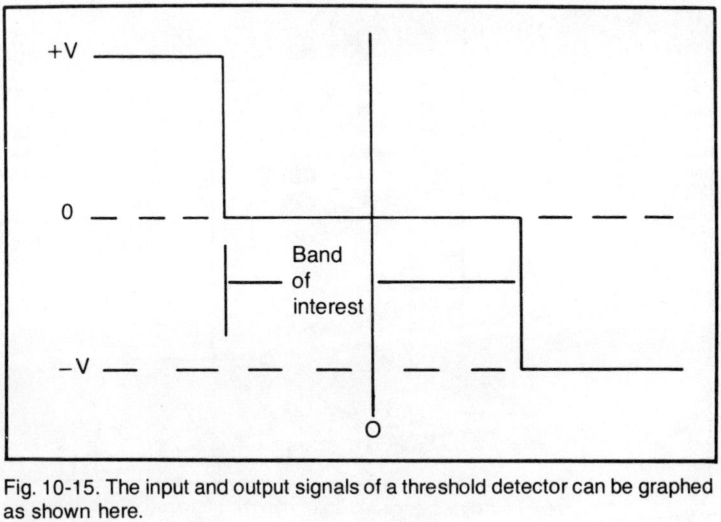

Fig. 10-15. The input and output signals of a threshold detector can be graphed as shown here.

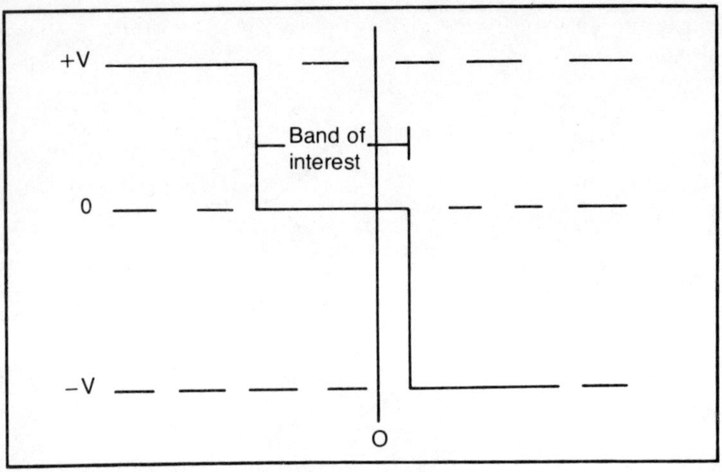

Fig. 10-16. Threshold detectors with nonsymmetrical bands of interest are also possible.

(+V and −V) dominate in these circuits which cause the diodes within the bridge to conduct. The low impedance path through the conducting diodes holds the op amp's output to a near zero value.

As the input signal is made more positive (or is moved further from zero), a point will be reached when the input signal current

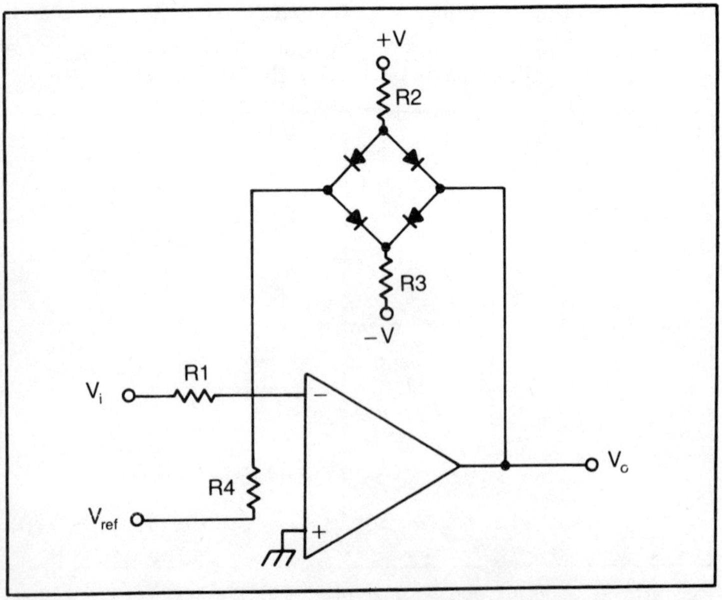

Fig. 10-17. This threshold detector circuit has a nonsymmetrical band of interest.

equals the forward biased diode current from the bridge voltage supplies. That is:

$$I_i = I_{d+}$$

I can use Ohm's Law to expand this equation so that I can work with the more obvious voltage and resistance values.

$$\frac{V_i}{R1} = \frac{+V}{R2}$$

Rearranging this formula, I find that the required input voltage for this switchover point must have a value according to the following formula:

$$V_i = \frac{R1(+V)}{R2}$$

When this point is reached, the diodes become reverse biased and stop conducting, which effectively opens up the feedback path. The high open-loop gain of the op amp causes it to go into saturation. Since the positive input voltage is fed into the inverting input, the output is the device's negative saturation voltage.

Now, if the input voltage begins to drop off, approaching zero (becoming less positive), when it drops below the critical value ($V_i = (R1 \times (+V))/R2$) the output returns to virtual zero. The input voltage is once again within the band of interest.

As the input voltage continues to drop, it passes through zero and goes negative. Soon the negative input current will equal the bridge supply current.

$$I_i = I_{d-}$$

$$\frac{V_i}{R1} = \frac{-V}{R3}$$

$$V_i = \frac{R1(-V)}{R3}$$

Once again, when this voltage is exceeded the diodes become reverse-biased, opening up the feedback path. The op amp's output is driven into saturation. Since the input voltage is negative and is fed into the inverting input, the output is negative.

If I graph the input voltage (starting negative and going positive) against the output voltage I get the graph shown in Fig. 10-15.

In most circuits, resistors R2 and R3 will have equal values. The bridge supply voltages will usually be symmetrical (equal voltages with opposite polarities). The bridge supply voltages may be tapped off of the dual-polarity power supply driving the op amp itself.

Let's take a closer look at a typical design. In my example, the bridge voltage supplies will be ±12 volts, resistor R1 will have a value of 2200 ohms, and R2 and R3 will be 10,000 ohms each.

The crossover point for a positive voltage occurs when the input current equals 1.2 mA.

$$I_i = I_{d+} = \frac{+V}{R2} = \frac{12}{10{,}000} = 0.0012 \text{ amps} = 1.2 \text{ mA}$$

In terms of voltage, it equals 2.64 volts.

$$V_i = \frac{R1(+V)}{R2} = \frac{2200 \times 12}{10{,}000} = \frac{26{,}400}{10{,}000} = 2.64 \text{ volts}$$

Below 2.64 volts, the output will be zero (or near zero). Raising the input voltage above 2.64 volts drives the op amp into negative saturation. If the op amp is powered by ±12 volts, this will be approximately equal to −11 volts. Since circuit values are symmetrical, I get identical values (with reversed polarities) when the input voltages go negative.

$$I_i = I_{-d} = \frac{-V}{R3} = \frac{-12}{10{,}000} = -0.0012 \text{ amps} = -1.2 \text{ mA}$$

$$V_i = \frac{R1(-V)}{R3} = \frac{2200 \times -12}{10{,}000} = \frac{-26{,}400}{10{,}000} = -2.64 \text{ volts}$$

If the input voltage is lower than −2.64 volts, the op amp will be driven into positive saturation, or about 11 volts.

I can summarize the action of this circuit by taking a look at a few typical input and output voltages.

V_i	V_o
−5	+11
−4	+11

−3	+11
−2	0
−1	0
−0	0
+1	0
+2	0
+3	−11
+4	−11
+5	−11

Compare this list to the output graph shown in Fig. 10-15.

The band of interest ranges from −2.64 volts to +2.64 volts. If I want a nonsymmetrical band of interest, I add a constant reference voltage to be added to the actual input voltage, as shown in Fig. 10-17. Generally, things are easiest if the V_{ref} input resistor has the same value as the V_i input resistor (R4 = R1).

To illustrate the effects of this added reference voltage, I will continue to use the same resistor and voltage values that I used in the original example.

R1 = 2.2 K
R2 = 10 K
R3 = 10 K
+V = +12 volts
−V = −12 volts

Since R4 is equal to R1, it has a value of 2.2 K, of course. I'll set my V_{ref} signal at +1.5 volts. As far as the circuit is concerned, it is seeing a single input voltage which is actually the sum of V_i and V_{ref}.

$$V_t = V_i + V_{ref}$$

Therefore, I replace V_i with V_t in the basic voltage equations.

$$V_t = \frac{R1(+V)}{R2} = V_i + V_{ref}$$

$$V_t = \frac{R1(-V)}{R3} = V_i + V_{ref}$$

Since V_{ref} is a constant, I need to rewrite the equations to solve for V_i.

$$V_i = \frac{R1(+V)}{R2} - V_{ref}$$

$$V_i = \frac{R1(-V)}{R3} - V_{ref}$$

So, for our example, the positive switchover point becomes 1.14 volts.

$$V_i = \frac{R1(+V)}{R2} - V_{ref} = \frac{2200 \times 12}{10,000} - 1.5$$

$$V_i = \frac{26,400}{10,000} - 1.5 = 2.64 - 1.5 = 1.14 \text{ volts}$$

Similarly, the negative switchover point is lowered to 4.14 volts.

$$V_i = \frac{R1(-V)}{R3} - V_{ref} = \frac{2200 \times -12}{10,000} -1.5$$

$$V_i = \frac{-26,400}{10,000} - 1.5 = -2.64 - 1.5 = -4.14 \text{ volts}$$

The band of interest now extends from −4.14 volts to +1.14 volts. It is no longer symmetrical around zero. The effect is best illustrated by again listing some typical input and output voltages.

V_i	V_o
−5	+11
−4	0
−3	0
−2	0
−1	0
0	0
+1	0
+2	−11
+3	−11
+4	−11
+5	−11

This corresponds to the output graph of Fig. 10-16.

Notice that the band of interest remains the same size. It is only shifted with respect to zero. To shift the band to interest in the opposite direction, use a negative reference voltage.

These circuits are functional and will work fine in many applications, but they are far from ideal. Accuracy is limited by the op amp and the diodes, in addition to the resistance tolerance.

The input bias current and input offset voltage of the IC can shift the switchover points which could be a problem in certain precision applications. Since a real world op amp does not have truly infinite open-loop gain, there will be some indefinite outputs when the input voltage is near one of the switchover values. In addition, the switchover points can also be affected by the voltage drops across the diodes themselves (typically about 0.7 volts each for silicon diodes). This last source of error can be overcome by replacing the bridge resistors and voltage supplies with constant current sources (see Chapter 9).

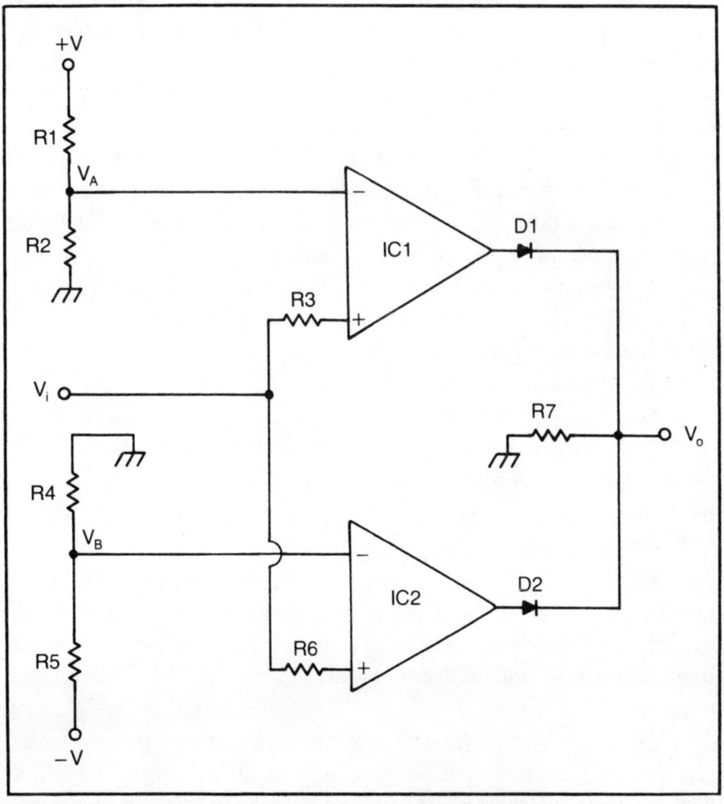

Fig. 10-18. An improved window comparator may be created using two separate op amps.

The speed at which the output can respond to changes in the input voltage will be limited by the inherent slew rate of the op amp used in the circuit.

Another window comparator circuit is shown in Fig. 10-18. Two separate op amps follow for improved operation. This circuit will directly reflect the polarity of the input signal. If the input voltage is more positive than the band of interest, the output will also be positive. Similarly, if the input voltage is more negative than the low end of the band of interest, the output of the circuit will be negative. Polarity is not inverted.

The upper end of the window (positive switching point) is set by the voltage divider made up of resistors R1 and R2. By the same token, resistors R4 and R5 form a voltage divider to set the lower end of the window (negative switching point).

Notice that with this approach, the two switchover points are independent, and there is no reason why the band of interest has to be symmetrical around zero. If you do want a symmetrical window, simply set the R1/R2 ratio equal to the R5/R4 ratio. The easiest way to do this would be to make R1 equal to R5 and R2 equal to R4.

Input resistors R3 and R6 should have equal values. Resistor R7 should be approximately equal to the resistance of the load circuit to be driven by the window comparator. Each of the op amps serves as a simple single-voltage comparator such as those described in the first part of this chapter.

IC1 compares the input signal (V_i) to the positive switchover voltage (V_A). If V_i is less than V_A, the output of IC1 will be its negative saturation voltage. If V_i is greater than V_A, however, IC1 goes into positive saturation.

At the same time, IC2 is comparing the input voltage (V_i) to the negative switchover voltage (V_B). If V_i is less than V_B then IC2 will be in negative saturation (V_i more negative than V_B). On the other hand, if V_i is greater than V_B (V_B more negative than V_i), the output of IC2 will be in positive saturation.

The real key to the functioning of this circuit is in the output diodes D1 and D2. These components are what make the circuit operate as a threshold detector. IC1 and diode D1 handle positive input signals, while IC2 and diode D2 are set up to deal with negative input signals.

The precision of this circuit can be greatly improved by replacing the diodes and R7 with a pair of op amp precision diode circuits (such as the one illustrated in Fig. 10-19) and an op amp summing amplifier (see Chapter 3).

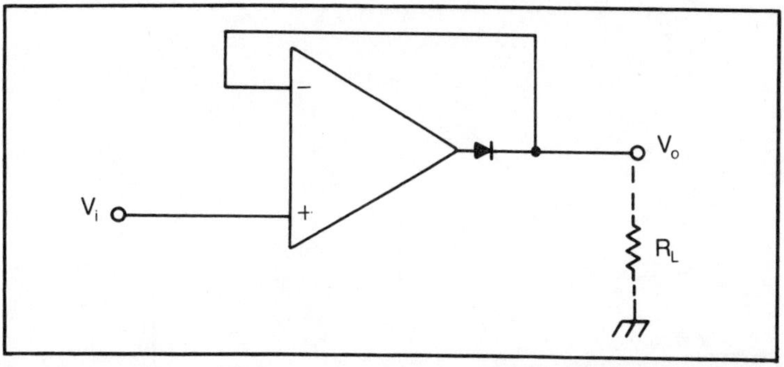

Fig. 10-19. An op amp can be made to serve as a precision diode.

Input bias currents can cause some dc error in the output signals, so the input resistor connected to the noninverting input of each op amp should be selected to equal the parallel combination of the appropriate voltage divider resistor.

$$R3 = \frac{R1 \times R2}{R1 + R2}$$

$$R6 = \frac{R4 \times R5}{R4 + R5}$$

MAGNITUDE COMPARATOR

Yet another op amp detector circuit is shown in Fig. 10-20. This is a magnitude detector.

Notice how much this circuit resembles the two op amp window detector circuit of Fig. 10-18. The main difference is the elimination of the two output diodes.

The magnitude detector has two output stages. One output stage handles input voltages within a specified range, and the other deals with out-of-range signals. The window range is set by the two voltage dividers—R1/R2 and R5/R4. These resistors bias the two parallel comparator stages.

When the input voltage is within the specified range, IC2 wants to go into positive saturation, and IC1 tries to go into negative saturation. These out-of-band signals cancel each other and keep the output at zero.

Now, let's say the input voltage becomes more positive and becomes greater than the upper reference voltage (V_A). Since V_i at the noninverting input is greater than V_A at the inverting input, the output of this op amp is the positive saturation voltage. At the same

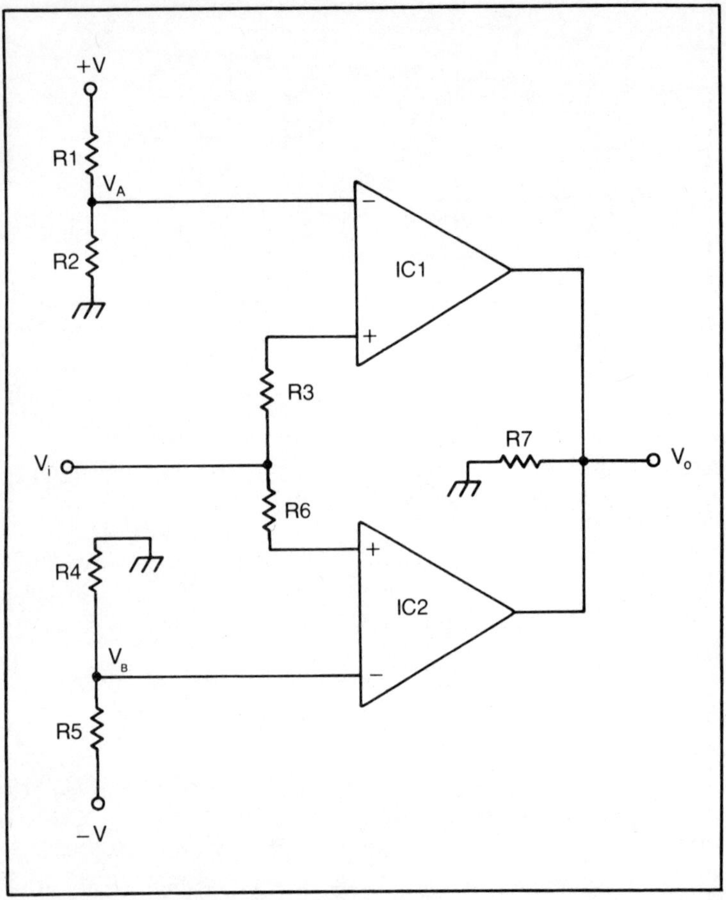

Fig. 10-20. A magnitude detector is similar to the window comparator circuit of Fig. 10-18, but without the output diodes.

time, V_i is also fed into the noninverting input of IC2. V_B, which is more negative than V_i is fed to the inverting input. This op amp remains in positive saturation.

The two op amps are now both in positive saturation, so their output voltages work together, instead of canceling each other out. Essentially the same thing occurs when the input voltage drops below the lower reference voltage (V_B).

As long as the input voltage is within the specified range, the output remains at zero. If the input voltage is outside the band of interest in either direction (either positively or negatively), both comparators switch into positive saturation. The output graph for this circuit is shown in Fig. 10-21.

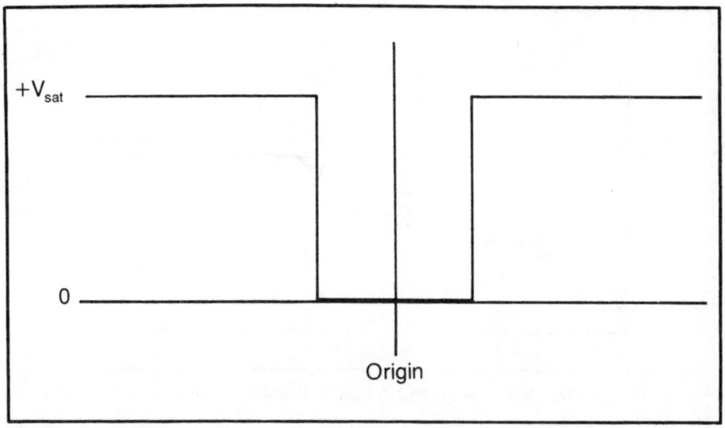

Fig. 10-21. Here is the output graph for the magnitude detector circuit of Fig. 10-20.

PEAK DETECTOR

It is sometimes necessary to determine the maximum value of a varying voltage. This can be done with a circuit known as a peak detector. The simplest possible peak detector consists of a diode and capacitor, as illustrated in Fig. 10-22.

The diode allows the voltage to flow in only one direction—from the input to the capacitor and onto the output. As the input voltage is increased, the capacitor is charged through the diode. Now, if the input voltage begins to drop off, the capacitor wants to discharge, but that would reverse bias the diode, so the capacitor holds its peak charged value, until the input voltage begins to increase again and exceeds the previous peak level, charging the capacitor further. The action of this circuit is illustrated in the output graph of Fig. 10-23.

The input voltage is tracked and followed until a maximum level is reached. This peak value is held by the peak detector circuit.

This simplified, idealized circuit isn't really very practical. The capacitor will soon discharge through the load circuit. Even

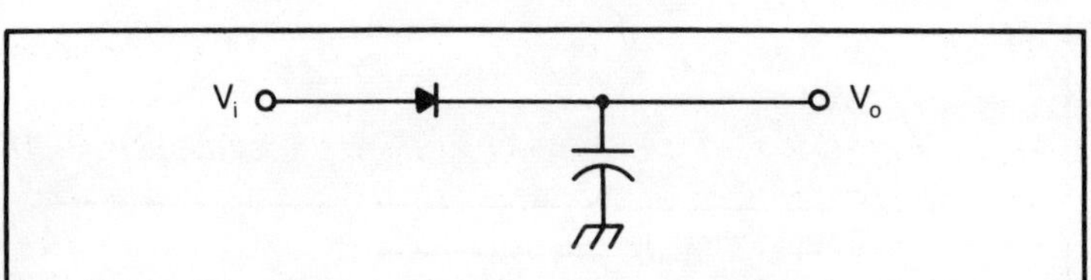

Fig. 10-22. A simple peak detector can be made from a diode and a capacitor.

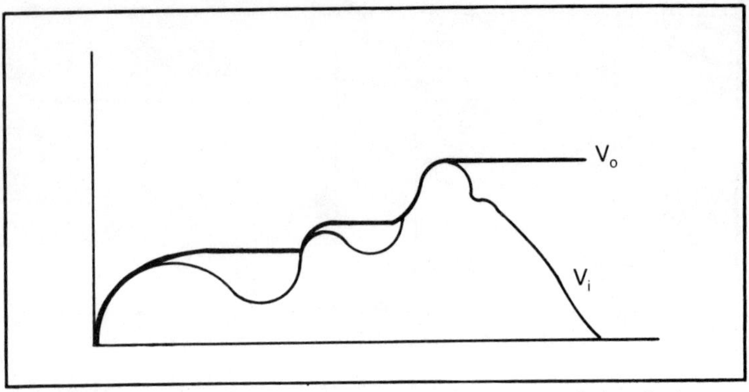

Fig. 10-23. A peak detector remembers the highest voltage fed to the input.

without a load, a capacitor cannot hold a charge indefinitely. It will begin to taper off due to leakage, as shown in Fig. 10-24.

Somewhat better results can be achieved by using an op amp precision diode in the place of the diode. This is shown in Fig. 10-25. The op amp in this circuit should be a type with a FET input stage. This type of device has an extremely high input resistance that will minimize the decay of the output voltage during the hold mode.

To prevent loading effects on the capacitor, an output buffer stage should be added, as shown in Fig. 10-26. This will prevent the capacitor from discharging too quickly through the load circuit. The

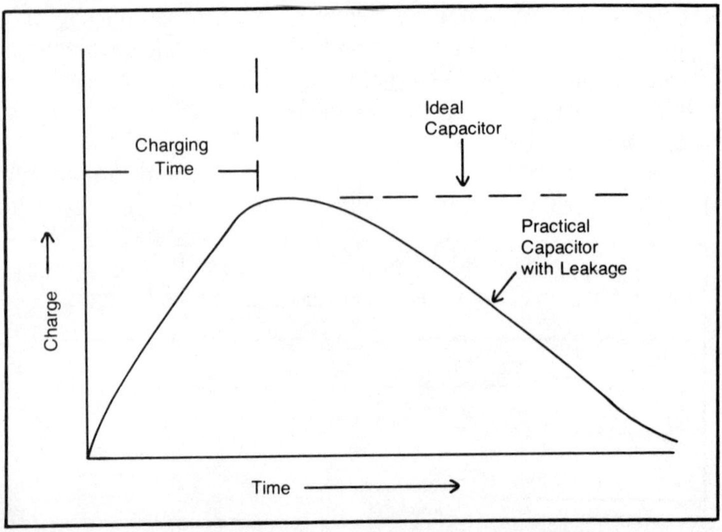

Fig. 10-24. Practical capacitors cannot hold a charge indefinitely.

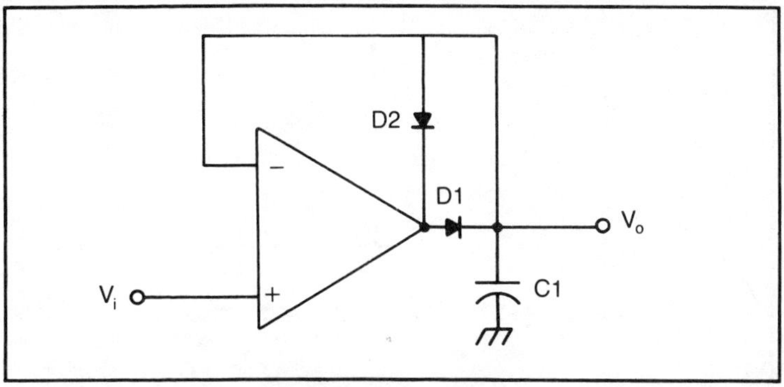

Fig. 10-25. A practical peak detector can be built around an op amp precision diode.

output will still eventually decay due to leakage within the capacitor itself.

For a practical peak detector, some means must be provided to manually reset the output to zero. The simplest method is to add a push-button switch to temporarily short out the capacitor and force it to quickly discharge.

For long holding times, a high-quality capacitor with a large capacitance value should be used. A large capacitor will respond to brief surges in the input voltage more sluggishly than a smaller value component. If the input peak is very brief, a large capacitor may not charge up to the full value before the input begins to drop off. Larger capacitances take longer to charge and to discharge.

Better performance can be achieved by using two op amps instead of just one. Figure 10-28 shows a noninverting peak detector. The output will have the same polarity as the input signal.

Capacitor C3 serves as the peak holding capacitor. A low-leakage capacitor should be used. The other two capacitors (C1 and C2) are not always needed. They are often included to improve

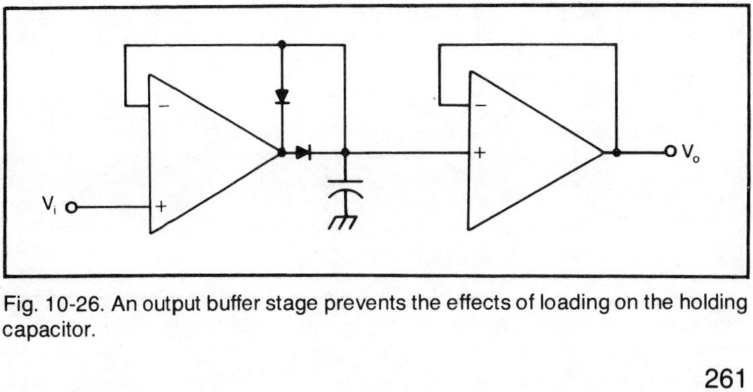

Fig. 10-26. An output buffer stage prevents the effects of loading on the holding capacitor.

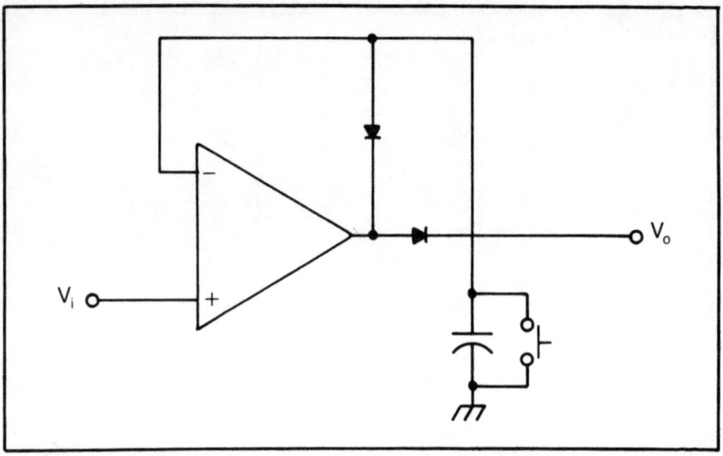

Fig. 10-27. Some means of resetting the output to zero is needed in a practical peak detector.

circuit stability and to prevent overshoot from a step input signal (see Fig. 10-29).

IC2 is simply an output buffer to prevent the holding capacitor (C3) from being discharged through the external load circuit.

One major advantage of this circuit is that IC1 cannot latchup (get stuck at a certain output voltage), as can happen with some peak detector circuits.

Another peak detector circuit is shown in Fig. 10-30. The primary difference between this and the circuit of Fig. 10-28 is that this circuit inverts the output polarity. The op amps are operated in the inverting mode.

IC1 serves as a precision diode while IC2 tracks and inverts the

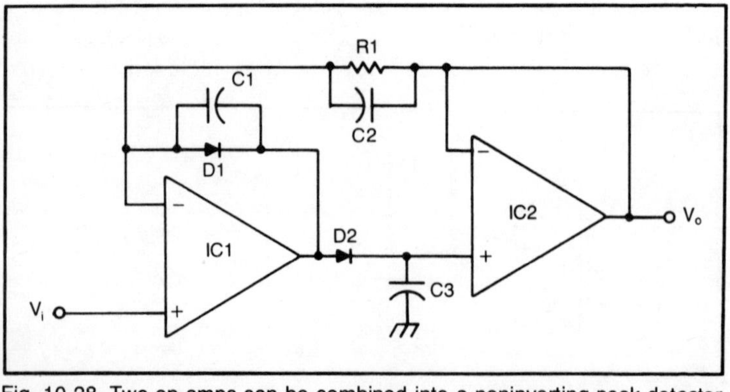

Fig. 10-28. Two op amps can be combined into a noninverting peak detector circuit.

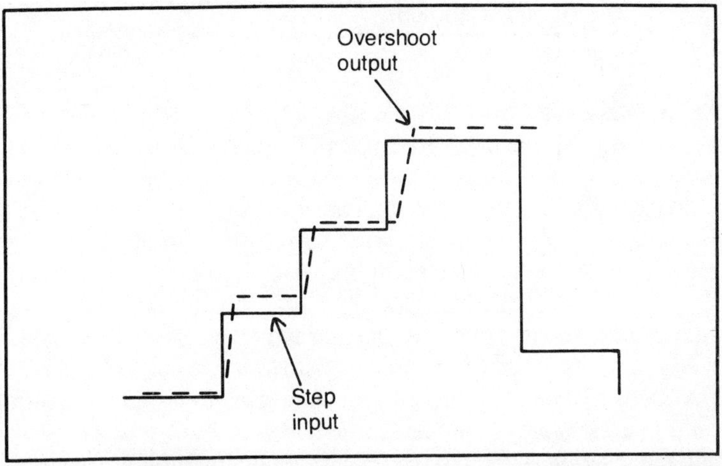

Fig. 10-29. Some peak detectors may tend to overshoot on step input signals.

signal. Capacitor C2 is the peak holding capacitor. Positive feedback from the output of IC2 is returned to the noninverting input of IC1 through resistor R4 and capacitor C3.

To minimize any dc errors within the circuit, the value of resistor R1 should be equal to the parallel combination of R2 and R4.

$$R1 = \frac{(R2 \times R4)}{(R2 + R4)}$$

For example, in a typical circuit, R2 and R4 may each be equal to 10 K. This means R1 should have a nominal value of 5000 ohms.

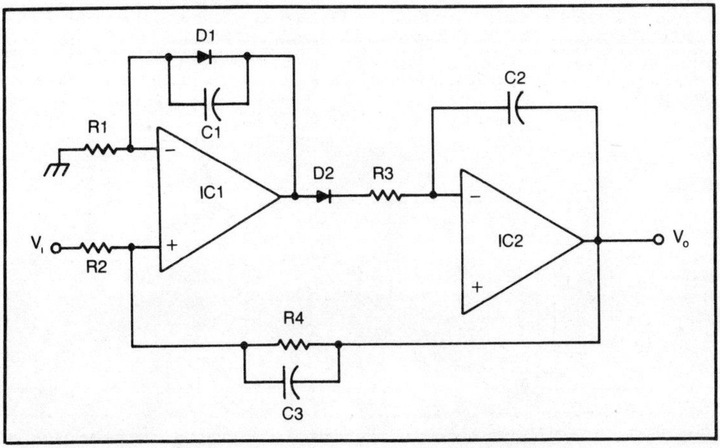

Fig. 10-30. This is an inverting peak detector circuit using two op amps.

$$R1 = \frac{(10,000 \times 10,000)}{(10,000 + 10,000)} = \frac{100,000,000}{20,000} = 5000 \text{ ohms}$$

Of course, a standard 4.7 K resistor could be used for resistor R1 without any significant degradation of the circuit's performance. A good value for R3 would be about 1 K. A 0.1 μF capacitor will do a pretty good job as the peak holding capacitor (C2).

Output buffering is performed by IC2 to prevent the holding capacitor from being affected by the load circuit.

The circuit shown in Fig. 10-31 is probably the simplest two op amp peak detector circuit. Momentarily closing switch S1 allows the capacitor to discharge and resets the circuit output to zero.

Let's examine just what happens in this circuit during operation. I will assume the output has just been reset to zero a positive voltage is being applied to the noninverting input of IC1. Since the output of this op amp is not inverted, it forward biases diode D1.

Even a very small input voltage will forward bias the diode, thanks to the high gain of the op amp. The effects of the voltage drop across the diode are negligible and can be ignored altogether in this application.

Since the diode is forward-biased, it allows current to flow back to capacitor C1, which begins to charge. When the charge on C1 equals the input voltage, the differential input voltage across the inputs of IC1 is zero. The output of this op amp drops to zero, and the capacitor does not charge any further. It can't discharge because that would reverse-bias the diode. If the input voltage is now increased, the capacitor will be charged up to this new maximum value.

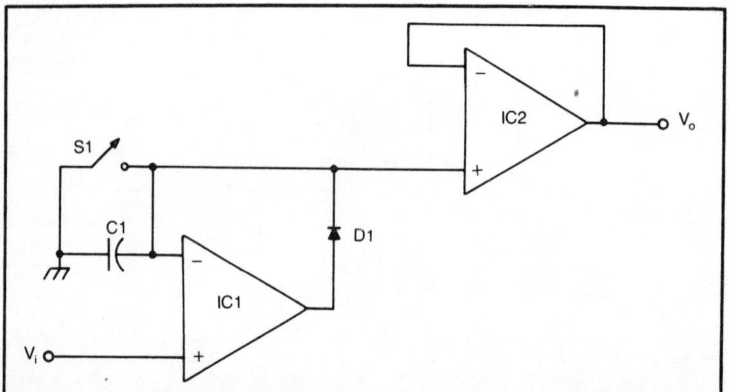

Fig. 10-31. This peak detector circuit features manual output resetting.

Let's say the input voltage begins to drop off, and becomes less positive. Since the voltage to the inverting input of IC1 (C1's charge) is greater than the voltage at the noninverting input (V_i), the output of the op amp goes negative. This reverse biases the diode, blocking current flow. Capacitor C1 continues to hold the peak charge value (ignoring leakage).

The voltage across the holding capacitor will always be equal to the largest positive voltage that has been presented to the circuit input since it was last reset.

IC2 is simply a noninverting voltage follower to act as an output buffer for the circuit to prevent loading effects.

The peak voltage can be held by the capacitor for periods lasting up to several minutes, but remember that any capacitor will have some internal leakage that will cause its charge to eventually drift off. The capacitor will also suffer some discharging through leakage current in the diode and the bias current required by the op amps. The output buffer will also place some slight loading on the holding capacitor, although considerably less than most external loading circuits would.

11

Analog Computation Circuits

Operational amplifiers were originally designed to perform mathematical operations for analog computers. An analog computer differs from the digital devices known as computers in a number of ways. A digital computer is a binary device. That is, there are just two output conditions—on and off, or high and low, or 1 and 0. Each binary signal is called a *bit.* More complex signals can be created by combining bits into *bytes* and *words.* Because of the binary structure, the output will always consist of a series of definite steps without intermediate values. This idea is illustrated in Fig. 11-1.

Analog devices, on the other hand, directly reflect all changes in the input signal. The output signal can vary continuously through an infinite number of intermediate values.

A digital circuit with an output that ranges from zero to five volts in one volt steps has only six possible output signals.

0 volts	3 volts
1 volt	4 volts
2 volts	5 volts

An analog circuit with the same range would have an infinite number of possible output voltages. An output of 2.5 would be no problem for the analog device. The digital circuit would have to round off 2.5 to either 2 volts or 3 volts.

Another important difference between analog and digital com-

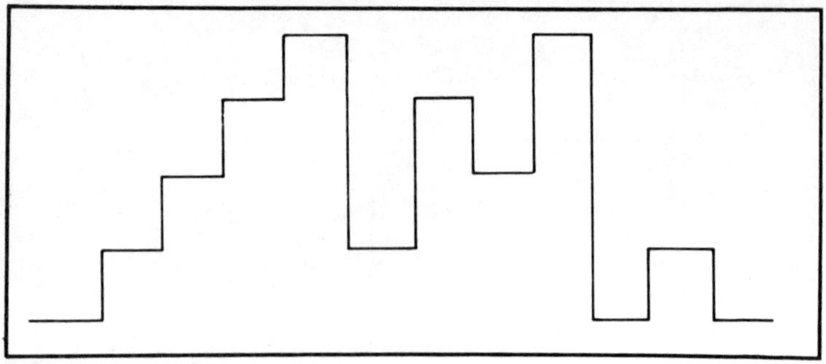

Fig. 11-1. Digital outputs consist of discrete steps.

puters is programmability. Analog computers are almost always dedicated devices. That is, they are hard-wired to perform a specific type of operation. A logarithmic amplifier, for example, can not calculate the sine of an angle.

Digital computers, however, are usually controlled by a program, or a series of instructions that tell a general purpose circuit which operations to perform. A single circuit can do addition, subtraction, multiplication, logarithms, and trigonometry.

In this chapter, I examine a few dedicated analog computation circuits built around op amps that perform some of the more frequently used mathematical operations.

ADDITION AND SUBTRACTION

I have already presented circuits that perform the mathematical operations of addition and subtraction. The circuit shown in Fig. 11-3 adds together the values applied to the three inputs.

This is simply a summing amplifier. For most applications as an adder, you will not want to weight the inputs. In other words, the

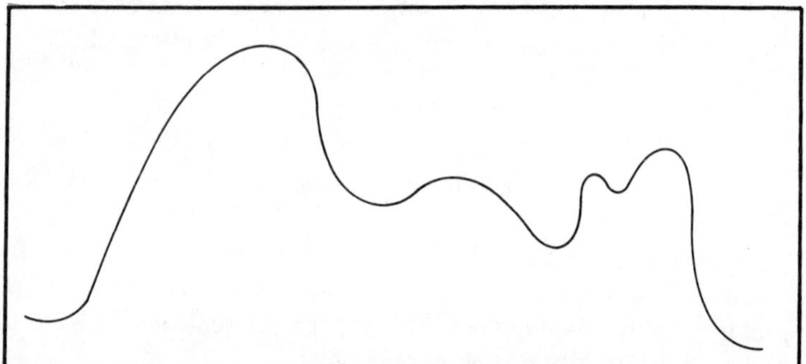

Fig. 11-2. Analog outputs can pass through a continuous range of values.

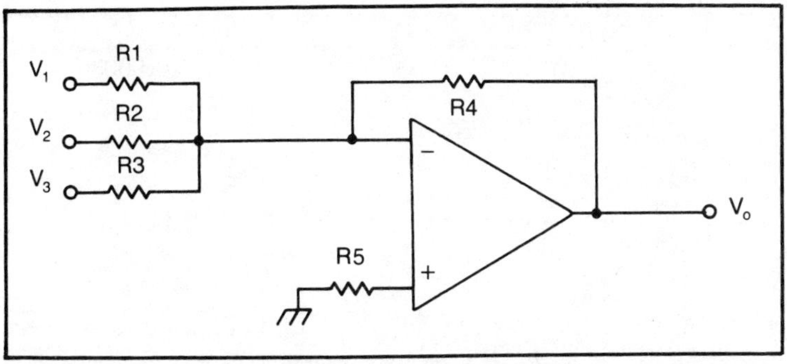

Fig. 11-3. Two or more voltages can be added together by a summing amplifier.

three input resistors should have equal values. For unity gain, feedback resistor R4 will have the same value.

$$R1 = R2 = R3 = R4$$

Resistor R5, which compensates for input bias errors, should have a value equal to the parallel combination of the input resistors.

$$R5 = \frac{1}{\left(\frac{1}{R1} + \frac{1}{R2} + \frac{1}{R3}\right)}$$

Since R1 = R2 = R3, I can simplify this equation to the following:

$$R5 = \frac{1}{\left(\frac{3}{R1}\right)} = \frac{R1}{3}$$

Let's say resistors R1 through R4 have a value of 10 K each. I will calculate the desired value of R5 using both forms of the equation given above to prove that the results are identical with either method.

$$R5 = \frac{1}{\left(\frac{1}{R1} + \frac{1}{R2} + \frac{1}{R3}\right)}$$

$$R5 = \frac{1}{\left(\frac{1}{10{,}000} + \frac{1}{10{,}000} + \frac{1}{10{,}000}\right)}$$

$$R5 = \frac{1}{(0.0001 + 0.0001 + 0.001)} = \frac{1}{0.0003}$$

$$R5 = 3333.333333 \text{ ohms}$$

$$R5 = \frac{R1}{3} = \frac{10{,}000}{3} = 3333.333333 \text{ ohms}$$

Yes, the same answer is obtained from either equation. I can round off the value for R5 to 3.3 K. That will be close enough.

Using these (or similar) component values, this circuit will add whatever voltages are fed into each of the inputs. Since the inverting input of the op amp is used, the polarity of the output is reversed. The output voltage is the inverted sum of the input voltages.

$$V_o = -(V_1 + V_2 + V_3)$$

For example, let's assume input V_1 is 2 volts, V_2 is 3 volts, and V_3 is 5 volts.

$$V_o = -(2 + 3 + 5) = -10 \text{ volts}$$

Even if some of the input voltages may be negative, circuit operation will not be affected. Assume the following input voltages:

$$V_1 = 3.5 \text{ volts}$$
$$V_2 = -2.25 \text{ volts}$$
$$V_3 = 4.35 \text{ volts}$$

The output voltage is the sum of the inputs.

$$V_o = -(3.5 + (-2.25) + 4.35) = -(3.5 - 2.25 + 4.35)$$
$$V_0 = -5.6 \text{ volts}$$

In the next example I will make the negative input voltage dominant. Assume the following input signals:

$$V_1 = 3 \text{ volts}$$
$$V_2 = -5.87 \text{ volts}$$
$$V_3 = 1.32 \text{ volts}$$

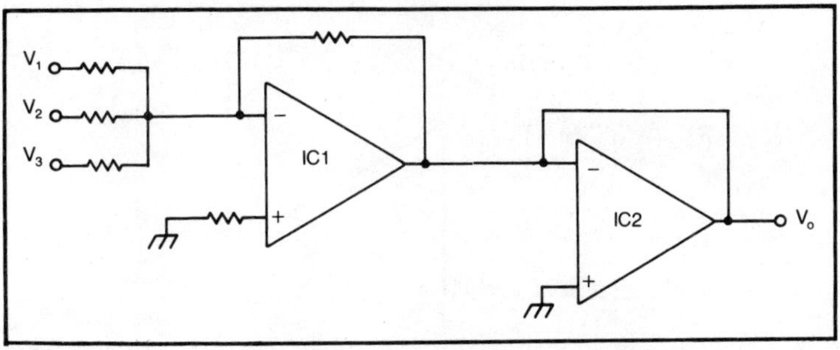

Fig. 11-4. An inverting voltage follower permits the output of the summing amplifier of Fig. 11-3 to maintain the original input polarity.

$$V_o = -(3 + (-5.87) + 1.32) = -(3 - 5.87 + 1.32) = -(-1.55) = +1.55 \text{ volts}$$

Remember the output polarity is inverted from its direct mathematical value. This may be a problem in some applications. One solution would be to follow the circuit with an inverting voltage follower stage, as shown in Fig. 11-4 to reinvert the resulting signal to its original mathematical polarity.

Subtraction can also be performed with an op amp by using the basic difference amplifier that is illustrated in Fig. 11-5. If all four of the resistors are given equal values, the circuit will have unity gain, and the output will directly indicate the difference between the two input signals:

$$V_o = V_2 - V_1$$

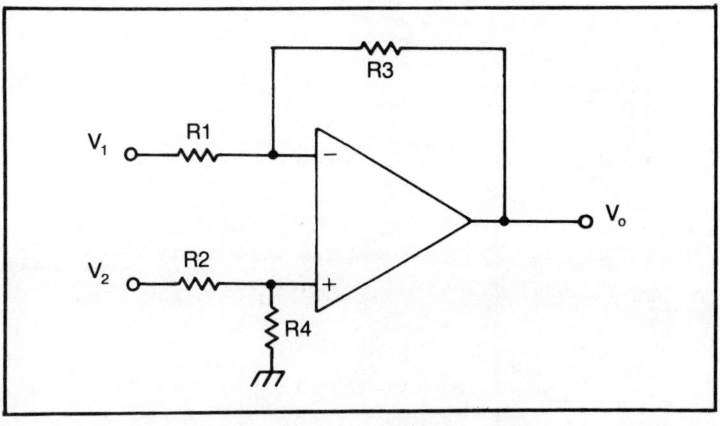

Fig. 11-5. A difference amplifier can be used to perform subtraction operations.

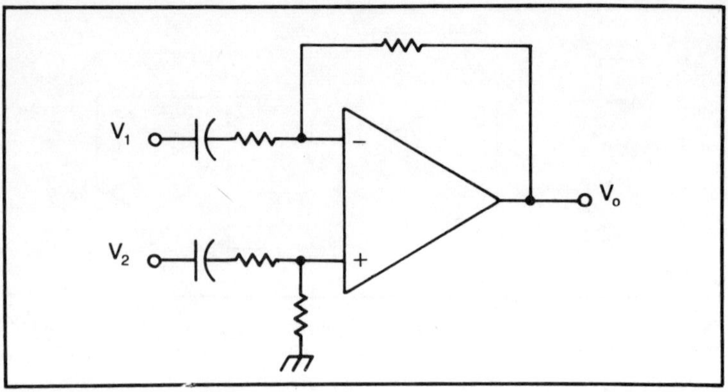

Fig. 11-6. By adding dc blocking capacitors to the inputs, a difference amplifier can be adapted for ac operation.

The signal fed to the inverting input (V_1) is subtracted from the signal fed to the noninverting input (V_2). Some typical input combinations and the resulting outputs are summarized in Table 11-1.

For ac signals you may want to adapt the circuit as shown in Fig. 11-6. The capacitors block any undesired dc components in the input signals.

LOGARITHMIC AMPLIFIERS

After addition and subtraction, the two simplest mathematical operations are multiplication and division. Unfortunately, these operations are somewhat difficult to perform electronically. Therefore, I will take a detour for now and examine circuits that produce an output proportional to the logarithm of the input signal.

V_1	V_2	V_o
0	0	0
0	2	2
2	0	−2
2	3	1
3	2	−1
3	−4	−7
−3	4	7
−2	−3	−1
−5	−2	3
3	3	0
−3	3	6
3	−3	−6
−3	−3	0

Table 11-1. Typical Inputs and Outputs for A Unity-Gain Difference Amplifier.

The logarithm of a number is the exponent that indicates the power to which a number must be raised to produce a given number. Natural logarithms use a base of 10. Here are a few numbers and their logarithms:

1	0	20	1.30103
2	0.30103	25	1.39794
3	0.04771213	30	1.4771213
4	0.60206	100	2
5	0.69897	1000	3
10	1	10000	4
12	1.0791812	100000	5

Raising ten to the logarithm produces the original number.

$$10^{0} = 1$$
$$10^{0.30103} = 2$$
$$10^{0.4771213} = 3$$
$$10^{1} = 10$$
$$10^{2} = 100$$
$$10^{3} = 1000$$

Raising any number to the power of zero (X^0) always results in 1, so the logarithm of 1 is 0, no matter what base is used. The logarithm of zero is infinity in the negative direction. Negative logarithms indicate numbers less than one.

$$\log 0.1 = -1$$
$$\log 0.2 = -0.69897$$
$$\log 0.005 = -2.30103$$

To work from the logarithm back to the original number is called finding the antilogarithm. For example, the antilog of 0.30103 is 2, the antilog of 1 is 10, and so forth.

So what? Well, logarithms can come in extremely handy in many applications. In fact, old fashioned slide rules in the precalculator days used the principle of logarithms to solve various equations.

Logarithms can be used to perform multiplication and division and convert values into decibels (dB) among other applications.

A logarithmic amplifier built around an op amp uses a resistor as the input component and a transistor as the feedback component.

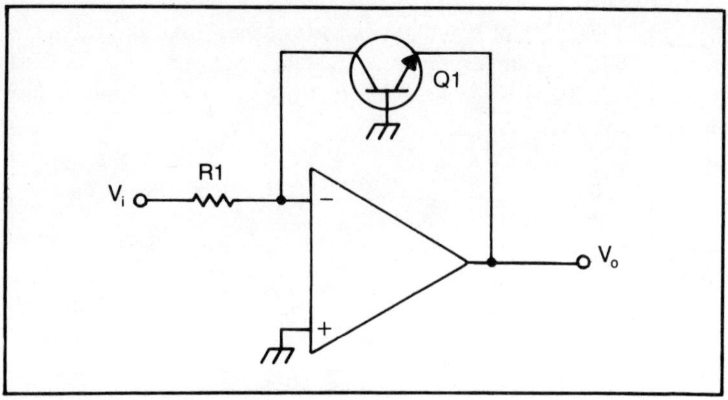

Fig. 11-7. A logarithmic amplifier uses a transistor as the feedback component.

This circuit is shown in Fig. 11-7. The base of the transistor is grounded.

The collector/emitter curve of many transistors is very close to a true logarithmic curve, making the devices ideal for use in logarithmic amplifiers. The collector/emitter current through the transistor will be equal to the input current, thanks to the nature of

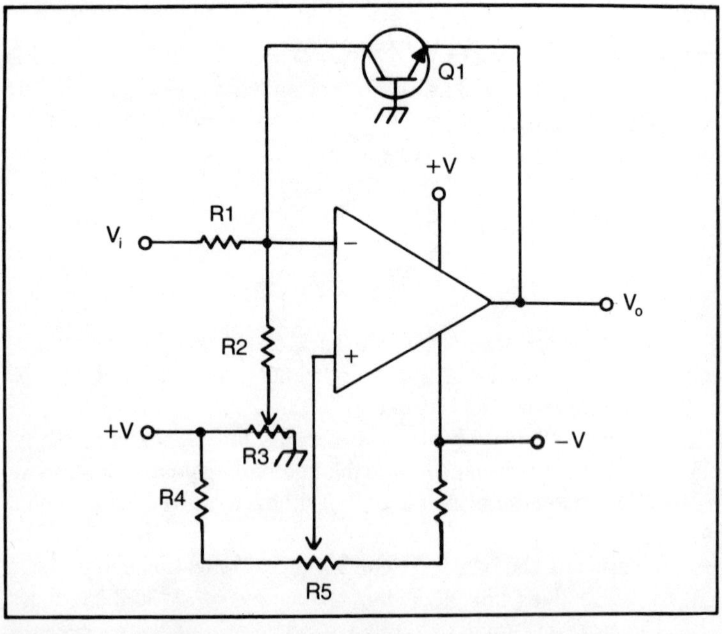

Fig. 11-8. A practical logarithmic amplifier circuit permits calibration of the offsets.

Table 11-2. Typical Value for Fig. 11-8.

Q1	2N2222 or similar npn transistor
R1	100 K resistor
R2	82 K resistor
R3	5 K potentiometer
R4	33 K resistor
R5	1.5 K potentiometer
R6	33 K resistor

the op amp. As a result, the output voltage will be proportional to the logarithm of the input voltage. A power transistor is typically used to reduce the series feedback resistance and improve the circuit's current handling capabilities.

A practical logarithmic amplifier circuit is shown in Fig. 11-8. Typical component values are listed in Table 11-2. The offset null is adjusted to zero for an input signal corresponding to one unit (the log of 1 is 0, remember). The size of this unit, which can be almost anything the designer desires, corresponds to the zero dB reference level used in the circuit.

Another, simplified variation on the basic logarithmic amplifier circuit is illustrated in Fig. 9-9.

The opposite of a logarithm is the antilogarithm. An antilogarithmic amplifier can be easily created simply by reversing the

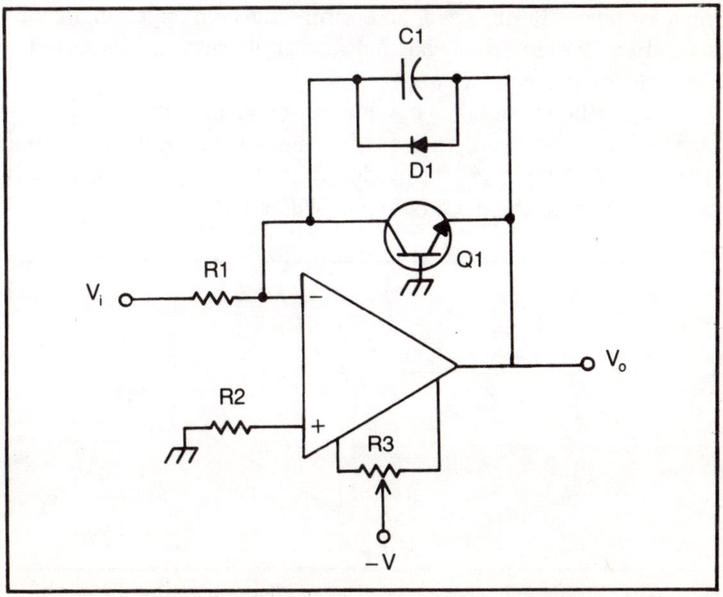

Fig. 11-9. Here is a simplified variation of the basic logarithmic amplifier.

input and feedback components of a logarithmic amplifier, as shown in Fig. 11-10. Now the input component is a grounded base transistor, and the feedback component is a resistor.

MULTIPLICATION AND DIVISION

Multiplying the input signal by a constant value is not particularly difficult. All you need is a noninverting amplifier circuit with the appropriate amount of gain. You should recall that the output of a noninverting amplifier equals the input voltage multiplied by the gain.

$$V_o = V_i \times G$$

Therefore, if I need a circuit to multiply the input signal by a constant factor of 10, I need only design a noninverting amplifier with a gain of 10.

This is functional, but severely limited, and hopelessly inadequate in many applications. A little more versatility can be achieved by using a potentiometer as the feedback resistance. In this case, the multiplication factor can be manually adjusted although calibration may be difficult.

In many practical applications, it is necessary to multiply two input voltages. In this case, an entirely different approach may be used. Here's where the logarithms and antilogarithms discussed in the last few pages come in.

Logarithms exhibit many interesting characteristics. For instance, if you add the logarithms of two numbers and take the antilog of their sum, you will get the product of the two original numbers. In algebraic terms, this idea is expressed as follows:

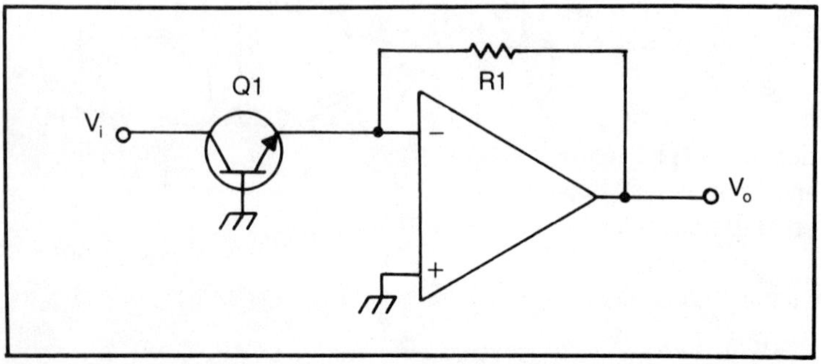

Fig. 11-10. An antilogarithmic amplifier is just the opposite of a logarithmic amplifier.

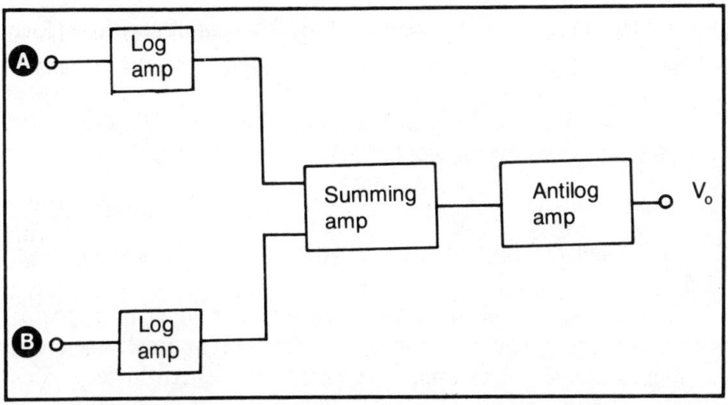

Fig. 11-11. Multiplication can be performed by adding logarithms.

$$\text{antilog}(\log A + \log B) = A \times B$$

The use of logs permits multiplication by simple addition. Let's try a simple example. I need to multiply 2 times 3. Of course, the answer should work out to be 6. Let's see if the logarithm method will get the desired result.

$$C = A \times B$$
$$C = \text{antilog}(\log A + \log B) = \text{antilog}(\log 2 + \log 3)$$
$$C = \text{antilog}(0.30103 + 0.4771213) = \text{antilog } 0.7781513$$
$$C = 6$$

This was a trivial example. The logarithm method is more useful in complex problems, such as 3607.814 × 90011.0043.

$$C = \text{antilog}(\log 3607.814 + \log 90011.0043)$$
$$C = \text{antilog}(3.5572441 + 4.9542956)$$
$$C = \text{antilog } 8.5115397 = 324742960.21$$

I can combine four basic op amp circuits to perform this operation for us. I need two logarithmic amplifiers, a summing amplifier, and an antilogarithmic amplifier. The block diagram for a complete op amp multiplier is illustrated in Fig. 11-11.

Similarly, division can be performed by taking the antilogarithm of the difference of the logarithms of the two original numbers.

$$A/B = \text{antilog}(\log A - \log B)$$

As an example, let's divide 15 by 3. The result should be 5, of course.

$$C = \text{antilog}(\log 15 - \log 3)$$
$$C = \text{antilog}(1.1760913 - 0.4771213)$$
$$C = \text{antilog } 0.69897 = 5$$

The block diagram for an op amp divider circuit is shown in Fig. 11-12.

Sometimes it is necessary to raise an input signal to a specific power. This is simply multiplying the number by itself a specific number of times. For example, 5 raised to the fourth power is 625.

$$5^4 = 5 \times 5 \times 5 \times 5 = 625$$

A logarithmic amp will also allow me to perform this function. To raise a number, A, to the nth power I can use this logarithmic formula:

$$A^n = \text{antilog}(\log A \times n)$$

Let's check this out by solving for 5 to the fourth power again.

$$5^4 = \text{antilog}(\log 5 \times 4) = \text{antilog } (0.69897 \times 4)$$
$$5^4 = \text{antilog } 2.79588 = 625$$

Since the power (n) will be a constant in most applications, I can use the gain of a noninverting amplifier to perform multiplication.

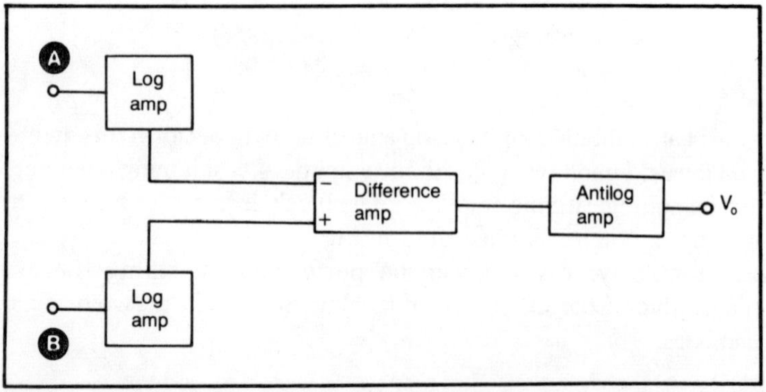

Fig. 11-12. Division can be performed by subtracting logarithms.

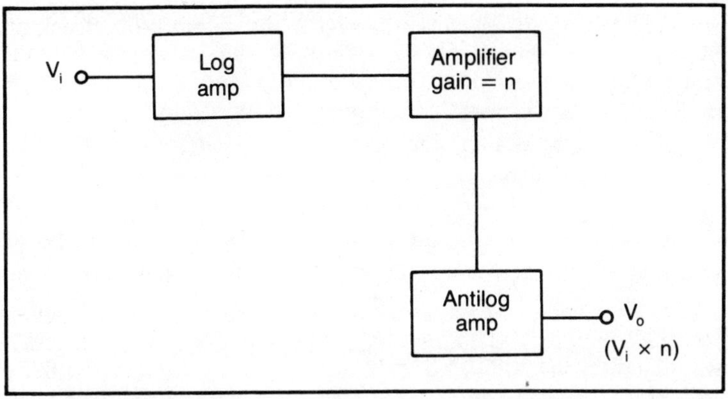

Fig. 11-13. Logarithm and antilogarithm amplifiers can be used to raise a number to a specific power.

The block diagram for an A^n amplilfier circuit is shown in Fig. 11-13. The amplifier at the second stage could be replaced with the multiplication network illustrated in Fig. 11-11. The output of the log amplifier is fed to input A of the multiplier, and a voltage corresponding to the desired power is fed to input B. The complete block diagram is shown in Fig. 11-14. Notice that three logarithmic amps, two antilogarithmic amps, and a summing amp are required.

SINE AMPLIFIERS

Op amps can also be used to perform trigonmetric functions.

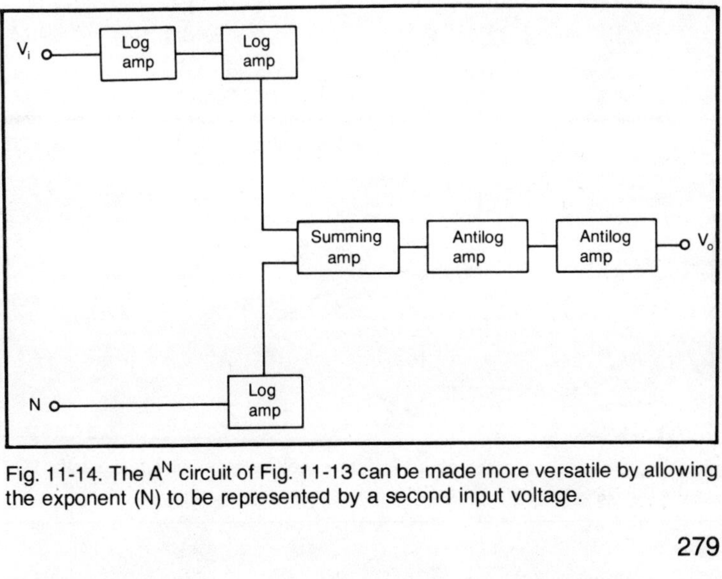

Fig. 11-14. The A^N circuit of Fig. 11-13 can be made more versatile by allowing the exponent (N) to be represented by a second input voltage.

All trigonmetric functions are simply ways of expressing the ratios between the sides of right triangles. If the input voltages to an op amp circuit can be scaled to represent the sides of a right triangle, a division circuit will yield that trigonmetric function.

A simple sine amplifier circuit is shown in Fig. 11-15. The output voltage will be proportional to the input voltage and the sine of the angle defined by the position of the potentiometer's shaft. A major drawback of this circuit is that a special sine potentiometer is required. The resistance at the wiper must be proportional to the sine of the wiper angle.

ABSOLUTE VALUE CIRCUITS

In a number of applications, the magnitude of a voltage is important, but not the polarity. For instance, I may need to detect five volt signals of either polarity. This could be done using two detectors—one for +5 volts, and one for −5 volts. This isn't always practical.

What is needed here is a circuit to take the absolute value of the input voltage. Essentially the polarity sign is discarded and ignored. Absolute value is written as two bars on either side of the number.

$$|X| = \text{absolute value of } X$$

The absolute value function always has a positive result, regardless of the polarity of the original number. Here are some typical examples:

$$|5| = 5 \qquad |19| = 19$$
$$|-7| = 7 \qquad |-19| = 19$$
$$|4.32| = 4.32 \qquad -|3| = -3$$
$$|-0.1| = 0.1$$

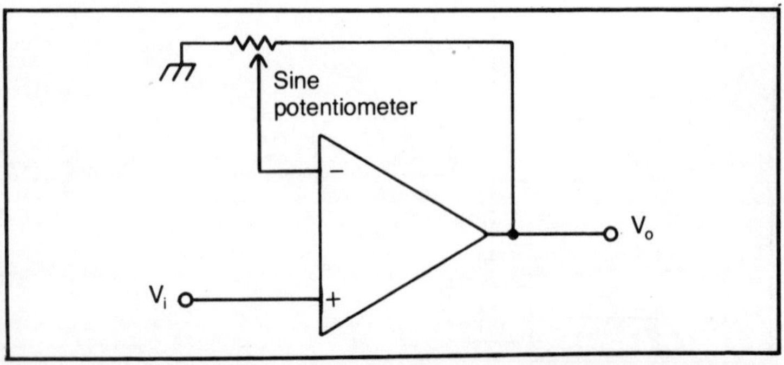

Fig. 11-15. The output of this circuit is determined by the input signal and the sine of the potentiometer shaft's angle.

Did the last one catch you off guard? If the sign is outside the bars it is not part of the function.

A circuit for extracting the absolute value from an input signal is shown in Fig. 11-16. Notice that IC1 is an inverting amplifier and IC2 is a noninverting amplifier. Let's assume that the input voltage, V_i, is positive. The output of IC1 will go negative and reverse bias diodes D1 and D2. The output of IC1 is blocked from the circuit output (V_o). Meanwhile, IC2 has a positive output. Since this is a unity gain amplifier, the output of IC2 is equal to V_i. Diodes D3 and D4 are forward biased by this positive signal, which is fed to the circuit output (V_o).

Now, if the input voltage (V_i) goes negative, just the opposite happens. IC1 puts out a positive signal equal to the inverted value of V_i that forward biases diodes D1 and D2 and allows this signal to reach the circuit output (V_o). IC2, on the other hand, has a negative

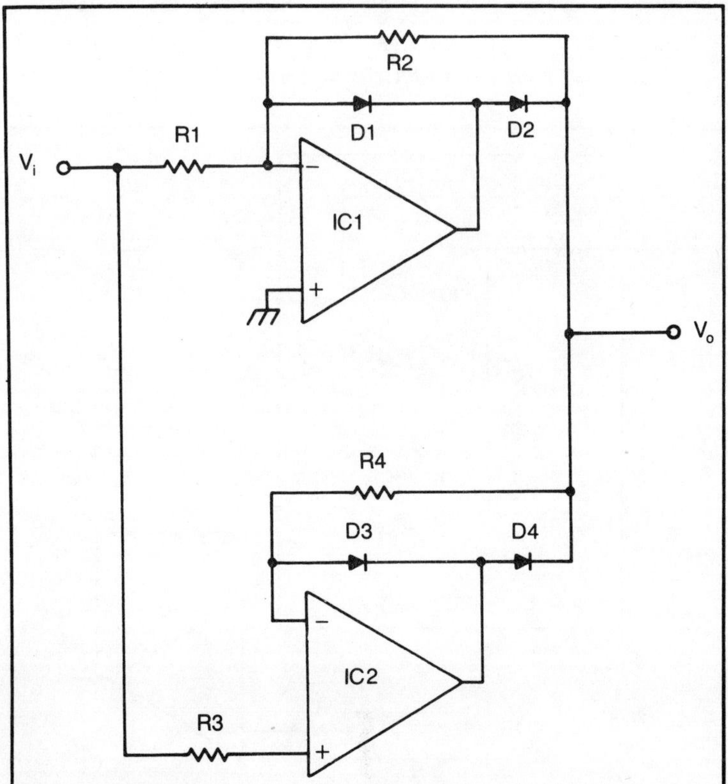

Fig. 11-16. This circuit can be used to extract the absolute value of a signal.

output that reverse biases diodes D3 and D4 and blocks this signal from the circuit output (V_o).

Regardless of the polarity of the input voltage, the output of this circuit will always be positive with a magnitude equal to that of the input signal.

Another approach for an absolute value circuit is shown in Fig. 11-17. Here IC1 passes and inverts positive inputs and blocks negative inputs. IC2 serves as a summing amplifier in which the original input signal (through R5) is added to two times the output of IC1 (through R4). This is accomplished by making R4 half the value of R5.

Let's see what happens with a couple of typical input signals. Assume the input voltage is +5 volts. IC1 passses and inverts this signal. The output of IC1 equals −5 volts. This is fed to the IC2 through R4, along with the original input signal (V_i) through R5. This gives us an output signal equal to +5 volts.

$$V_o = -(V_i + 2V_{o1}) = -(5 + 2 \times -5) = -(5 + -10)$$
$$= -(-5) = +5$$

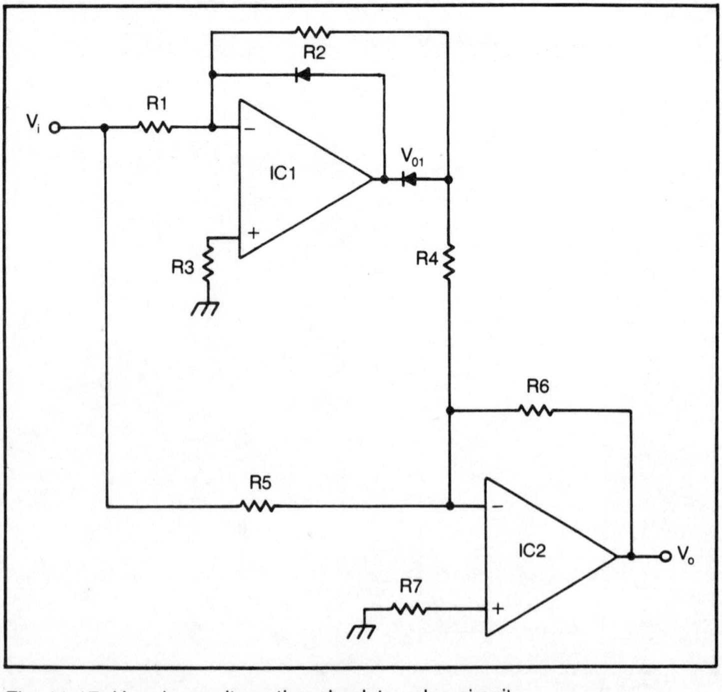

Fig. 11-17. Here is an alternative absolute value circuit.

If the input voltage is −4 volts, this signal will be blocked by IC1 and the diodes. V_{o1} therefore has a value of 0, making the output signal equal to $-V_i$.

$$V_o = -(-4 + 2 \times 0) = -(-4 + 0) = -(-4) = +4$$

The output of this circuit is always positive, regardless of the polarity of the original input signal. Since unity gain is used throughout, the magnitude of the signal is not altered.

Absolute value circuits are also used for precision full-wave rectification.

For the circuit of Fig. 11-17 to work properly, the following value relationships must be maintained in the resistors:

$$R1 = R2 = R5 = R6$$

$$R4 = \frac{R1}{2}$$

$$R3 = R4$$

$$R7 = \frac{R1}{4}$$

12

Digital Applications

Op amps are basically analog devices. The output may generate any of an infinite series of continuously changing values. No one involved with modern electronics can totally ignore digital circuitry, which deals only with whole numbers (integers). No fractional outputs are possible.

Op amps can be used in digital circuitry, especially in applications where digital and analog circuits must be interfaced. Before I begin to discuss op amp digital circuits, I want to review the basics of digital electronics.

THE BINARY SYSTEM

We are used to counting in the decimal numbering system, which is comprised of 10 digits.

0 5
1 6
2 7
3 8
4 9

If I need to express a quantity greater than nine, I start a new column to the left of the first. This second column indicates how many units equal to the base I need to add. In the decimal system the base is 10. For example, the number 27 represents two 10s plus seven ones. Additional columns represent higher powers of the base. Let's break down the number 473921.

$$4 \times 10^5 + 7 \times 10^4 + 3 \times 10^3 + 9 \times 10^2 + 2 \times 10^1 + 1 \times 10^0 =$$
$$4 \times 100,000 + 7 \times 10,000 + 3 \times 1000 + 9 \times 100 + 2 \times 10 + 1 \times 1$$

Ordinarily, this breakdown is performed mentally without really thinking about it. We are used to the system, and we work with it automatically.

Ten is not the only possible base for a numbering system. For electronics circuitry it isn't even particularly convenient, no matter how comfortable humans may feel with it. For electronic circuitry, a binary system turns out to be the most convenient. The binary numbering system consists of just two digits.

0
1

It is very easy to represent these two digits unambiguously by the presence or absence of a voltage. 1 and 0 can be represented by on and off, or high and low. Of course, single digit binary numbers can't communicate very much information. If I want to express numbers higher than one, I need to add columns to the left. Each column represents a higher power of 2.

first column	$2^0 = 1$
second column	$2^1 = 2$
third column	$2^2 = 4$
fourth column	$2^3 = 8$
fifth column	$2^4 = 16$
sixth column	$2^5 = 32$
seventh column	$2^6 = 64$
eighth column	$2^7 = 128$

Converting from one numbering system to another is simply a matter of multiplying each column by the appropriate value. For instance, let's convert the binary number 10011101 into the decimal system.

$$10011101_2 = 1 \times 2^7 + 0 \times 2^6 + 0 \times 2^5 + 1 \times 2^4 + 1 \times 2^3 + 1 \times 2^2 + 0 \times 2^1 + 1 \times 2^0$$
$$= 1 \times 128 + 0 \times 64 + 0 \times 32 + 1 \times 16 + 1 \times 8 + 1 \times 4 + 0 \times 2 + 1 \times 1$$
$$= 128 + 0 + 0 + 16 + 8 + 4 + 0 + 1$$
$$= 157$$

Converting from decimal to binary is a little more complex. I start with the decimal number, say 140, and subtract the highest possible power of two. In this case, this would be 2^7, or 128. I repeat this process until I get down to 2^0, or 1. If a given power of two can't be divided into the number, a 0 is inserted in the appropriate column.

140		(Original Decimal Number)
−128	$2^7 \times 1$	
12	$2^6 \times 0$	(64)
	$2^5 \times 0$	(32)
	$2^4 \times 0$	(16)
−8	$2^3 \times 1$	
4	$2^2 \times 1$	
−4		
0		
	$2^1 \times 0$	(2)
	$2^0 \times 0$	(1)

Decimal 140 equals binary 10001100. Fortunately, in practical design work, you probably won't have to do much conversion between numbering systems. However, you should have a fair understanding of what's involved.

DIGITAL GATES

Digital circuits are generally built around subcircuits known as gates. A gate produces a specific digital output in response to the digital values of its input or inputs.

The simplest digital gates have one input and one output. There are four possible ways for a digital output to respond to the single input. Two of these are entirely trivial and useless.

In one case, the output is always 0, regardless of the value of the input. That is, if the input is 0, the output will be 0, and if the input is 1, the output will be 0. The action of a digital gate can be summarized in a chart that compares all possible input conditions to the appropriate outputs. The *truth table* for the gate we've just described would look like this:

Input	Output
0	0
1	0

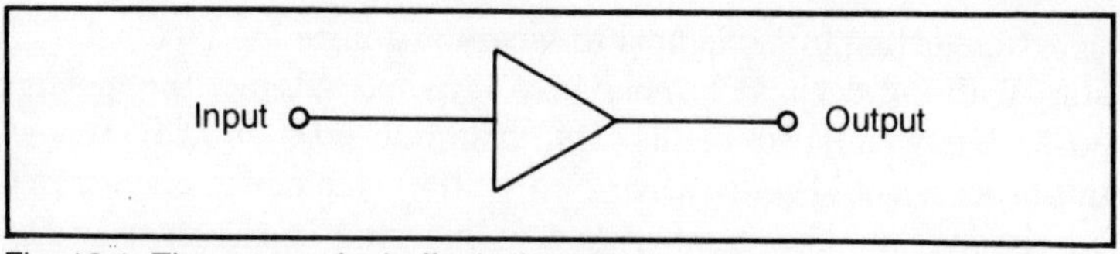

Fig. 12-1. The output of a buffer is the same as the input.

Another trivial possibility would have the output equal to 1 regardless of the input condition.

Input	Output
0	1
1	1

Both of these possibilities are useless, of course, since the input has absolutely no effect on the output. Another possibility would be for the output to always be equal to the input.

Input	Output
0	0
1	1

This one might also seem to be trivial and pointless, but it isn't. This type of gate is used for buffering and interfacing. It is similar to the noninverting voltage followers used in the earlier chapters of this book. In fact, this type of single input gate is called a buffer. Its schematic symbol is shown in Fig. 12-1.

The final possibility for a single input/single output digital gate would have the output always in the opposite state as the input.

Input	Output
0	1
1	0

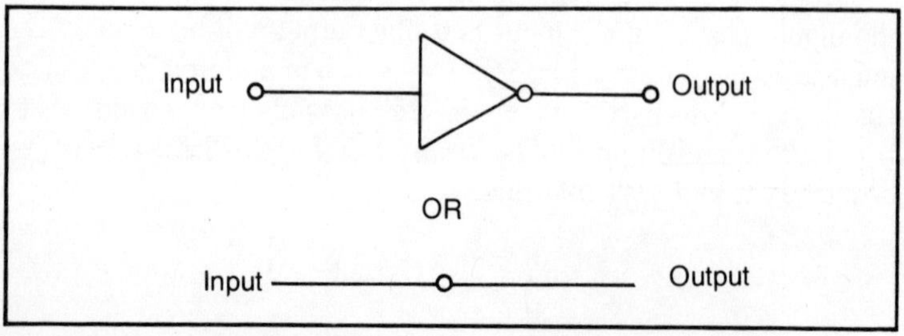

Fig. 12-2. An inverter changes a logic 1 to a logic 0, and vice versa.

This is conceptually similar to the inverting voltage followers I have been dealing with throughout this book.

The schematic symbol for this type of gate is shown in Fig. 12-2. The small circle is used in digital schematics to represent inversion. Occasionally the triangle may be eliminated, leaving just the circle, but this can make diagrams less clear, so it is not preferred. This type of gate is called an inverter or a NOT gate. Digital inversion is often written as a bar over the value to be inverted. For example, if the input of an inverter is identified as A, the output can be written as $\overline{A}$.

While buffers and inverters are useful, their applications are limited. More powerful circuitry combines two or more signals to determine the output.

There are sixteen possible output patterns for gates with two inputs and one output. The truth tables for each possibility are shown in Table 12-1. Notice that some are trivial. A and P ignore both inputs and hold constant output values. F simply reflects input

Table 12-1. Truth Tables for Two-Input, One-Output Digital Gates.

a	b	out A	out B	out C	out D	out E	out F	out G	out H	out I	out J	out K	out L	out M	out N	out O	out P
0	0	0	0	0	0	1	0	0	1	0	1	1	0	1	1	1	1
0	1	0	0	0	1	0	0	1	0	1	0	1	1	0	1	1	1
1	0	0	0	1	0	0	1	0	0	1	1	0	1	1	0	1	1
1	1	0	1	0	0	0	1	1	1	0	0	0	1	1	1	0	1

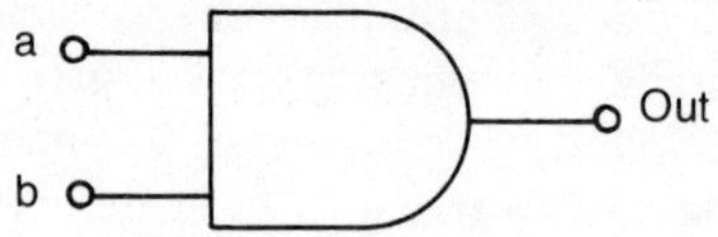

Fig. 12-3. An AND gate has an output of 1 only when all inputs are 1s.

a and ignores input b, and K inverts a, while continuing to ignore b. Similarly, F and K ignore input a.

Pattern B is an important pattern in digital circuitry. The output is a 1 if and only if both of the inputs are 1s. If either or both of the inputs are a 0, the output will be a 0.

Since the output is a 1 only when input is a 1 *and* input b is a 1, this type of gate is called an *AND* gate. The schematic symbol for this device is shown in Fig. 12-3. AND gates with more than two inputs are also possible, although not as common. The schematic symbol and truth table for a three input AND gate is shown in Fig. 12-4. The action of an AND gate can be reversed by inverting the output as shown in Fig. 12-5. In this case the truth table becomes:

a	b	c	$\overline{c}$
0	0	0	1
0	1	0	1
1	0	0	1
1	1	1	0

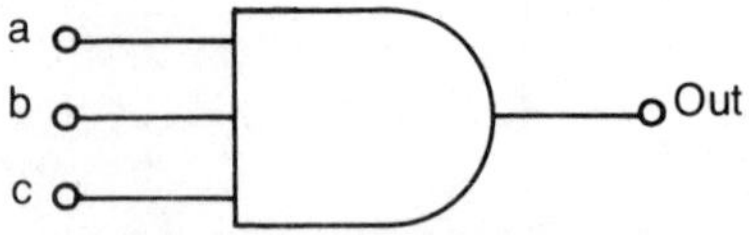

a	b	c	Out
0	0	0	0
0	0	1	0
0	1	0	0
0	1	1	0
1	0	0	0
1	0	1	0
1	1	0	0
1	1	1	1

Fig. 12-4. An AND gate can be designed with more than two inputs.

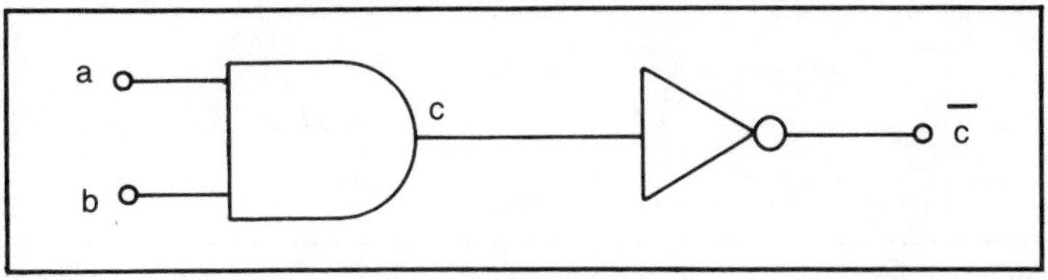

Fig. 12-5. The action of an AND gate can be reversed by adding an inverter at the output.

Output $\overline{c}$ (NOT c) always has the opposite value of c. Notice that the output pattern is the same as truth table O in Table 12-1. Since the output is a 1 unless input a and input b are both 1s, this can be called a NOT AND gate. It is usually shortened to NAND gate. The AND gate and the inverter can be combined into a single device. The schematic symbol for a NAND gate is illustrated in Fig. 12-6.

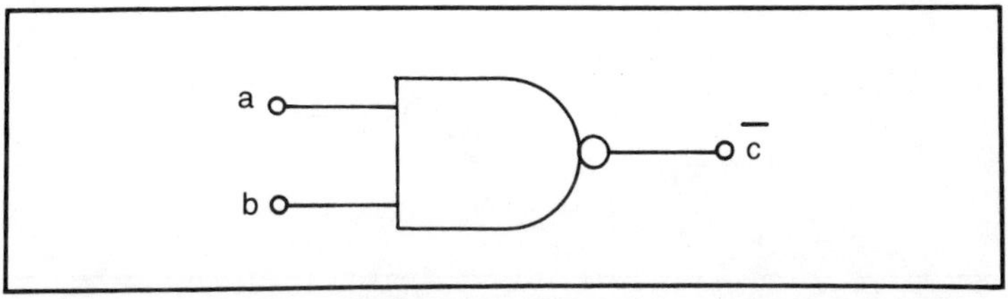

Fig. 12-6. A NAND gate has a logic 1 output unless all inputs are at logic 1.

The next important type of two input/one output digital gate uses truth table L in Table 12-1.

As long as input a *or* b is a 1, the output will be a 1. Not surprisingly, this is called an *OR* gate. The schematic symbol is shown in Fig. 12-7. A variation on the basic OR gate is the exclusive-OR, or XOR gate, shown in Fig. 12-8. In this device, the output is a 1 if input a is a 1 or if input b is a 1, but not both. Another way to phrase this would be that the output is a 1 when the two

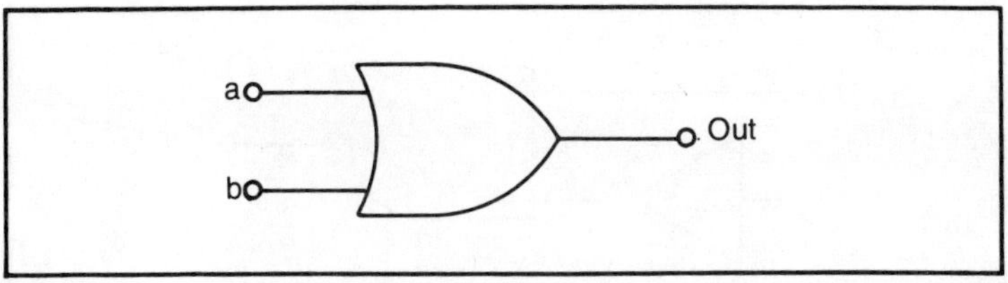

Fig. 12-7. A logic 1 at either or both of the inputs of an OR gate will produce a logic 1 output.

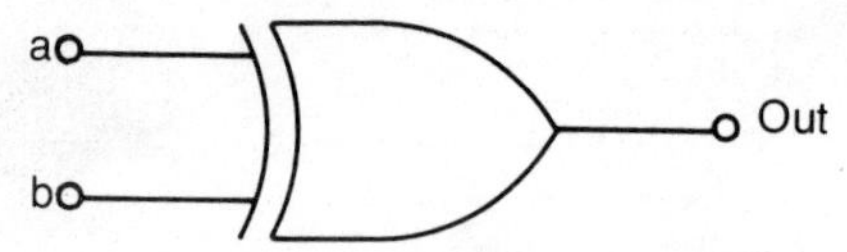

Fig. 12-8. A variation on the basic OR gate is the XOR (exclusive-OR) gate.

inputs have differing values. If a and b are both 0s or both 1s, the output will be a 0. This is the same as truth table I from Table 12-1.

Inverting the output of an OR gate produces a NOT OR, or *NOR* gate. The output is a 1 if and only if neither input a *nor* input b is a 1. This is truth table pattern E in Table 12-1. The schematic symbol for a NOR gate is shown in Fig. 12-9.

The other truth table patterns listed in Table 12-1 can be created by combining the basic gates. For example, pattern N can

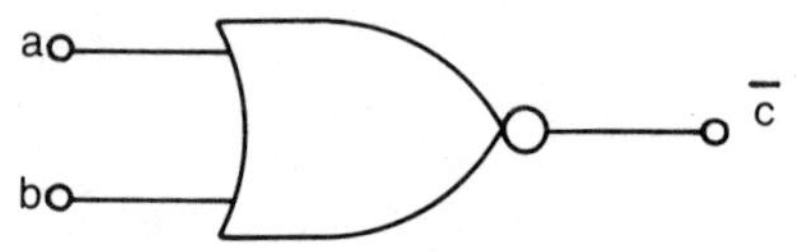

Fig. 12-9. A NOR gate is the opposite of an OR gate.

be created with an inverter and a NAND gate, as shown in Fig. 12-10.

The basic gates can be combined in countless ways to create new patterns. For instance, the five input/three output circuit of Fig. 12-11 will generate the truth table in Table 12-2.

OP AMP LOGIC LEVEL SHIFTER

Digital gates are generally in IC form. There are several basic

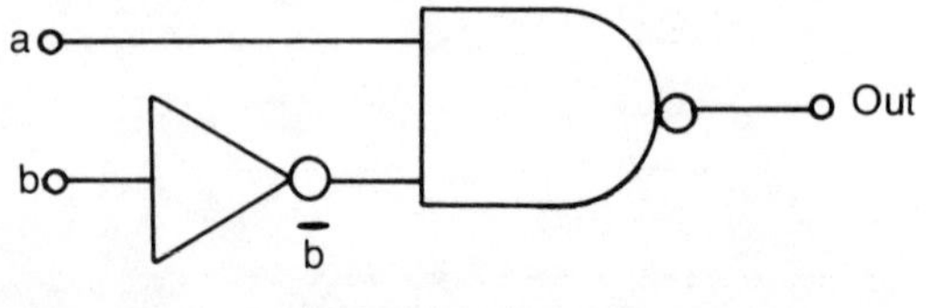

Fig. 12-10. Gates can be combined to generate new truth tables.

Table 12-2. Truth Table for Fig. 12-11.

Inputs					Outputs			Inputs					Outputs		
a	b	c	d	e	f	g	h	a	b	c	d	e	f	g	h
0	0	0	0	0	0	0	1	1	0	0	0	0	1	1	1
0	0	0	0	1	0	0	0	1	0	0	0	1	1	1	0
0	0	0	1	0	0	0	0	1	0	0	1	0	1	1	0
0	0	0	1	1	0	0	1	1	0	0	1	1	1	1	1
0	0	1	0	0	0	0	1	1	0	1	0	0	1	1	1
0	0	1	0	1	0	0	0	1	0	1	0	1	1	1	0
0	0	1	1	0	0	0	0	1	0	1	1	0	1	0	0
0	0	1	1	1	0	0	1	1	0	1	1	1	1	0	1
0	1	0	0	0	0	0	1	1	1	0	0	0	0	0	1
0	1	0	0	1	0	0	0	1	1	0	0	1	0	0	0
0	1	0	1	0	0	0	0	1	1	0	1	0	0	0	0
0	1	0	1	1	0	0	1	1	1	0	1	1	0	0	1
0	1	1	0	0	0	0	1	1	1	1	0	0	0	0	1
0	1	1	0	1	0	0	0	1	1	1	0	1	0	0	0
0	1	1	1	0	0	0	0	1	1	1	1	0	0	0	0
0	1	1	1	1	0	0	1	1	1	1	1	1	0	0	1

classes, or families, each family has its own specific characteristics. For example, TTL (transistor-transistor logic) gates use 0 volts to represent a 0, and +5 volts to represent a 1. In some applications it may be necessary to change these logic levels to some other voltage. This may be necessary when interfacing differing logic families, or when driving analog devices such as relays.

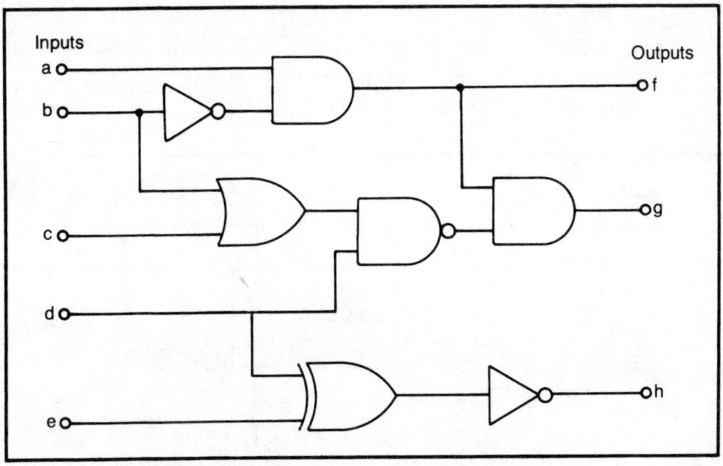

Fig. 12-11. This moderately complex gating circuit combines five input signals into three output signals.

Figure 12-12 shows the circuit for an op amp logic level shifter. Notice that there are two inputs. One (V_i) accepts the digital signal to be shifted. The other input (V_c) is a control signal input, that turns the circuit on and off.

Assume V_c is at logic 1, and V_i is a 0. The inputs to AND gate A are V_i and V_c. For an AND gate this combination results in a 0 output. V_i is also inverted to V_i before being combined with V_c by AND gate B. Since $V_i = 0$ then V_i must equal 1. The inputs to this gate are therefore both 1s, producing a 1 at the output.

The op amp is wired as a differential amplifier with its gain determined by the R2/R1 ratio. R3 should equal R1, and R4 should have the same value as R2. For purposes of discussion, let's say R2 has twice the value of R1. This makes the gain equal to 2. I will also assume the digital input signal uses 0 volts for logic 0 and +5 volts for logic 1.

AND gate A is putting out a 0. This means 0 volts is being fed through resistor R1 to the inverting input of the op amp. At the same time, the output of AND gate B is a logic 1, or 5 volts. This signal is fed through R3 to the noninverting input. The differential voltage is 5 volts. Since the noninverting input has the large signal, the output polarity is not inverted. The output equals 5 volts times the gain. Assuming the gain is 2, the output when $V_i = 0$ and $V_c = 1$ is +10 volts.

Now let's change V_i to a logic 1. V_c will remain unchanged at logic 1. Both inputs (V_i and V_c) for AND gate A are at logic 1, making its output a 1. This means a +5 volts signal is fed through resistor R1 to the inverting input of the op amp. The inverter changes $V_i = 1$ to $\overline{V_i} = 0$, making the output of gate B (and the input to the noninverting side of the op amp) equal to 0. Again the absolute differential input to the op amp is 5 volts, but since the inverting

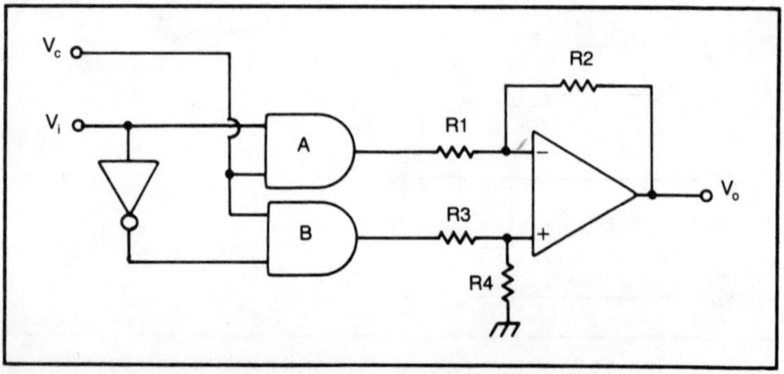

Fig. 12-12. This circuit is an op amp logic level shifter.

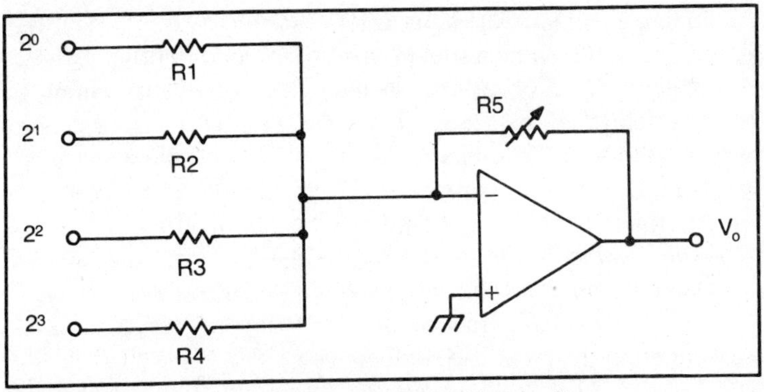

Fig. 12-13. Here is a simple op amp D/A converter circuit.

input is now dominant, the output polarity will be reversed, making the output equal to −10 volts.

The original digital input (V_i) switches from 0 to +5 volts. The output of this circuit switches between −10 volts and +10 volts.

What happens if V_c is changed to a logic 0? The output of both AND gates will have to be 0s, since an AND gate can only go to logic 1, if all inputs equal 1. The differential input voltage to the op amp is 0, so the output is 0 volts. I can summarize the action of this circuit with a modified truth table.

Inputs		Output
V_i	V_c	
x	0	0 volts
0	1	+10 volts
1	1	−10 volts

The x in the V_i column represents *don't care*. The output will not be affected by this input value. It may be either a 0 or a 1. Don't care conditions will often show up in complex gating circuits.

D/A CONVERTERS

It is often necessary to use a digital signal to operate some analog circuit. Some means of translating the digital numbers into continuously varying analog voltages or currents is needed. A circuit that performs this translation is called a digital-to-analog converter, or D/A converter, for short.

Op amps are frequently used in D/A conversion applications. A simple op amp D/A converter is shown in Fig. 12-13. This circuit should certainly look familiar to you by now. It is nothing more than a summing amplifier. The input resistors are scaled to weight the inputs. The 2^1 (2) column should have twice the weighting as the 2^0 (1) column. The 2^2 (4) column should have twice the weighting as the 2^1 (2), and four times the weighting as the 2^0 (1) columns, and so forth. This can be extended as far as needed to allow one input resistor for each binary digit, or bit. Binary numbers with more than 8 digits (bits) would be awkward for this simple circuit.

The diagram in Fig. 12-13 accepts a 4-bit binary number and translates it into a single analog voltage at the output.

For proper D/A translation, the following ratios must be maintained in the resistance values:

$$R2 = \frac{R1}{2}$$

$$R3 = \frac{R1}{4} = \frac{R2}{2}$$

$$R4 = \frac{R1}{8} = \frac{R2}{4} = \frac{R3}{2}$$

The feedback resistor (R5) is shown as a variable resistance. It is selected to give the desired gain.

If I select a 100 K resistor for R1, the other three resistors will take on the following values:

$$R2 = \frac{R1}{2} = \frac{100,000}{2} = 50,000 = 50 \text{ K}$$

$$R3 = \frac{R1}{4} = \frac{100,000}{4} = 25,000 = 25 \text{ K}$$

$$R4 = \frac{R1}{8} = \frac{100,000}{8} = 12,500 = 12.5 \text{ K}$$

High quality precision resistors with low tolerances should be used. Standard 5 percent resistors probably won't be close enough for any but the least critical of applications.

Let's say I have set feedback resistor R5 so that a 1 at 2^0 input and 0s at the other three inputs will produce an output of exactly -1 volt, observe what happens with other input combinations.

The binary number is written in descending order.

$$2^3 \quad 2^2 \quad 2^1 \quad 2^0$$

The sample input I just mentioned ($2^0 = 1$, $2^1 = 0$, $2^2 = 0$, $2^3 = 0$) could be written as 0001. If all of the inputs are 0 (0000), the output will also be 0 volts, of course. Now let's consider what happens when the 2^1 input is the only input at logic 1.

$$\begin{aligned} 0010 &= 0 \times 2 + 0 \times 2^2 + 1 \times 2^1 + 0 \times 2^0 \\ &= 0 \times 8 + 0 \times 4 + 1 \times 2 + 0 \times 1 \\ &= 2 \end{aligned}$$

Input resistor R2 has half the resistance as input resistor R1, so naturally it has only half the voltage drop. This means twice as much signal reaches the input of the op amp, so the output should logically be twice the value obtained when the output is 0001. Sure enough, the output in this case is −2 volts.

The next consecutive binary number sets both 2^0 and 2^1 to logic 1, while leaving 2^2 and 2^3 at 0.

$$\begin{aligned} 0011 &= 0 \times 2^3 + 0 \times 2^2 + 1 \times 2^1 + 1 \times 2^0 \\ &= 0 \times 8 + 0 \times 4 + 1 \times 2 + 1 \times 1 \\ &= 2 + 1 \\ &= 3 \end{aligned}$$

A logic 1 at 2^0 produces a −1-volt output. A logic 1 at 2^1 produces a −2-volt output. Since this is a summing amplifier, these two simply add together, producing an output voltage of −1 volt plus −2 volts, or −3 volts. The same sort of thing occurs for all possible input combinations up to 1111.

$$\begin{aligned} 1111 &= 1 \times 2^3 + 1 \times 2^2 + 1 \times 2^1 + 1 \times 2^0 \\ &= 1 \times 8 + 1 \times 4 + 1 \times 2 + 1 \times 1 \\ &= 8 + 4 + 2 + 1 \\ &= 15 \end{aligned}$$

A 1111 input results in a −15 volt output signal. A complete summary of the output voltage for all possible output combinations is listed in Table 12-3. The output voltage will change in a stepwise fashion, as shown in Fig. 12-14, rather than in a smooth continuous manner. The output will never be at an intermediate level, such as 3.5 volts.

For some applications it may be desirable to smooth out the

Table 12-3. Input/Output Values for Fig. 12-13.

Inputs 2^3	2^2	2^1	2^0	Output	
0	0	0	0	0 volts	−(0 + 0 + 0 + 0)
0	0	0	1	−1 volt	−(0 + 0 + 0 + 1)
0	0	1	0	−2 volts	−(0 + 0 + 2 + 0)
0	0	1	1	−3 volts	−(0 + 0 + 2 + 1)
0	1	0	0	−4 volts	−(0 + 4 + 0 + 0)
0	1	0	1	−5 volts	−(0 + 4 + 0 + 1)
0	1	1	0	−6 volts	−(0 + 4 + 2 + 0)
0	1	1	1	−7 volts	−(0 + 4 + 2 + 1)
1	0	0	0	−8 volts	−(8 + 0 + 0 + 0)
1	0	0	1	−9 volts	−(8 + 0 + 0 + 1)
1	0	1	0	−10 volts	−(8 + 0 + 2 + 0)
1	0	1	1	−11 volts	−(8 + 0 + 2 + 1)
1	1	0	0	−12 volts	−(8 + 4 + 0 + 0)
1	1	0	1	−13 volts	−(8 + 4 + 0 + 1)
1	1	1	0	−14 volts	−(8 + 4 + 2 + 0)
1	1	1	1	−15 volts	−(8 + 4 + 2 + 1)

stepwise motion of the output with a filter capacitor, causing the output to look something like Fig. 12-15. The modified circuit is shown in Fig. 12-16. Since the inverting input is used, the output will always be negative. A noninverting version of the D/A converter circuit is illustrated in Fig. 12-17. It works in basically the same manner as the inverting version of Fig. 12-13. The gain is determined by the R2/R1 ratio in the normal manner for a noninverting amplifier. The other resistors should maintain the following relationships.

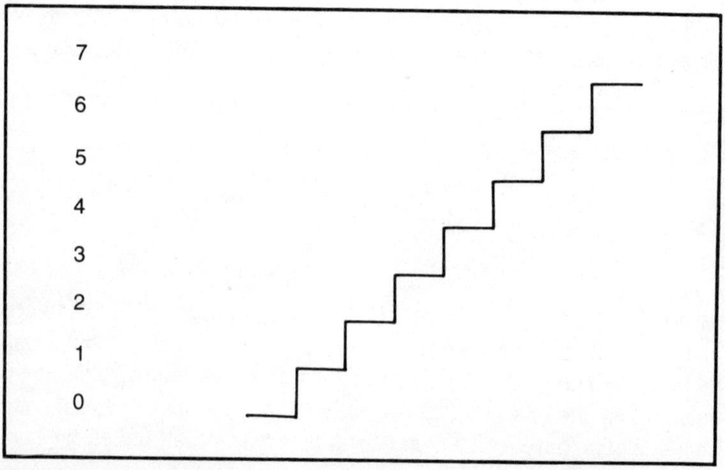

Fig. 12-14. The output of a D/A converter will move in a stepwise fashion between specific values.

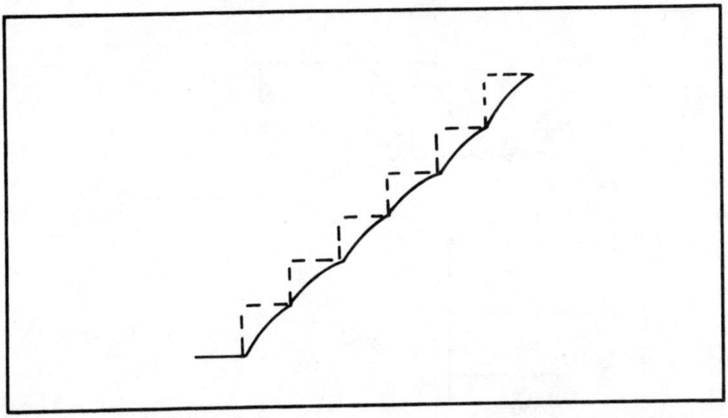

Fig. 12-15. The output of a D/A converter can be smoothed out somewhat by a filter capacitor.

$$R4 = \frac{R3}{2}$$

$$R5 = \frac{R3}{4} = \frac{R4}{2}$$

$$R6 = \frac{R3}{8} = \frac{R4}{4} = \frac{R5}{2}$$

$$R7 = R3$$

Both of these circuits (Fig. 12-13 and Fig. 12-17) suffer from some common defects. First off, the halving of each consecutive

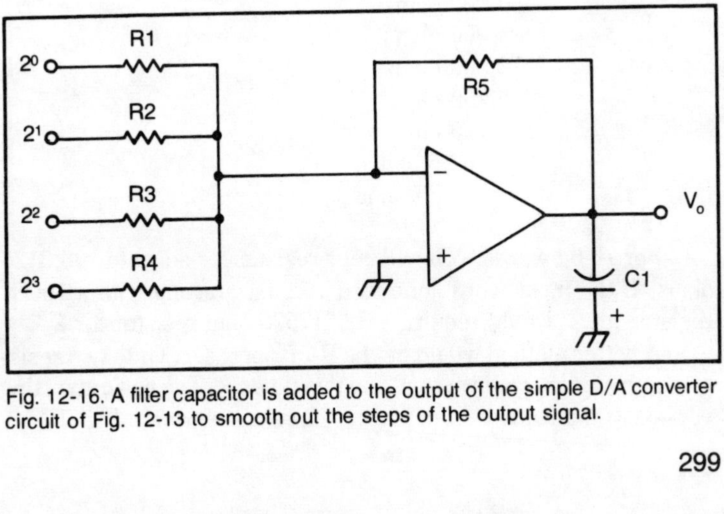

Fig. 12-16. A filter capacitor is added to the output of the simple D/A converter circuit of Fig. 12-13 to smooth out the steps of the output signal.

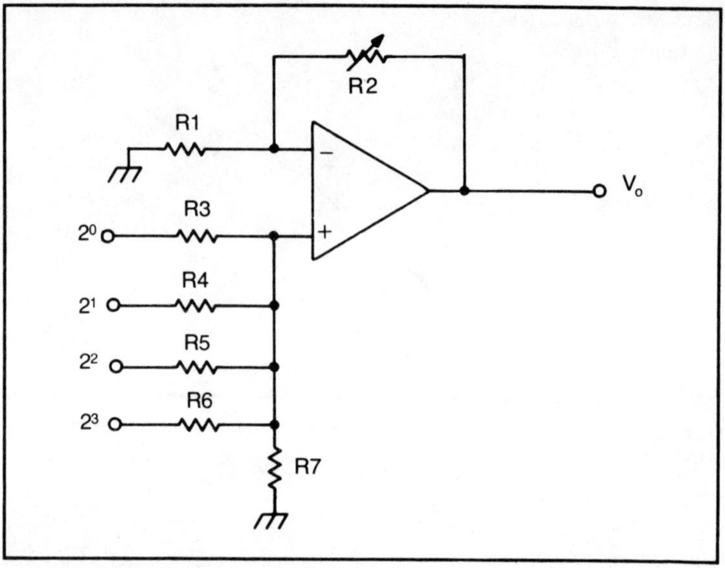

Fig. 12-17. This D/A converter will provide positive output voltages.

input resistor results in some oddball values which may be difficult to find. Exact values are needed, or the results will be imprecise.

Moreover, if the input signals consist of more than just a few bits, the range of required resistance values quickly becomes excessive. For example, if you need to translate 8-bit words (sometimes called bytes) and if you use a 100 K resistor for the 2^0, you will need the following resistance values:

100,000 ohms	2^0
50,000 ohms	2^1
25,000 ohms	2^2
12,500 ohms	2^3
6,250 ohms	2^4
3,125 ohms	2^5
1,562.5 ohms	2^6
781.25 ohms	2^7

For 10-bit words, you will need resistances down to 195.3125 ohms. If the input word contained 16 bits, starting with a 100 K resistor for 2^0 would require a 3.0517578 ohm resistor for 2^{15}!

A better method would be the R-2R approach. Only two resistances values are required—R and two times R. The basic inverting version of the R-2R D/A converter circuit is shown in Fig. 12-18.

The following relationships should be maintained between the resistance values:

$$R4 = R6 = R8 = R11 = R$$
$$R1 = R2 = R5 = R7 = R9 = R10 = 2R$$

Feedback resistor R3 sets the circuit gain. Except for R3, all of the resistors in this circuit have one of two values. If R equaled 12 K then 2R would equal 24 K. No other resistance values will be needed.

The R-2R ladder can be extended indefinitely to accommodate as many input bits as are required. Larger binary numbers (with more bits) offer finer resolution. A 4-bit binary word can take on just 16 values from 0000 to 1111 (15). An 8-bit word can go as high as 255 (1111 1111), and a 16-bit word can cover 65536 possible combinations (1111 1111 1111 1111).

A noninverting version of the R-2R D/A converter circuit is illustrated in Fig. 12-19. The relationships between the resistor values in this circuit are as follows:

$$R1 = R4 = R6 = R8 = R$$
$$R3 = R5 = R7 = R9 = R10 = 2R$$

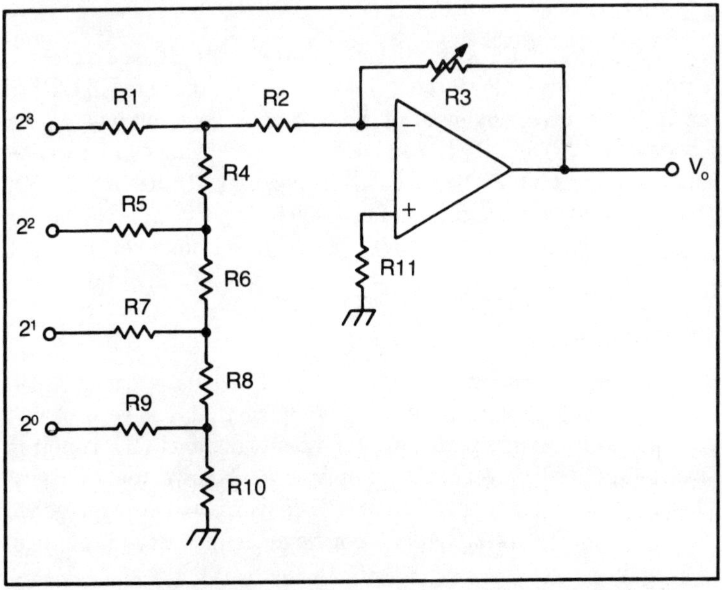

Fig. 12-18. A better approach to D/A conversion is to use an R-2R ladder.

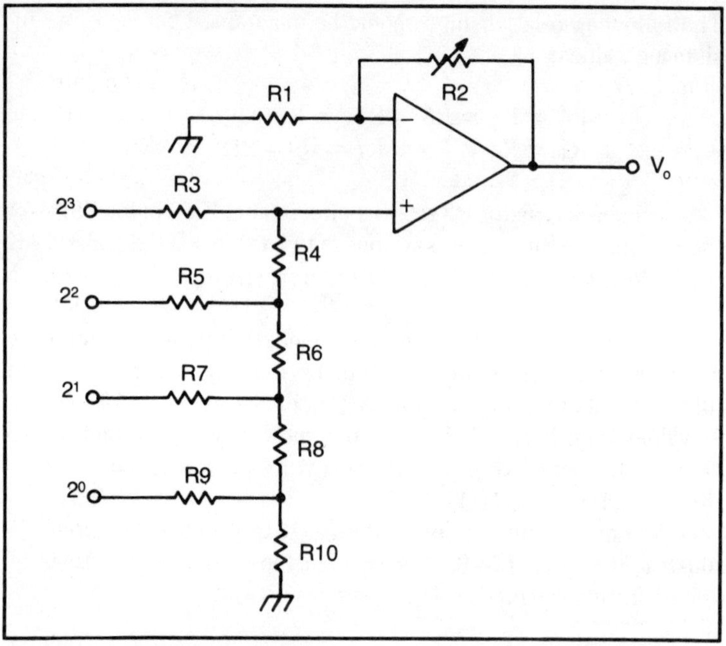

Fig. 12-19. This is the noninverting version of the R-2R D/A converter circuit.

Feedback resistor R2 is selected to produce the desired output gain.

The R-2R circuit works because lower order bits see a greater resistance. In Fig. 12-19 the 2^3-bit only passes through R3 (2R) to reach the noninverting input of the op amp. The 2^2-bit has to pass through R5 (2R) and R4 (R) and sees a total resistance of 3R. The least significant bit (2^0) has to pass through R9 (2R), R8 (R), R6 (R), and R4 (R), for a grand total of 5R. Each bit's position in the ladder forms a weighted voltage divider between the noninverting input and ground.

A/D CONVERTERS

Converting analog signals into digital signals is trickier than converting digital signals into analog signals. As you've probably guessed, a circuit that performs this translation function is called an analog-to-digital converter, or simply an A/D converter, for short. There are two basic approaches to A/D conversion. One approach is illustrated in Fig. 12-20. A D/A converter is included in the feedback path.

The op amp is wired as a basic comparator. When the circuit is reset the output of the comparator is 0, and the counter begins counting upwards from 0000. Each digital count is converted back to an analog voltage by the D/A converter. This voltage is compared to the original input voltage. If the input voltage is greater than the D/A output voltage, the count continues. Finally, when the D/A converter produces an output voltage that is equal to or greater than the input voltage, the output of the comparator goes high, stopping the digital counter.

A binary word now appears at the output , corresponding to the analog input voltage. The counter is reset and the process begins anew. Some kind of latching circuitry is required at the outputs so that only the desired final count values are allowed to reach the output. I don't want to output the intermediate count values.

The other basic approach to A/D conversion is known as the dual-slope integrator method. A typical circuit is shown in Fig. 12-21. This circuit is made up of a number of stages—an integrator, a zero-crossing detector, a NAND gate, a resettable binary counter, and a switch-driving comparator. With the exception of the counter, these are all circuit types that have been discussed earlier in this book.

The binary counter must output 1-bit higher than the desired digital output. If I want to output 4-bit binary words, I will need a 5-bit counter. The extra bit is fed to IC3.

In following the action of this circuit, I will start at the reset point. The counter is outputting all zeroes (00000). The output signal from the comparator pulls the electronic switch into the V_i position shown in the schematic. The output of the integrator (IC1) is about +0.7 volts (due to D1), and the output of the zero-crossing

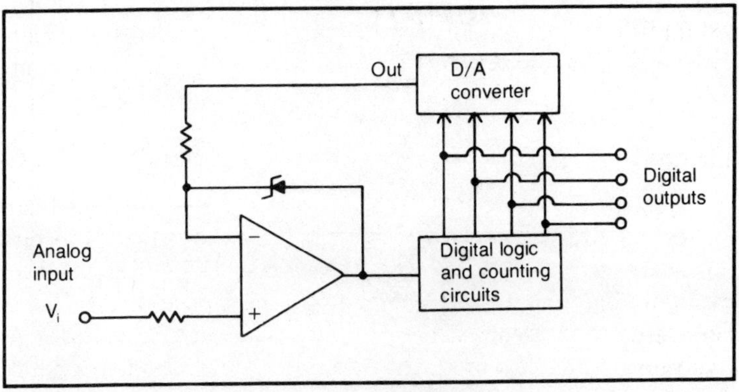

Fig. 12-20. One method of A/D conversion uses a D/A converter in the feedback path.

detector (IC2) is at about −0.7 volts (due to D2). This low signal at one of the inputs of the NAND gate forces its output to logic 1, inhibiting the counter from responding to incoming clock pulses.

The input signal (V_i) starts to charge capacitor C1, causing the output of IC1 to pass through 0 and become increasingly negative. When this signal passes through zero the detector (IC2) fires. Zener diode D2 will clamp the output of op amp B at a high positive voltage, allowing the NAND gate to start responding to the incoming clock pulse signals. The output of the gate goes high when the clock pulse goes low, triggering the counter.

Midway through the counting cycle, the most significant bit (which is connected to the input of IC3) will go high. This will trigger the electronic switch so that V_i is replaced by a known value—V_{ref}. The reference voltage (V_{ref}) has the opposite polarity as the unknown input voltage (V_i). This means capacitor C1 begins to discharge, causing the output of integrator IC1 to become less

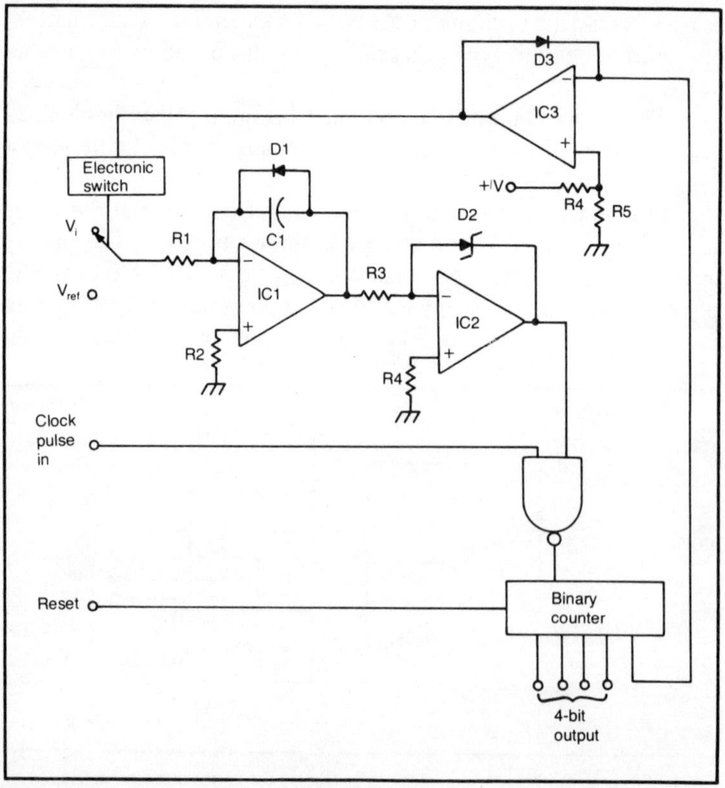

Fig. 12-21. This circuit uses the dual-slope integrator method of A/D conversion.

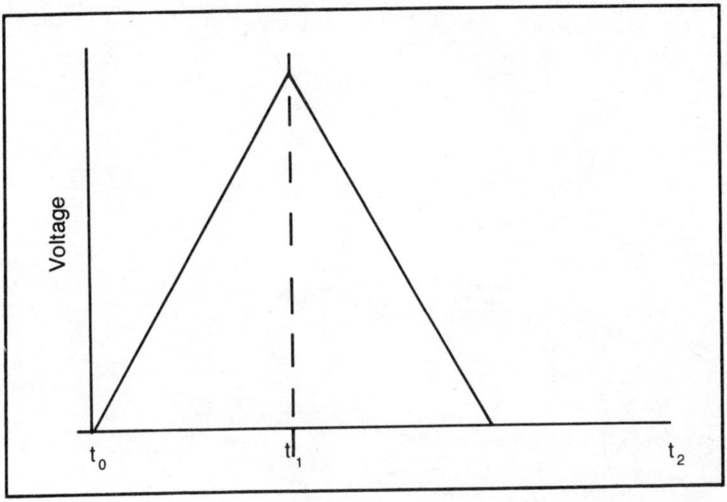

Fig. 12-22. In a dual-slope integrator A/D converter the charging and discharging rates of a capacitor are compared.

negative. When the integrator output passes back up through zero, the zero-crossing detector (IC2) fires again, cutting off the NAND gate, and stopping further clock pulses from incrementing the counter.

The output of the binary counter (excluding the extra bit going to the comparator) will now be a binary number that is proportionate to the input voltage (V_i). This value is stored in external latches and the converter circuit is reset, and the cycle starts over.

This method works by comparing the time V_{ref} takes to discharge the integration capacitor at a known rate.

The greater the input voltage (V_i), the higher the charge on the capacitor will be. The higher the charge on the capacitor, the longer it will take V_{ref} to discharge it. Since the clock pulses arrive at a steady rate, the longer it takes V_{ref} to discharge C1, the higher the count on the binary counter will be.

This is called the dual-slope integration method because graphing the voltage across the capacitor will show a charging slope and a discharging slope, as illustrated in Fig. 12-22. The time at which the circuit is reset is t_o. The capacitor is charged by V_i until time t_1 is reached. This is the point when the counter overflows and triggers the comparator and electronic switch. V_{ref} discharges the capacitor from time t_1 to time t_2 at which point final count is fed to the output. The ratio of t_1 to t_2 compared with t_0 to t_1 is proportional to the value of V_i.

13

Metering Circuits

In Chapter 10, I covered a number of op amp circuits that could detect specific input conditions. Since op amps can generate a continuous range of values at their outputs they can do more than simple yes/no indications. In this chapter, I will look at various ways op amps can be used in measuring various electrical characteristics, such as voltage, current, and resistance. By combining these circuits, you can even design and build your own op amp VOM (volt-ohm-milliammeter).

BASIC DC VOLTMETERS

A super simple op amp dc voltmeter circuit is shown in Fig. 13-1. This circuit consists of just a noninverting voltage follower and a regular analog voltmeter. What's the point of this circuit? I could just apply the voltage directly to the voltmeter, and skip the op amp stage altogether.

True, but most voltmeters have a fairly low input impedance that can load down the circuit being measured, affecting the voltage reading. Let's say the voltmeter has an input impedance of 1 K, and I need to measure the voltage drop across R2 in Fig. 13-2. The battery voltage is 6 volts. Assume I have the following resistor values in the circuit:

R1 = 470 ohms
R2 = 100 ohms
R3 = 820 ohms

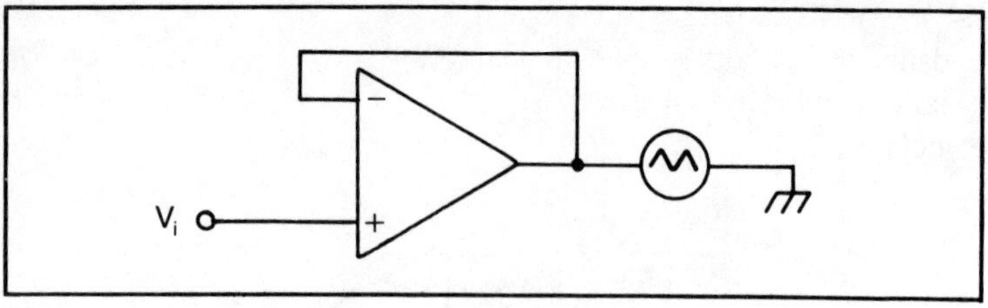

Fig. 13-1. A basic op amp voltmeter serves primarily as an impedance multiplier.

The total circuit resistance is 470 + 100 + 820 ohms, or 1390 ohms. The current through the circuit can be found with Ohm's Law.

$$I = \frac{E}{R} = \frac{6}{1390} = 0.0043165 \text{ amp} = 4.3 \text{ mA}$$

The current through each resistor is equal. Therefore, I can find the nominal voltage drop through each resistor:

$$E_{R1} = IR = 0.0043 \times 470 = 2.03 \text{ volts}$$
$$E_{R2} = 0.0043 \times 100 = 0.43$$
$$E_{R3} = 0.0043 \times 820 = 3.54$$

The total voltage drop in the circuit should equal the original source voltage (6 volts):

$$E_t = E_{R1} + E_{R2} + E_{R3} = 2.03 + 0.43 + 3.54 = 6 \text{ volts}$$

When the voltmeter is connected across R2 the effective resistance changes due to the internal resistance of the meter itself

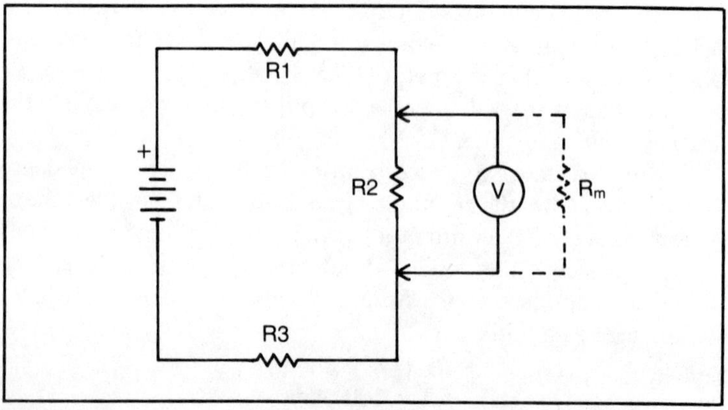

Fig. 13-2. Low meter impedance can affect the voltage being measured.

(R_m). I have already stated that the voltmeter has an input impedance of 1 K. I am actually measuring the voltage drop across the parallel combination of R2 and R_m. The effective resistance can be calculated as follows:

$$R_e = \frac{R2 \times R_m}{R2 + R_m} = \frac{100 \times 1000}{100 + 1000} = \frac{10,000}{1100}$$

$$R_e = 90.9090 \text{ ohms} \cong 91 \text{ ohms}$$

This changes the current flow through the circuit.

$$I = \frac{E}{R} = \frac{6}{470 + 91 + 820} = \frac{6}{1,381} =$$

$$I = 0.0043447 \text{ amps} \cong 0.043 \text{ amps} = 4.3 \text{ mA}$$

The change in current flow is negligible. The actual calculated voltage drop across R2 should be 0.4316547. The voltage drop across the parallel combination of R2 and R_m will be 0.3949707. Component tolerances are likely to introduce that much error.

Since the meter resistance is large compared to the circuit resistance, the introduced error is minor. But what if the meter resistance is relatively small compared to the circuit resistance?

This time I assign the following resistance values to the circuit and calculate the current and voltage drops.

$$R1 = 22 \text{ K}$$
$$R2 = 56 \text{ K}$$
$$R3 = 33 \text{ K}$$
$$Rt = 111 \text{ K}$$

$$I = \frac{E}{R} = \frac{6}{111,000} = 0.0000541 \text{ amps} = 0.054 \text{ mA}$$

$$E_{R1} = IR = 0.0000541 \times 22,000 = 1.19 \text{ volts}$$
$$E_{R2} = 0.0000541 \times 56,000 = 3.03 \text{ volts}$$
$$E_{R3} = 0.0000541 \times 33,000 = 1.78 \text{ volts}$$
$$E_t = E_{R1} + E_{R2} + E_{R3} = 1.19 + 3.03 + 1.78 = 6 \text{ volts}$$

However, when I try to measure the voltage drop across R2, I actually measure the voltage drop across the parallel combination of R2 and R_m:

$$R_e = \frac{R2 \times R_m}{R2 + R_m} = \frac{56{,}000 \times 1000}{56{,}000 + 1{,}000} = \frac{56{,}000{,}000}{57{,}000} = 982.5 \text{ ohms}$$

That's quite a difference from the nominal value of 56,000 ohms! Total circuit resistance has now been reduced.

$$R_t = R1 + R_e + R7 = 22{,}000 + 982.5 + 33{,}000 = 55982.5 \text{ ohms}$$

This dramatically changes the current flow through the circuit.

$$I = \frac{E}{R} = \frac{6}{55982.5} = 0.0001072 \text{ amps} = 0.11 \text{ mA}$$

The voltage drop measured across R2 is about .1 volt. That is a major error when the voltage I should be measuring is 1.19 volt.

$$E = IR_3 = 0.0001072 \times 982.5 = 0.1053 \text{ volt}$$

By placing a noninverting voltage follower between the circuit being measured and the voltmeter, the loading resistance becomes the input impedance of the op amp, which is very high. For the 741 op amp IC, this is typically about 1 megohm, or 1,000,000 ohms. The parallel combination with a 56 K (56,000 ohms) resistor works out as follows:

$$R_e = \frac{56{,}000 \times 1{,}000{,}000}{56{,}000 + 1{,}000{,}000} = \frac{560{,}000{,}000{,}000}{1{,}056{,}000} = 53{,}030 \text{ ohms}$$

The resistor's tolerance rating could introduce that much error.

Raising the input impedance of a voltmeter increases its accuracy by reducing the loading errors. The circuit of Fig. 13-1 is also known as an impedance multiplier.

Another advantage of the op amp input stage is that I can add gain to measure small voltages that may be too weak to register directly on the voltmeter. The circuit is shown in Fig. 13-3.

I can take this a step further and add a range switch, as shown in Fig. 13-4. Each feedback resistor should be factor of ten larger than the previous one.

R2a = 2.2 K
R2b = 22 K
R2c = 220 K

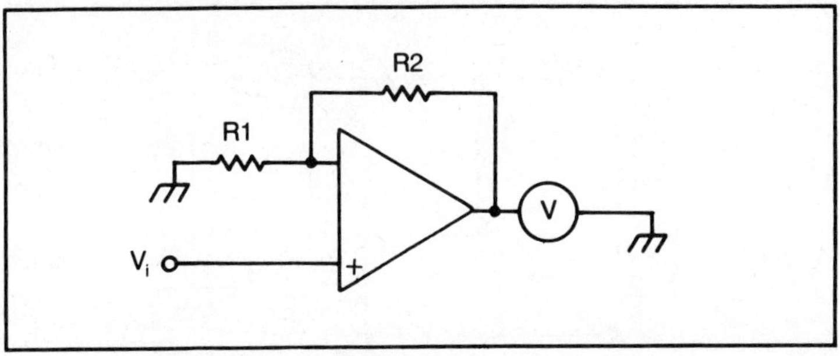

Fig. 13-3. Small voltages can be measured by adding gain to a basic op amp voltmeter.

This allows a single calibration scale to be used on the voltmeter. If I call the R2a position ×1, the R2b position would be ×10, and R2c would be selected for a ×100 range.

A voltmeter reading of 3.45 mV on the ×10 scale would indicate that the input voltage is 34.5 mV.

An even better op amp voltmeter circuit is shown in Fig. 13-5. Here the meter is placed in the feedback path. The exact value of the meter's internal resistance (R_m) will tend to vary somewhat from unit to unit. This will change the amount of current flow, and thus introduce some degree of error in the reading.

This circuit sidesteps the problem quite neatly. The current through the feedback loop of an op amp is determined by the input resistance. The feedback resistance is negligible as far as current flow is concerned. This is one of the natural characteristics of op amps. This means the current flowing through the meter in Fig. 13-5 is equal to the following relationship.

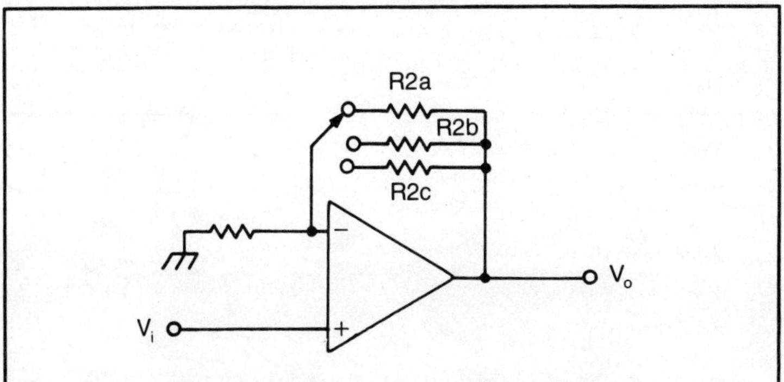

Fig. 13-4. Using switch selectable resistors adds the advantage of range selection.

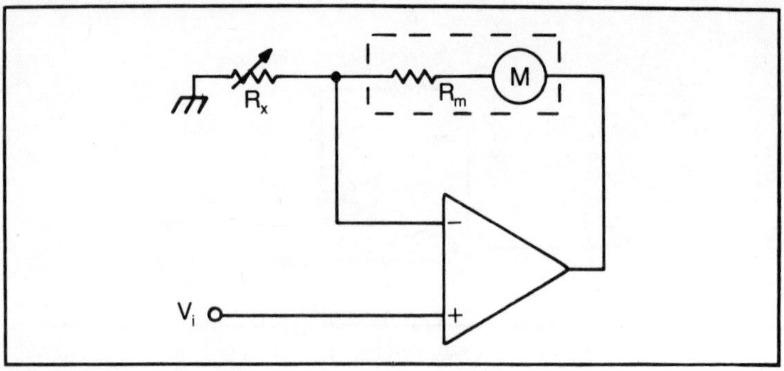

Fig. 13-5. This improved op amp voltmeter has the actual meter in the feedback path. The effects of its internal resistance become negligible.

$$I_m = \frac{V_i}{R_x}$$

The value of R_m is irrelevant. This circuit offers greater unit to unit consistency. Input resistor R_x is made variable to adjust the gain and to calibrate the meter reading.

AC VOLTMETER

Measuring dc voltages is fairly easy to measure, but once you start working with ac signals, things start to get complicated. For one thing, the unit of measurement must be made clear. 10 volts dc is clear and unambiguous. The meaning is simple—there is a potential of 10 volts between the point being measured and a common reference point (usually ground).

But just what is meant by 10 volts ac? The problem is that the exact voltage of an ac signal changes from instant to instant in a regular pattern, as shown in Fig. 13-6. When I say 10 volts ac, do I

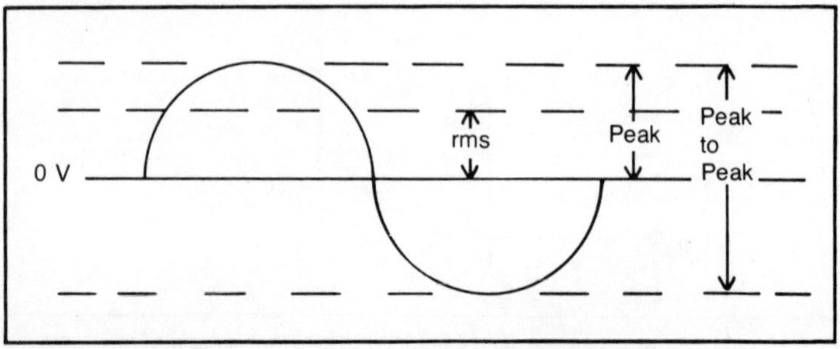

Fig. 13-6. There are several ways to measure ac voltages.

mean the maximum voltage in the cycle is 10 volts away from common (0)? If so, I should say 10 volts ac peak. Or perhaps I mean the cycle passes through a 10 volt range. For example, the positive peak reaches +5 volts, and the negative peak goes as low as −5 volts. The distance from the negative peak to the positive peak is 10 volts. In this case, I should say ten volts ac peak-to-peak.

Probably the most common and useful way of measuring ac voltages is also the least obvious. This method is called the rms (root-mean-square) method. The rms voltage can be found with a little math and some averaging. It's usefulness lies in the fact that the rms voltage is the effective voltage as compared to dc. A 10 volt ac rms signal produces the same amount of heating capacity as a 10 volt dc signal.

For sine waves the rms voltage is 0.707 times the peak voltage. For instance, a 10 volt ac peak signal is the same as a 7.07 volt ac rms signal. The power companies supplying house current generate sinusoidal signals. The sine waveform is maintained through most power supply circuits until rectification.

Most ac voltmeters read out in rms units. This is the most useful form of measurement because dc equations can be used. Ohm's Law applies to ac signals if rms voltages are used. To be useful, an ac voltmeter must maintain its accuracy over the widest possible frequency range.

A basic op amp ac voltmeter circuit is shown in Fig. 13-7. In most cases ordinary silicon diodes can be used in the bridge. If very high accuracy is important, you could substitute precision diode circuits. This circuit will measure the rms voltage of sine wave signals. Other waveshapes will produce inaccurate readings. This is true of most ac voltmeters. One way around this problem is to convert the ac signals into their equivalent dc voltage which can be easily and unambiguously measured. The circuit illustrated in Fig.

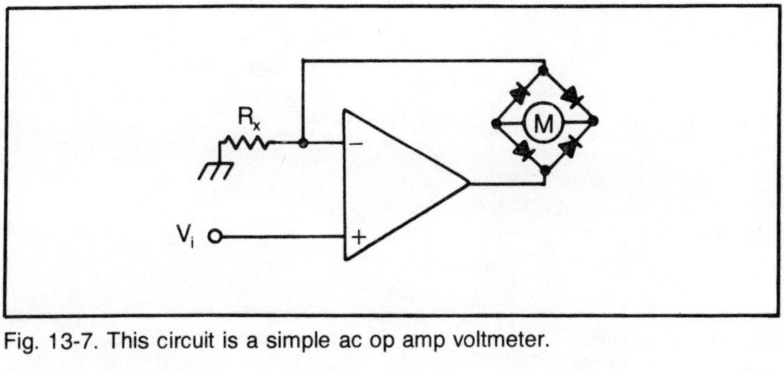

Fig. 13-7. This circuit is a simple ac op amp voltmeter.

13-8 is an rms to dc voltage converter. It uses the temperature conversion method.

Remember that X volts rms has the same heating capacity as X volts dc. This circuit makes use of this fact by comparison of a pair of heating elements (R1 and R2). R1 and R2 are specialized components known as thermocouples. A thermocouple produces an output voltage that is linearly proportional to the temperature of the resistance element (within a specific range).

IC1 is simply a noninverting voltage follower used to isolate this circuit from the circuit being measured. The output of this op amp is fed to thermocouple R1. The resistance element is heated by the rms voltage. A dc voltage (V_1) that is proportional to the temperature of R1 is generated by the thermocouple.

This voltage is fed into IC2, which is connected as a difference amplifier. The output of IC2 is fed back through the resistance element of thermocouple R2. This thermocouple produces a voltage (V_2) that is proportional to the temperature of its resistance element.

The thermocouple voltages (V_1 and V_2) are fed to the inputs of the difference amplifier which will adjust its output until both input voltages are equal. When V_2 equals V_1 I know resistance elements R1 and R2 are at the same temperature. This means there is the same amount of power across the resistance elements. The formula for power is as follows:

$$P = \frac{V^2}{R}$$

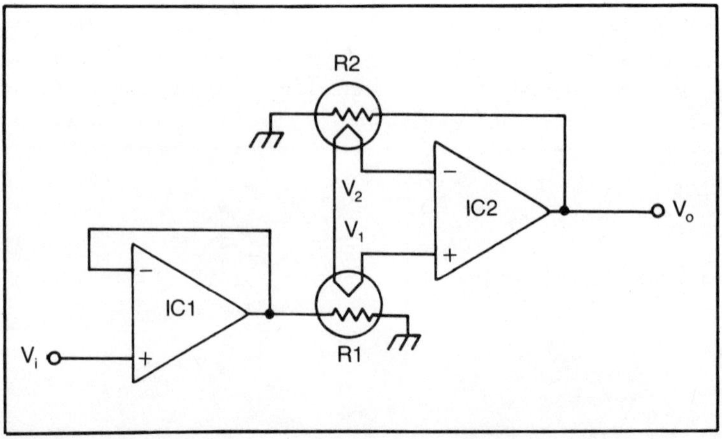

Fig. 13-8. This circuit can be used for rms-to-dc voltage conversion.

Since the two heating elements are at the same temperature, they are using the same amount of power.

$$P1 = P2$$

Substituting the appropriate voltage and resistance values give the following results:

$$\frac{V_i^2}{R1} = \frac{V_o^2}{R2}$$

I can rearrange the equation to solve for the output voltage (V_o).

$$V_o = \sqrt{\frac{R2 \times V_i^2}{R1}}$$

Assuming the resistance values are equal (R1 = R2), I can cancel these things out, simplifying the formula.

$$V_o = \sqrt{V_i^2} = V_i$$

V_i is the applied ac voltage in rms, and V_o is the dc output voltage of IC2. They are effectively equal. This method will work for any ac waveshape, since it is the heating power of the signal being measured and not the ac voltage itself.

CURRENT METERS

Voltage is probably the most commonly measured electrical quantity, but it is not the only thing you might need to find. Often you need to measure the current flowing through a circuit. This is done with a current meter. Where voltage meters are placed in parallel across the circuit element being measured, current meters must be inserted in series with the circuit itself.

An alternative approach is shown in Fig. 13-9. The current flowing through the meter is equal to the input voltage divided by the sum of the resistances in the op amp's output path.

$$I_m = \frac{V_i}{(R_x + R_m)}$$

R_m is the internal resistance of the current meter itself, and R_x

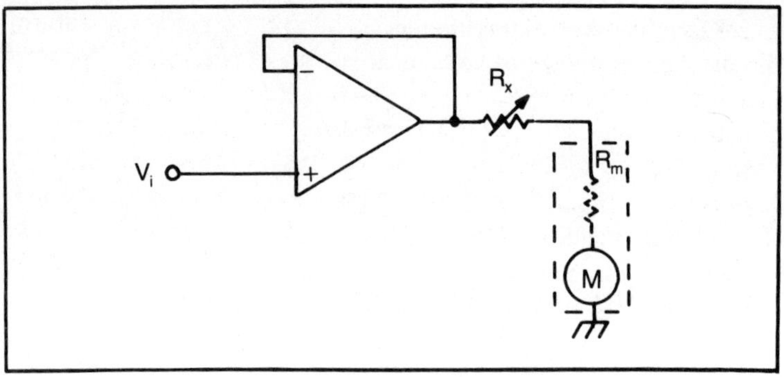

Fig. 13-9. Dc current can be measured using this circuit.

is an external multiplier resistance used to calibrate the meter readings.

OHMMETERS

One of the important features of op amp circuitry is that it allows the use of very small currents (in the microamp range) at voltage levels that are too small to forward bias any semiconductor junctions in the circuit. This characteristic permits us to design a low-power ohmmeter.

Ohmmeters place a voltage source across the component to be measured. If I am measuring a semiconductor device, I must be careful that the applied voltage is not too large or the semiconductor junction might be damaged. Moreover, I don't want high currents to pass through the component being tested if it happens to exhibit a very small amount of resistance.

The op amp ohmmeter circuit shown in Fig. 13-10 does not suffer from these problems. A constant current source insures that

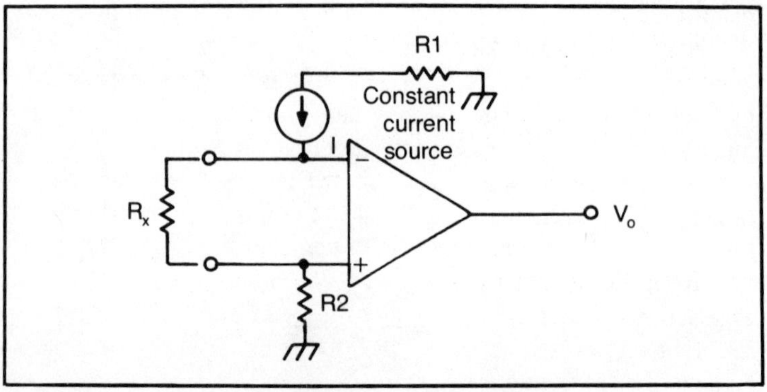

Fig. 13-10. Here is a circuit for an op amp ohmmeter.

excessive current will not flow through the device under test in any circumstances. Moreover, the applied voltage is very low, and thus safe for all semiconductor devices. The secret of this circuit is the constant current source. You can obtain a constant current source with a properly biased FET or a 1.36 volt mercury cell fed into a high resistance.

The op amp measures the voltage drop across the unknown resistance. According to Ohm's Law the voltage drop will equal the current flow (a known constant) times the unknown resistance.

$$V_i = IR_x$$

Therefore, the input voltage will vary directly with changes in the unknown resistance. The output of the op amp is fed to a properly calibrated voltmeter for measurement.

DC DIFFERENTIAL VOLTMETER

The circuits presented so far in this chapter will handle 90 percent of the basic measurement situations that an electronics technician is likely to encounter. There are some exceptions requiring specialized circuitry, however.

For example, the previous circuits all measured the unknown voltage with reference to ground. This is not always possible. Some circuits feature floating or balanced voltages that are not referenced directly to ground. In this case, measurements must be made with a differential amplifier that will determine the difference between two voltages which are individually referenced to ground (true 0).

A typical circuit for this situation is shown in Fig. 13-11.

A pair of op amps are connected as noninverting voltage followers to isolate the circuit being tested from the meter. For maximum accuracy, high input impedance devices should be used. Op amps with FET or Darlington input stages would be good choices.

The current flowing through the meter works out as follows:

$$I_m = \frac{V_i}{R_x R_m}$$

R_m is the internal resistance of the meter itself and R_x is an external resistance used for calibration. Because of the effects of R_m, this circuit will only be accurate for relatively large input signals. As the input voltage drops, so does the accuracy of the meter reading.

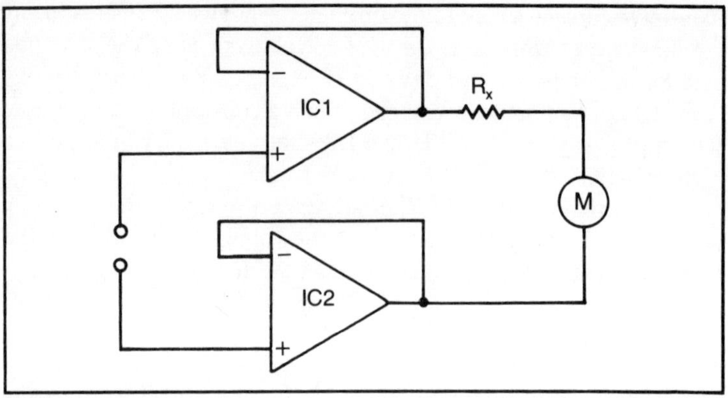

Fig. 13-11. Sometimes it is necessary to measure differential voltages that are not referenced to ground.

An improved dc differential voltmeter circuit is illustrated in Fig. 13-12. Because the meter is in the feedback path of IC5, its internal resistance will have virtually no effect on the amount of current flowing through it (determined by the input current being fed to IC5), and therefore not affecting the readout.

This circuit can be set up for different measurement ranges by changing the gain of the inverting amplifier IC4. If R1 and R2 equal 10 K and R4 equals 100 K, the gain of the entire circuit can be found with the following equation:

$$G = 10 \times \frac{R6}{R5}$$

Offset correction resistor R7 should be selected to match the parallel combination of R5 and R7.

NULL VOLTMETER

In some applications it is necessary to tune a voltage to a specific level. This is most easily done with a null voltmeter. This is a device that will tell you if you are above or below the desired level. Zero is in the middle of the range. This is not necessarily true 0 volts. It represents zero error. The meter pointer can move in either direction from this point. See Fig. 13-13.

A simple op amp null voltmeter circuit is shown in Fig. 13-14. The meter will give a reading of zero when V_i equals V_{ref}. A variation on this circuit is illustrated in Fig. 13-15. In this case, LEDs are used as the indicating devices. At null ($V_i = V_{ref}$) both LEDs should be completely dark, or glowing very dimly.

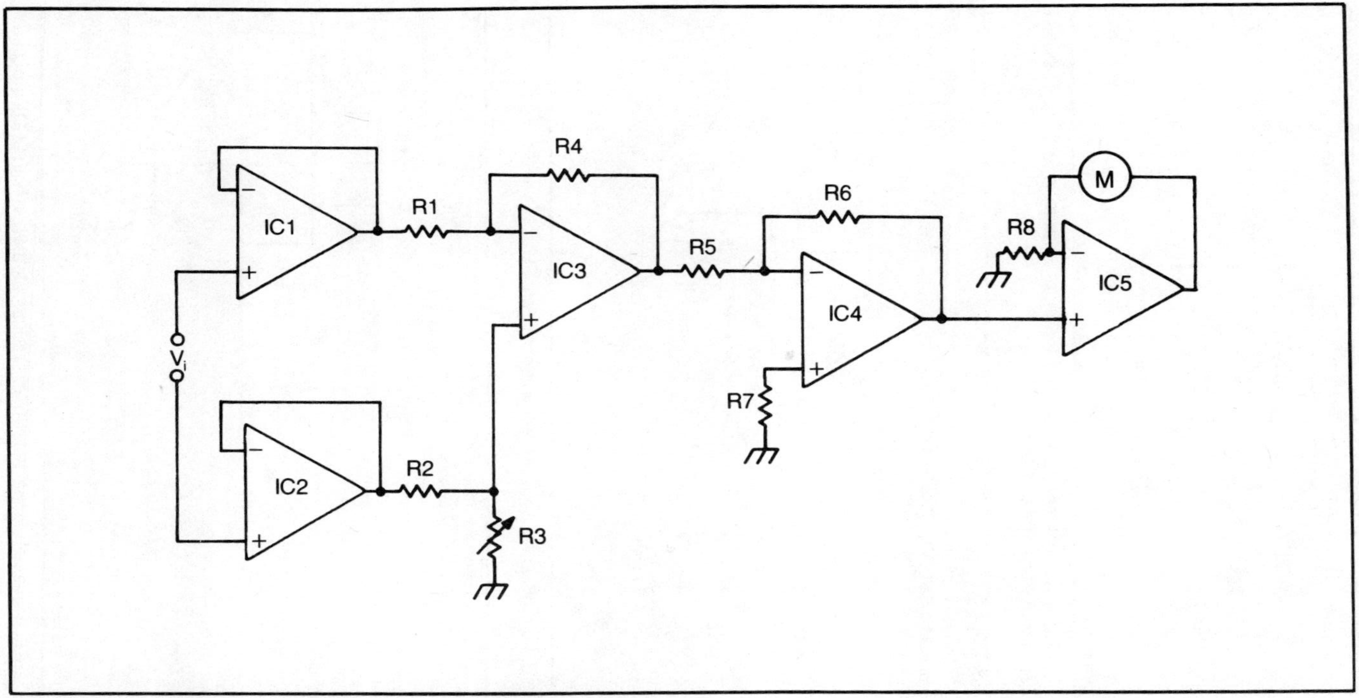

Fig. 13-12. This is an improved dc differential voltmeter circuit.

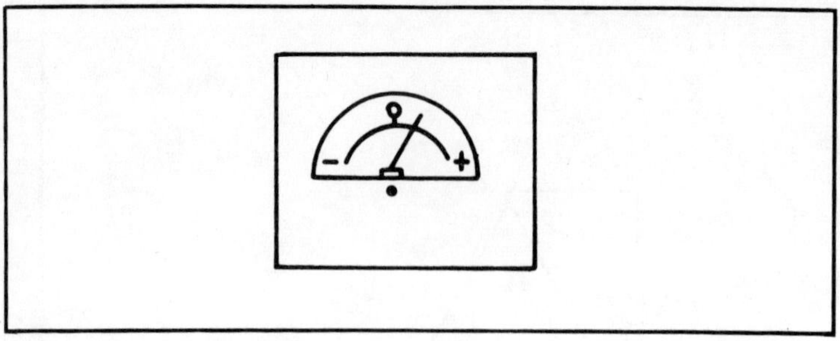

Fig. 13-13. A null voltmeter indicates if the input voltage is above or below a centered reference voltage.

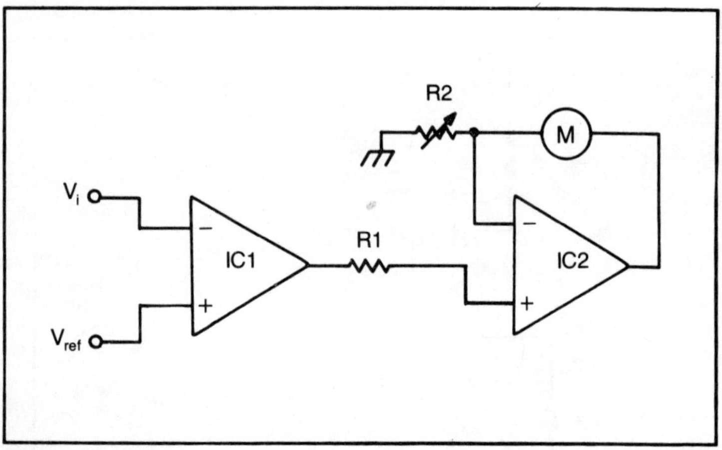

Fig. 13-14. This is an op amp null voltmeter circuit.

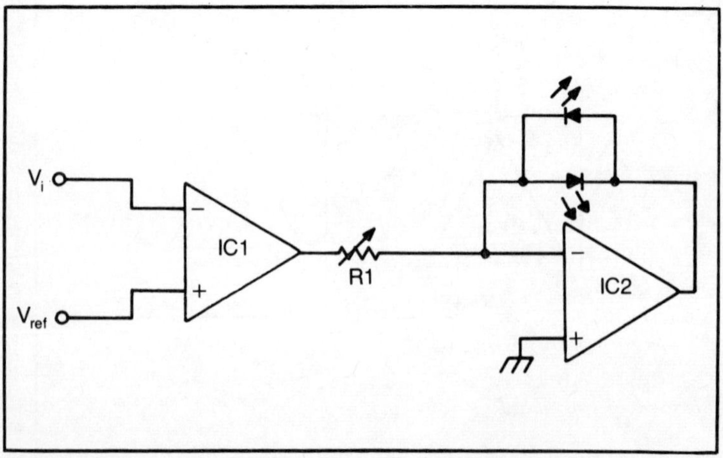

Fig. 13-15. In this circuit, the null condition is indicated when both LEDs are dark.

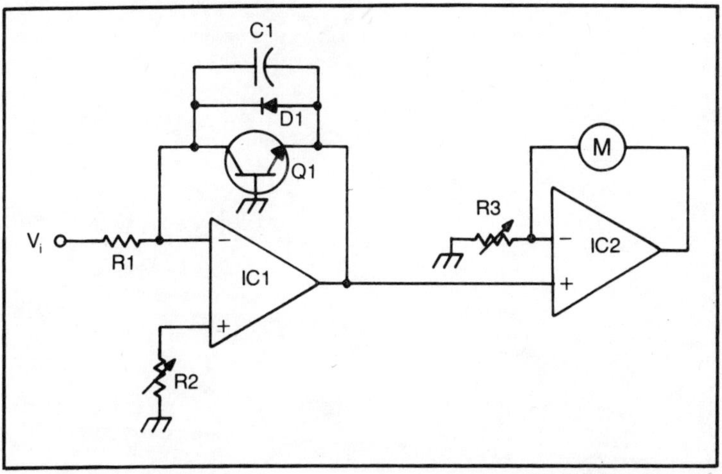

Fig. 13-16. A decibel meter is made up of a logarithmic amplifier and a dc voltmeter.

As the input voltage (V_i) moves away from the reference voltage (V_{ref}) one of the LEDs will start to glow brightly, and the order will go completely dark. The further V_i is from V_{ref}, the brighter the appropriate LED will glow (within the limitations of the device, of course). If the error is in the opposite direction (polarity reversed), the other LED will light, and the one that lit up in the previous example will be extinguished.

These circuits are quite simple and inexpensive, but they give quite good accuracy, and case to case repeatability. Of course, the exact difference voltage is not indicated by the LED circuit (Fig. 13-15), except in the most general way, but this is rarely required with a null meter. All I am concerned with is the question of whether V_i equals V_{ref}, is greater than V_{ref}, or is less than V_{ref}.

Obviously, the null voltmeter is basically a variation of the comparator circuits introduced in Chapter 10.

DECIBEL METER

Most meter circuits (including those discussed in this chapter) offer linear measurements. In most cases, that is highly desirable but not always. In many applications, measurements are needed in decibels (dB).

A circuit to accomplish this is shown in Fig. 13-16. IC1 is a logarithmic amplifier (see Chapter 11) and IC2 is a noninverting buffer/voltmeter.

14

Other Applications

I have covered dozens of applications for operational amplifiers but I am not done yet. The versatility of this device is truly remarkable. In this chapter, I will cover some miscellaneous applications that don't really fit into any of the major categories that were already covered.

AGC AMPLIFIER

The circuit shown in Fig. 14-1 is a basic inverting amplifier with an important difference. Notice that a second voltage is input through a FET (field effect transistor) into the feedback path. This allows automatic adjustment of the amplifier's gain.

A JFET (junction-FET) can be made to behave as a voltage controlled resistor over a limited range. The control voltage (V_c) input changes the effective value of the resistance between the transistor's source and drain. As the V_c is increased, this resistance (R_{DS}) decreases, increasing the effective gain of the amplifier. An equivalent circuit is shown in Fig. 14-2. Notice that R2, R3, and R_{DS} are essentially a voltage divider network. Changing the value of R_{DS} alters the amount of signal that makes it back to the inverting input of the op amp. The gain formula is as follows:

$$G = \frac{-\,R2 + R3 + \dfrac{(R2 \times R3)}{R_{DS}}}{R1}$$

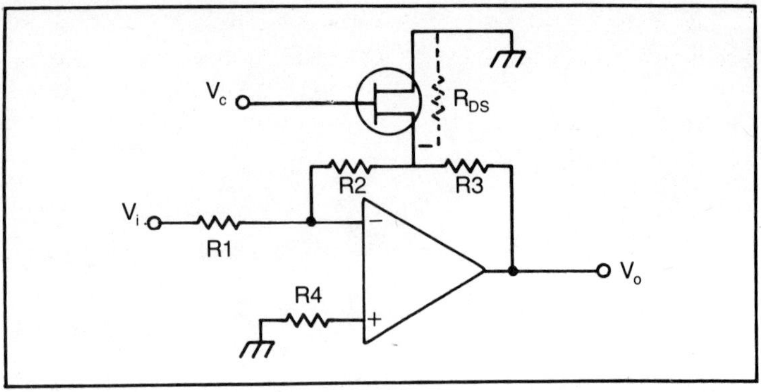

Fig. 14-1. In an agc amplifier the gain is set by an external control voltage.

The gain is negative because this is an inverting amplifier. The output polarity will be the opposite of the input polarity.

Let's try some practical examples. Assume the following component values are being used in the circuit:

$$R1 = 10\text{ K}$$
$$R2 = 22\text{ K}$$
$$R3 = 33\text{ K}$$

Ignoring the effect of R_{DS}, the nominal gain of this circuit would be as follows:

$$G = \frac{-R2 + R3}{R1} = \frac{-22{,}000 + 33{,}000}{10{,}000} = \frac{-55{,}000}{10{,}000}$$

$$G = -5.5$$

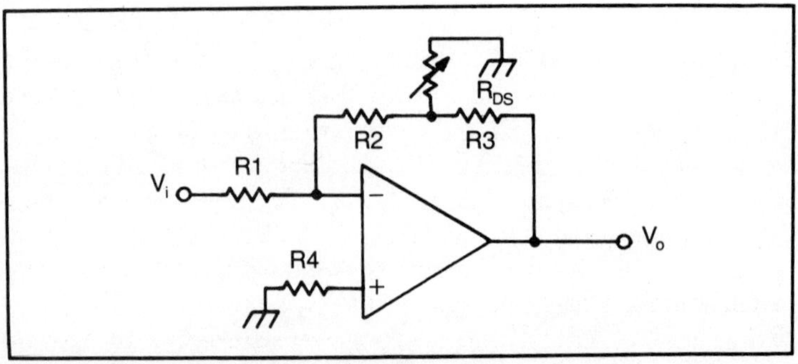

Fig. 14-2. The control voltage in an agc amplifier behaves like an electrically variable voltage divider.

Now, let's assume V_c has a value that makes R_{DS} equal 75 K. The actual gain under these circumstances works out as follows:

$$G = \frac{-R2 + R3 + \frac{(R2 \times R3)}{R_{DS}}}{R1}$$

$$= -\frac{22{,}000 + 33{,}000 + \frac{(22{,}000 \times 33{,}000)}{75{,}000}}{10{,}000}$$

$$= -\frac{55{,}000 + \frac{726{,}000{,}000}{75{,}000}}{10{,}000} = -\frac{55{,}000 + 9680}{10{,}000}$$

$$= -\frac{64{,}680}{10{,}000} = -6.468 \cong -6.5$$

If the control voltage is increased, the value of R_{DS} will decrease. Let's say it drops to 12,000 ohms. This time the gain becomes equal to:

$$G = -\frac{22{,}000 + 33{,}000 + \frac{(22{,}000 \times 33{,}000)}{12{,}000}}{10{,}000}$$

$$= \frac{-55{,}000 + \frac{726{,}000{,}000}{12{,}000}}{10{,}000} = \frac{--55{,}000 + 60{,}500}{10{,}000}$$

$$= \frac{-115{,}500}{10{,}000} = -11.55$$

Increasing the control voltage (V_c) decreases R_{DS}, and increases the circuit gain. This approach will only work properly over a fairly limited range because of the characteristics of the JFET being used as the voltage dependent resistance. If the voltage across the drain and source of the JFET is too large, distortion will result.

Because the gain of this circuit is determined by an external voltage, it is often called a *voltage-controlled amplifier* (or vca). In other applications, it may be referred to as an automatic gain control (agc) amplifier. Agc amps are often used in radio receivers to prevent fluctuations in the signal due to fading effects in the received signal. If the signal grows weaker the gain is increased. If the signal gets stronger, the gain is reduced.

VOLTAGE-TO-FREQUENCY CONVERSION

In many applications, it becomes necessary to transmit electrical information, such as voltages, from one place to another, or to record them for later recovery. This is fairly easy to do with ac signals in the audible range (approximately 50 Hz to 15 kHz). Such signals can be transmitted directly over telephone lines, or recorded using standard audio tape recorders.

For dc, or near dc signals, the normal limitations of audio circuits get in the way. One common approach to solving this problem is to convert dc or very low-frequency signals into higher frequency signals that can be recorded and/or transmitted directly. The new frequency at any instant will correspond to the level of the dc or low-frequency signal.

This can be done with an oscillator circuit that adjusts its output frequency in response to an input voltage. For obvious reasons such a device is called a *voltage-controlled oscillator*, or vco. Since the control voltage modulates the oscillator frequency the process is called frequency modulation, or FM.

When vcos are used to encode dc or low-frequency data, they are called voltage-to-frequency converters or V/F converters. One of the simplest approaches to V/F conversion is the single slope method. The circuit is illustrated in Fig. 14-3. The circuit's primary elements are an op amp comparator and an AND gate.

An external ramp or sawtooth signal is required for the functioning of this circuit. The voltage to be converted is fed to the inverting input of the op amp comparator. The reference ramp signal is applied to the noninverting input.

In practical circuits, switch S1 will be electronically controlled. It is shown here as a simple manual switch for clarity. At time t_0 the switch is in the ground position. Zero volts is applied to the inverting input. The ramp voltage begins at zero volts too, so, since the comparator inputs are equal, the output will be zero. This means the AND gate will be closed (output = 0), regardless of the value of the incoming clock pulses.

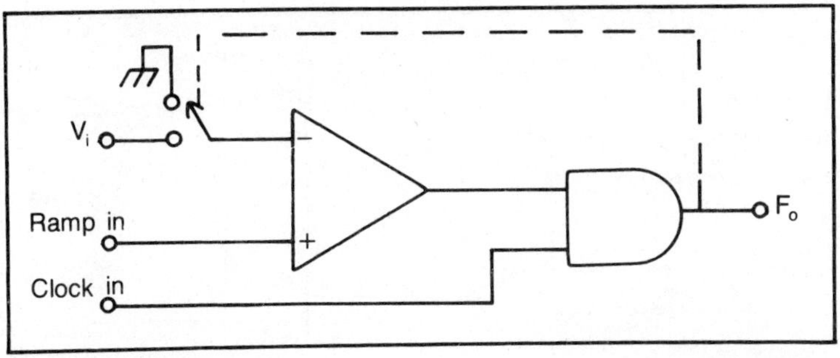

Fig. 14-3. This block diagram illustrates the single slope method of voltage-to-frequency conversion.

The ramp begins to rise from zero. As soon as it is a few millivolts above zero, the comparator's output snaps to the positive saturation state. This applies a logic 1 to the AND gate, allowing the clock pulses to pass through to the output. At the same time, switch S1 reverses its position, and starts feeding V_i to the inverting input of the comparator.

When the ramp voltage equals V_i, the comparator's output goes back to zero, cutting off the AND gate and the stream of clock pulses to the output.

S1 moves back to its grounded position, and the ramp voltage drops back to zero, and the cycle begins again. A typical cycle is shown in Fig. 14-4. The number of output pulses between times t_0 and t_1 is proportional to the input voltage.

Actually, this circuit does not perform true voltage-to-frequency conversion. The output frequency is constant, equal to the incoming clock pulses. The higher the level of V_i, the longer the AND gate will be open, permitting more clock pulses to the output.

The signal can be decoded with a digital counting circuit. A D/A converter (see Chapter 12) can be used to convert the digital count back to an analog voltage. This single slope approach is limited by the accuracy and linearity of the ramp waveform. The accuracy of the clock pulse frequency is also critical to reliable operation. In addition, this circuit is somewhat susceptible to interference errors due to noise.

Better results can be achieved by using dual slope V/F conversion. A typical circuit is shown in Fig. 14-5. Once again this is only pseudo-V/F conversion. The level of the input voltage controls the number of constant frequency output pulses. Notice that two op amp stages are required in this circuit. IC1 is an integrator, and IC2 is wired to perform as a comparator.

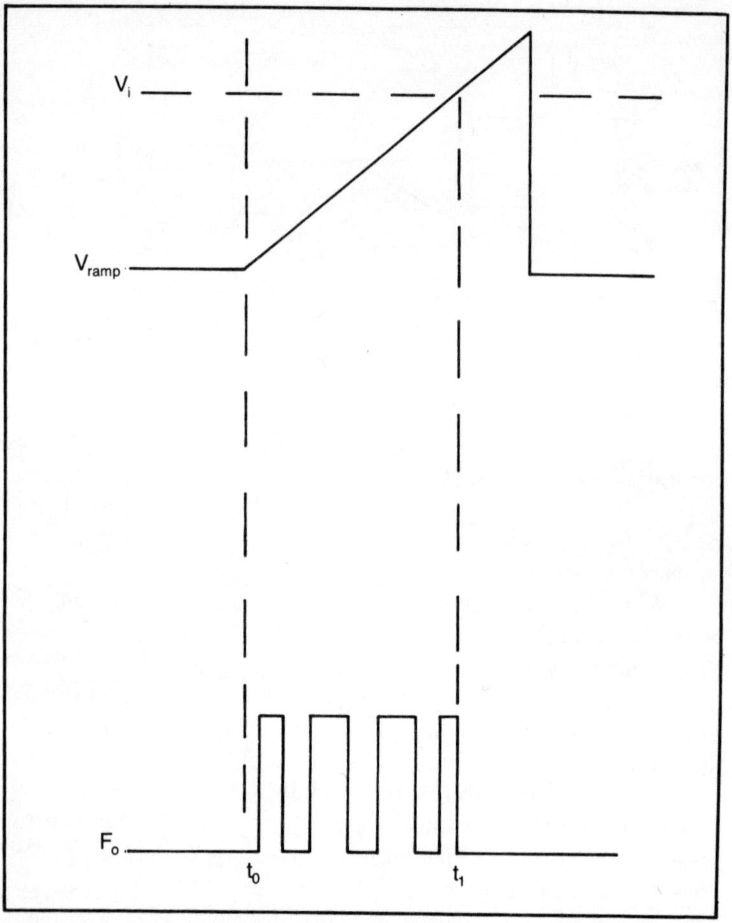

Fig. 14-4. The input and output signals illustrate how the single slope voltage-to-frequency converter functions.

At time t_0, the input voltage to be measured is applied to the integrator (IC1) through S1. The integrator capacitor is charged at a fairly linear rate until V_i is removed or reversed in polarity. The instant that the voltage across the capacitor rises above zero, the output of IC2 snaps to its high level and turns on the AND gate and permits the clock pulses to pass through. This will occur almost immediately.

A digital counter is incremented by the train of clock pulses at the output of the AND gate. Eventually its maximum count will be exceeded. The overflow signal is fed back to electronic switch S1, toggling it to the V_{ref} position. This is time t_1.

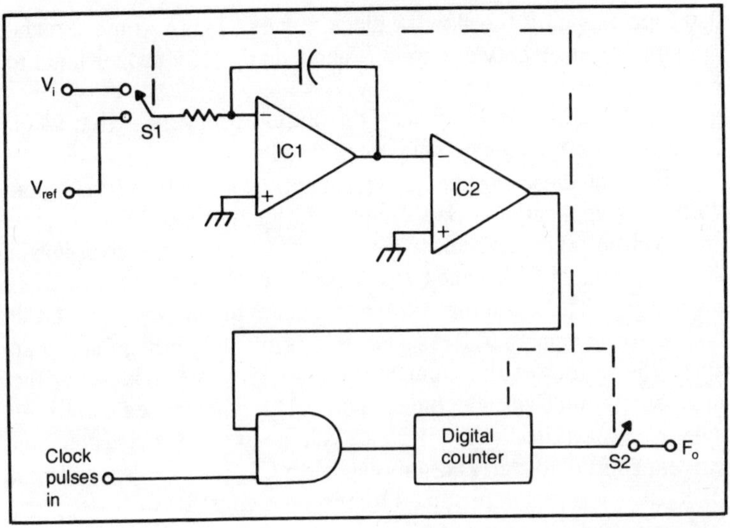

Fig. 14-5. Better results can be achieved with the dual slope voltage-to-frequency converter.

The reference voltage comes from a constant current voltage source with a polarity opposite of that of V_i. This will begin discharging the capacitor in the integrator circuit. The clock pulses will continue incrementing the counter (it starts over at 0000 when it overflows).

When the capacitor is completely discharged, the voltage across it will be zero once again (or slightly negative). The output of the comparator (op amp B) drops back to zero, turning off the AND

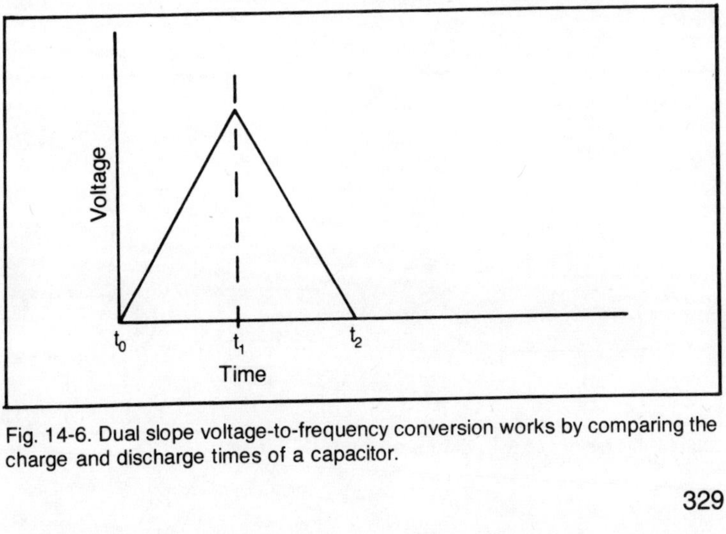

Fig. 14-6. Dual slope voltage-to-frequency conversion works by comparing the charge and discharge times of a capacitor.

gate and stopping the flow of clock pulses. This is time t_2. The number of pulses between time t_1 and t_2 is directly proportional to input voltage V_i.

Switch S2 is used to isolate the output from time t_0 to t_1 when the preliminary count is being performed.

Both of these circuits are frequently encountered in digital voltmeter applications. See Chapter 13 for more details.

A circuit that performs true voltage-to-frequency conversion is shown in Fig. 14-7. At time t_0, switch S1 is open, and capacitor C1 is discharged. The incoming voltage to be converted (V_i) is fed to the noninverting input of IC1. As soon as V_i is even slightly greater than zero, the output of the comparator goes positive, triggering the monostable multivibrator (see Chapter 12). This closes electronic switch S1, permitting the reference voltage (V_{ref}, a precise constant current source) to start charging capacitor C1. S1 is held closed for a fixed time interval determined by the time constant of resistor R2 and capacitor C2.

After the monostable multivibrator times out, one of two things can happen. If the charge across capacitor C1 is greater than the input voltage (V_i), the comparator's output will go negative, and the monostable multivibrator will not be retriggered. If, on the other hand, V_i is still greater than the charge on the capacitor, the comparator output will remain positive, immediately retriggering the monostable multivibrator for another fixed time period, giving V_{ref} more time to charge capacitor C1. This will continue until C1 is charged to a higher voltage than V_i.

The effect of all this is to create an output pulse whose width is proportional to the input voltage (V_i). This is illustrated in Fig. 14-8.

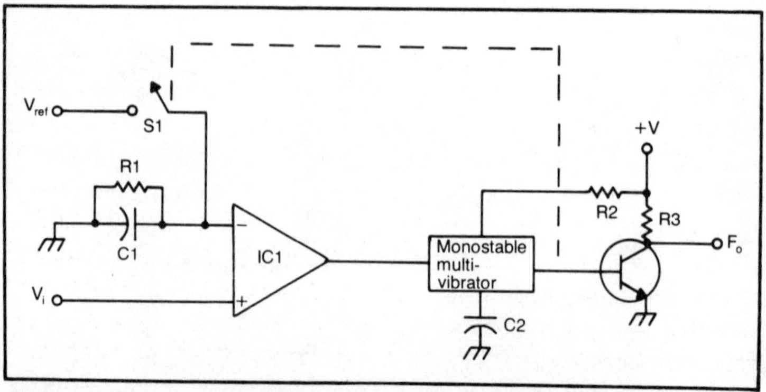

Fig. 14-7. This circuit will perform true voltage-to-frequency conversion.

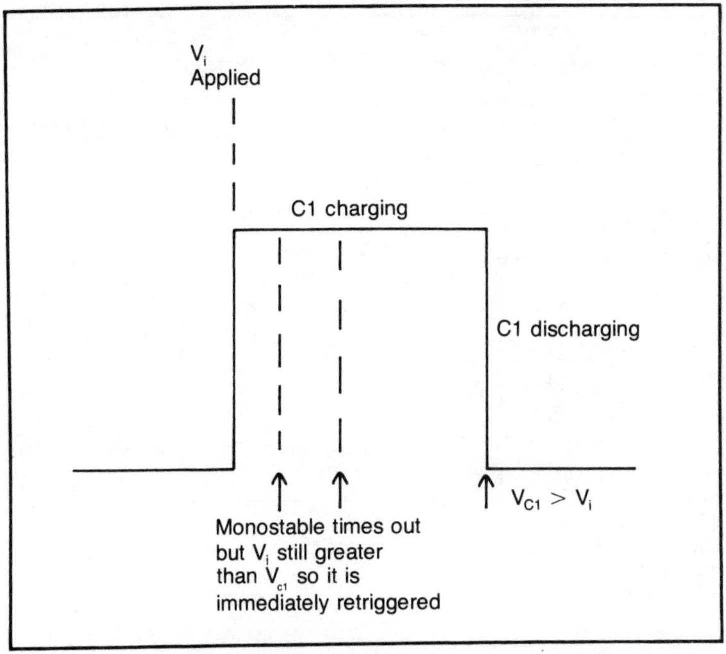

Fig. 14-8. The circuit shown in Fig. 14-7 produces an output pulse with a width that is dependent on the value of V_i.

When the monostable is no longer being retriggered, the output goes to its low state, and capacitor C1 begins to discharge through resistor R1. At some point, the voltage across the capacitor will drop below that of the input voltage (V_i, causing the charging cycle to begin again.

The length of each charging cycle, and the interval between them will be dependent on the amplitude of V_i. The higher the voltage, the longer it will take the capacitor to charge and discharge. In other words, increasing the input pulse will reduce the number of cycles that will occur per second. Raising the input voltage decreases the output frequency.

This method of V/F conversion is known as *automatic tracking* or *charge balancing*. It offers very high accuracy at fairly low cost.

Another voltage-to-frequency conversion circuit is illustrated in Fig. 14-9. Actually this circuit is a voltage-controlled oscillator, or vco. The differences are rather subtle and can be ignored for my purposes here. This circuit generates two output signals whose frequencies correspond directly to the amplitude of the input voltage (V_i). The output of IC1 is a triangle wave. IC2 converts this

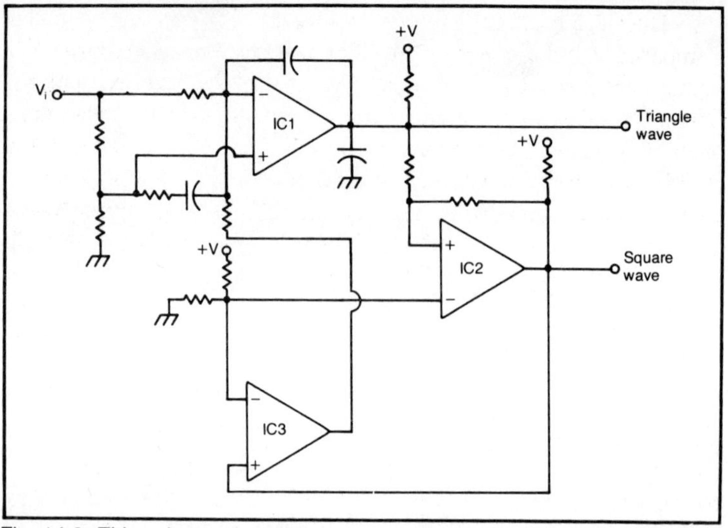

Fig. 14-9. This voltage-controlled oscillator offers simultaneous triangle wave and square wave output signals.

signal into a square wave. Both output signals will always have exactly the same frequency.

FREQUENCY-TO-VOLTAGE CONVERSION

Since voltage-to-frequency conversion is often used for recording and/or transmitting data, it follows that some means of transmitting the frequency encoded information back into dc (or low-frequency) signals will be required at the receiving end. This is called frequency-to-voltage, or F/V conversion.

For the pseudo-V/F converters shown in Fig. 14-3 and Fig. 14-5, the translation back into dc signals can be accomplished with a digital-to-analog (D/A) converter like those discussed in Chapter 12. Since these V/F converter circuits actually spew out a string of pulses whose number is proportional to the original signal (V_i), the receiving circuit only has to count the pulses received, and convert the number into an appropriate analog voltage.

True V/F to F/V systems are a bit more complicated. Figure 14-10 shows a block diagram for a F/V converter circuit. Compare this circuit with the V/F circuit illustrated in Fig. 14-7. Notice the great similarity between these diagrams. The major difference is the placement of the reference voltage (V_{ref}) source and the capacitor it charges.

Once again IC1 is a simple comparator. The incoming signal is compared to a fixed voltage set by a voltage divider comprised of resistors R1 and R2. The input signal will be a string of pulses. When the input pulses go high, the comparator's output also goes high, triggering the monostable multivibrator. The monostable multivibrator holds switch S1 closed for a time period determined by the values of resistor R3 and capacitor C1. As long as this switch is closed, capacitor C2 is charged by V_{ref} (once more a precise constant current source). The longer S1 is closed, the higher the voltage that C2 will be charged to.

When the input pulse goes low, the monostable multivibrator is no longer triggered, and it is permitted to time out and open switch S1, preventing further charging of capacitor C2. Resistor R4 allows the capacitor to discharge between cycles. The output voltage, tapped off across the holding capacitor (C2) will be proportional to the width of the incoming pulses. Wider pulses last longer, so there can be fewer per second. This implies that the output is also proportional to the frequency of the input signal.

DEAD-SPACE CIRCUIT

Many of the circuits discussed in this book work by comparing

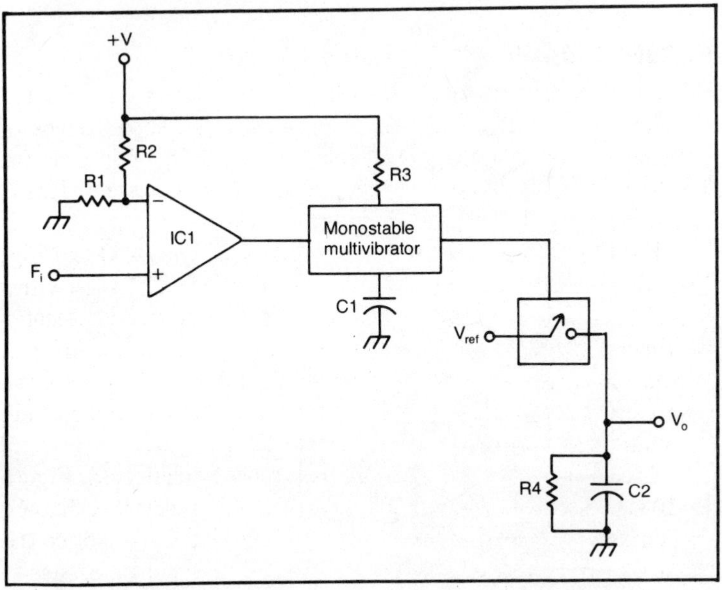

Fig. 14-10. Op amps can also be used to perform frequency-to-voltage conversion.

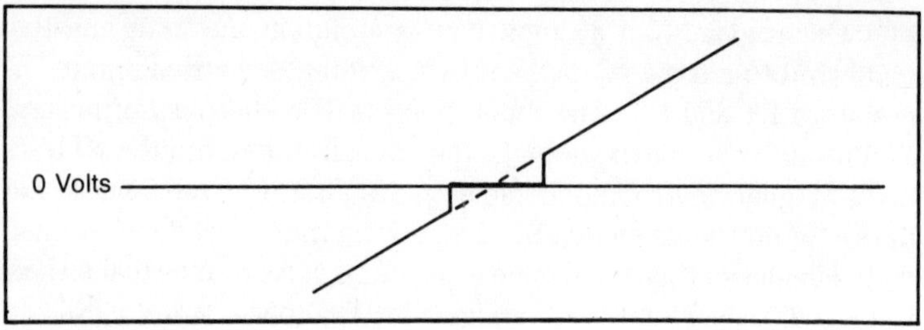

Fig. 14-11. A dead-space circuit forces near zero signals to zero.

an ac input signal to zero, or ground potential. Many waveforms pass through zero so quickly the circuit may not have time to respond properly. Moreover, some signals may approach very close to zero, but not reach it exactly.

Zero response can be improved by using a dead-space circuit. For most signals these circuits will function like an ordinary amplifier. However, signals of just a few millivolts (either positive or negative) are pulled to zero, creating a small *dead* space in the output graph, as shown in Fig. 14-11.

A typical dead space circuit is shown in Fig. 14-12. IC1 elimi-

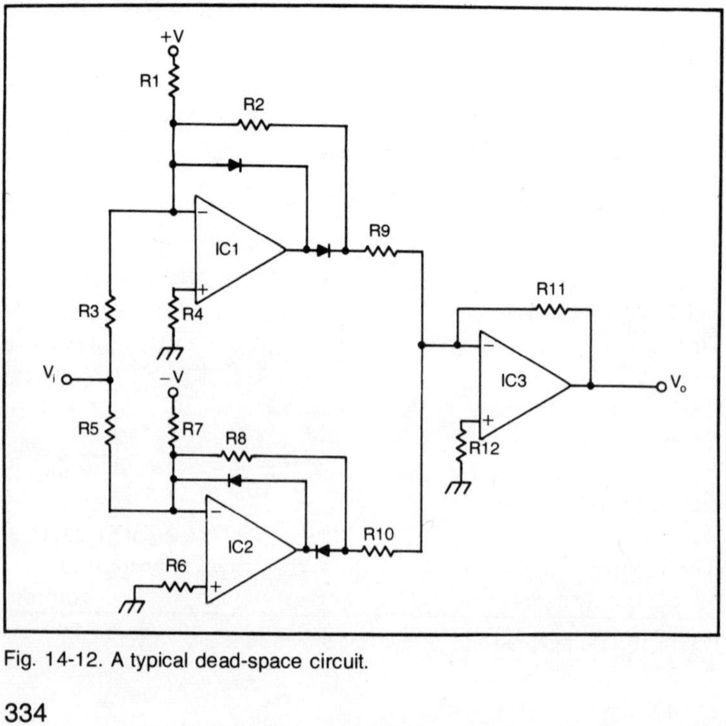

Fig. 14-12. A typical dead-space circuit.

nates small negative signals, and IC2 eliminates small positive signals. The outputs of these inverting amplifiers are fed into the summing amplifier (IC3) through R9 and R10. The gain for positive voltages is determined by the R1/R10 ratio. Similarly, the R11/R9 ratio sets the gain for negative voltages. These two gains do not necessarily have to be equal.

The dead space does not have to be made symmetrical either. For a dead space that is symmetrical around zero, resistor R2 and resistor R8 should have equal values. For asymmetrical response different values may be used for these components.

SERIES LIMITER

By combining an inverting amplifier with a dead-space circuit, I can limit or clip a signal either symmetrically or asymmetrically.

A basic series limiter circuit is illustrated in Fig. 14-13. A graph of the output voltage is shown in Fig. 14-14. IC1 and IC2 essentially make up a series dead-space circuit. If V_i is within the dead space defined by these circuit stages, IC3 will function like an ordinary inverting amplifier. If the input signal moves outside the defined dead space, either IC1 or IC2 will feed a signal of the opposite polarity to IC3. This will prevent the output voltage (V_o) from further increases.

The dead space may be symmetrical or asymmetrical with respect to zero. The limiting points are set by R1 and R3 for IC1 and by R5 and R7 for IC2. For reliable operation, resistors R9 and R10 should have equal values.

SAMPLE-AND-HOLD CIRCUITS

Another popular op amp circuit is the *sample-and-hold*, or S/H circuit. This is a circuit that is used to hold an instantaneous input value. This may be called a track-and-hold in some texts. S/H circuits are used in a number of signal storage, transmission, and measurement applications that require a continuous input signal. They are also at the heart of most analog-to-digital converters (see Chapter 12).

A basic S/H circuit is shown in Fig. 14-15. The op amp is connected as a simple voltage follower with unity gain. When switch S1 is closed, the output (V_o) will simply follow the input signal (V_i). The input voltage will also charge the holding capacitor, C1.

If the switch is opened, the circuit will be isolated from the input signal. The input to the op amp will now be just the voltage across the holding capacitor. The capacitor will hold the charge it

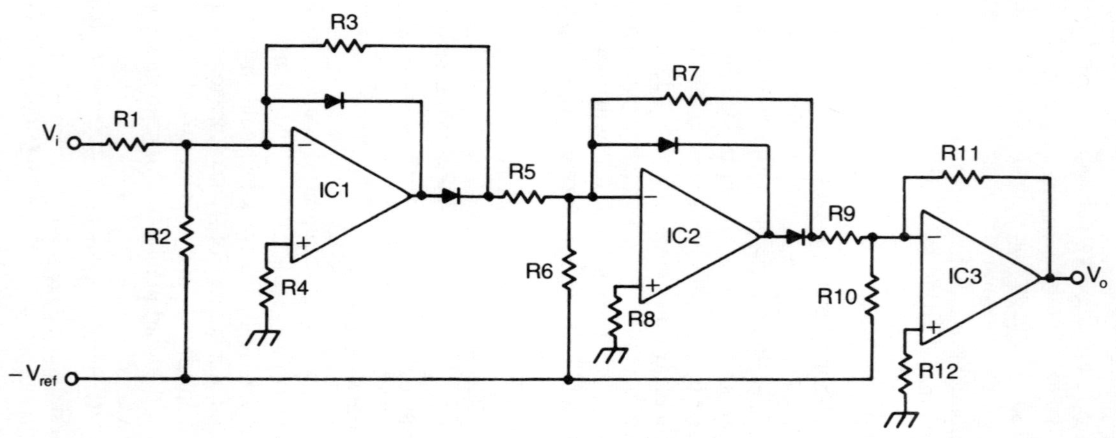

Fig. 14-13. The series-limiter circuit is similar to the dead-space circuit.

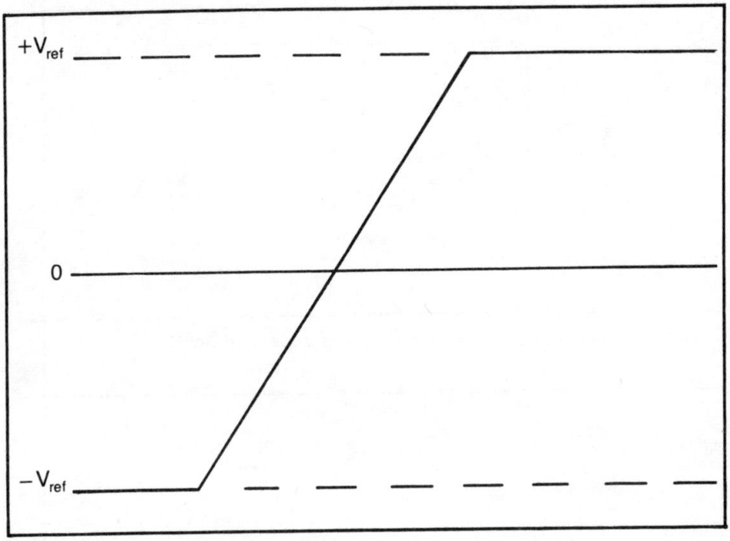

Fig. 14-14. A series limiter forces controllable clipping of the output voltage.

had just before the switch is opened. This value will be held until S1 is closed again, reapplying the input signal (or until the charge leaks off of the capacitor).

An inverting version of this circuit is shown in Fig. 14-16. It works in the same way, except the output polarity is inverted.

The functioning of a basic S/H circuit is illustrated in Fig. 14-17. The output will track the input signal until the switch is opened. The instantaneous value is then held constant, regardless of further changes in the input signal.

Practical S/H circuits have two inputs. Besides the input signal to be sampled (V_i), there is a clock signal (V_c) to trigger an elec-

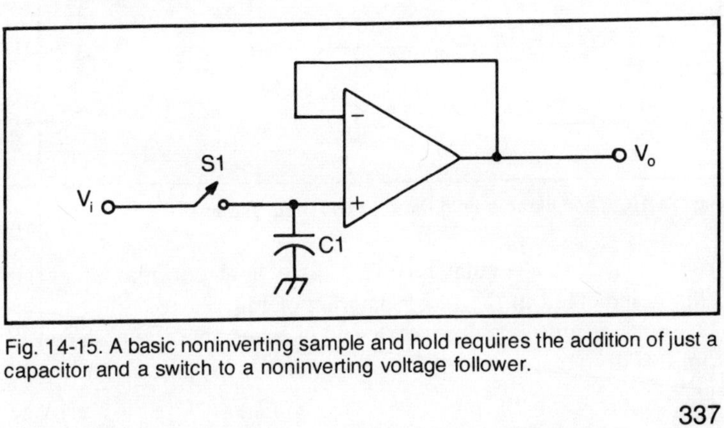

Fig. 14-15. A basic noninverting sample and hold requires the addition of just a capacitor and a switch to a noninverting voltage follower.

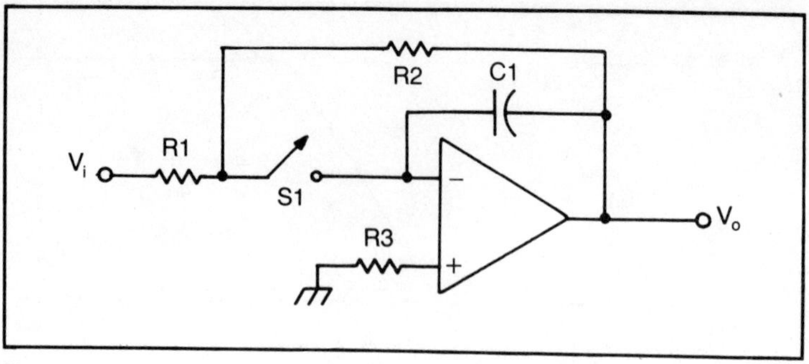

Fig. 14-16. This circuit is a basic inverting sample and hold.

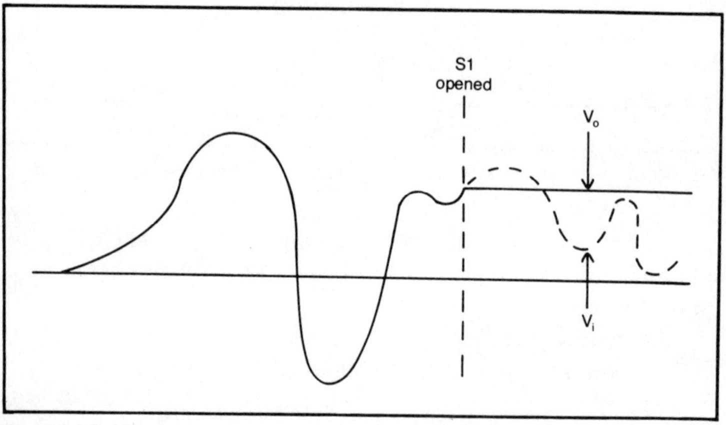

Fig. 14-17. A sample and hold remembers the last input voltage.

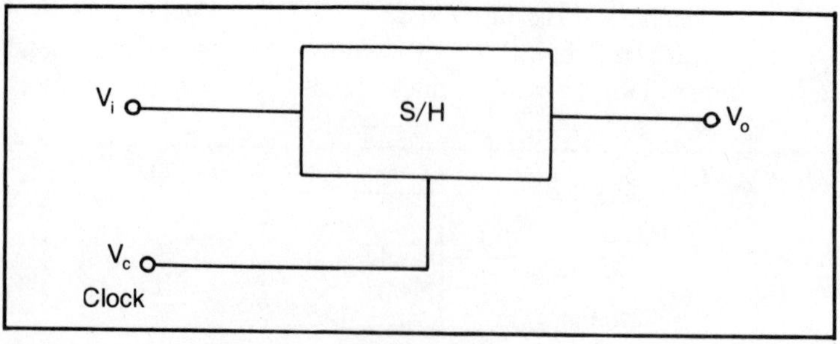

Fig. 14-18. A sample and hold requires two input signals.

tronic switch at a regular rate to sample V_i at periodic intervals. This is indicated in the block diagram of Fig. 14-18.

Periodically sampling a waveform gives the effect shown in Fig. 14-19. This series of instantaneous, held values can be fed

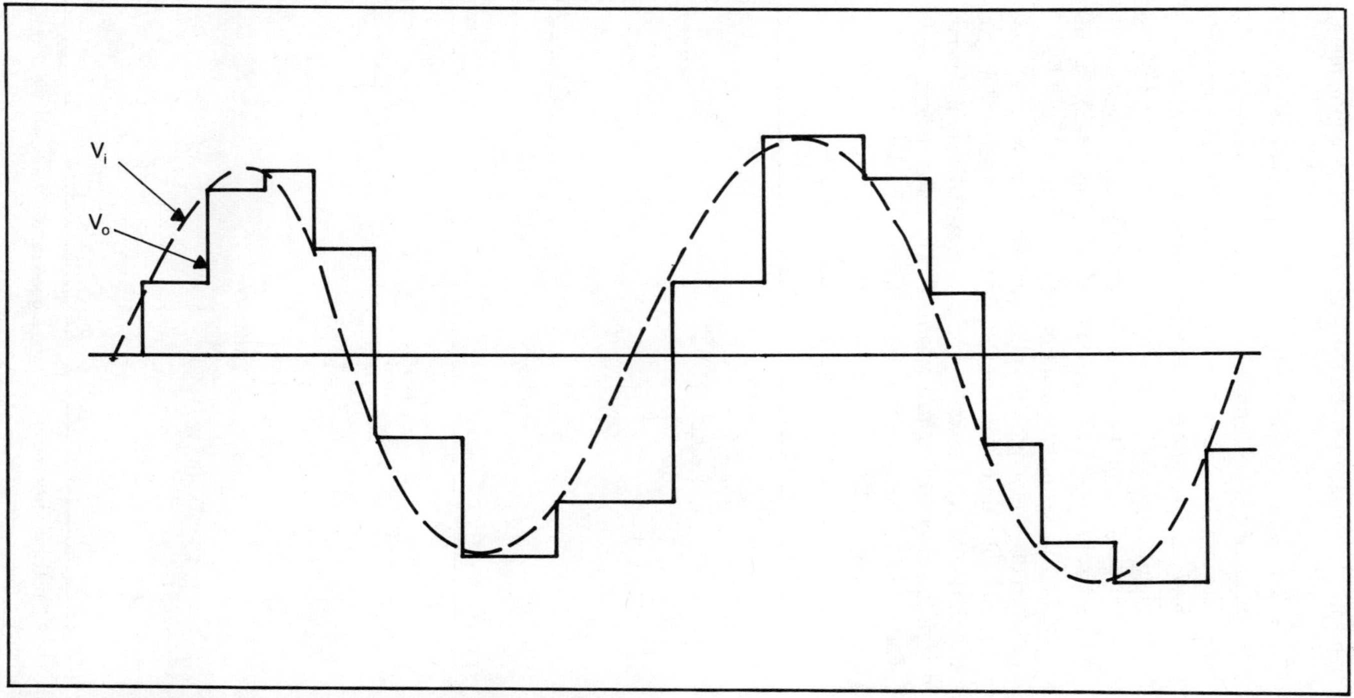

Fig. 14-19. Periodic sampling can indicate the input waveshape.

through an A/D converter to be transmitted or recorded in digital form.

There are two important factors that the designer should consider when setting up a sample-and-hold circuit. These are the sampling rate and the size of the holding capacitor. The sampling rate, of course, is determined by the frequency of the clock pulses. Lower frequencies will result in fewer samples, allowing less expensive recording and/or transmission, but at the cost of less detail.

A sine wave sampled 17 times, as shown in Fig. 14-20, creates a reasonable replica of the original waveform. But in Fig. 14-21, I have reduced the sampling rate so that the same sine wave cycle is sampled only four times. If you just saw this output, would you guess the original input signal was a sine wave? Generally it is a good idea to select the highest affordable clock frequency.

The holding capacitor value requires more of a decision on the part of the designer. Practical values may range from about 100 pF to 10 μF. Large capacitance values will hold the sampled value long

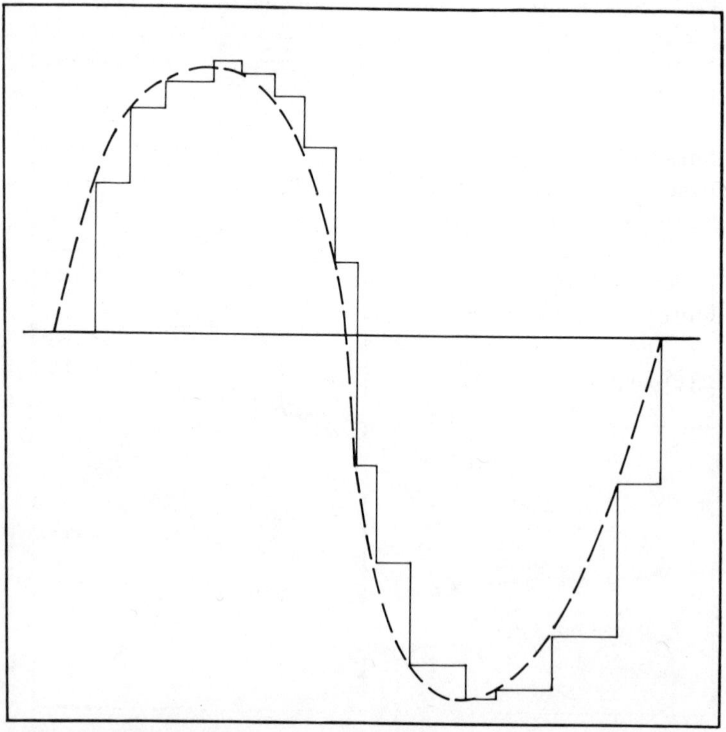

Fig. 14-20. Frequent sampling can provide a good representation of the original input waveform.

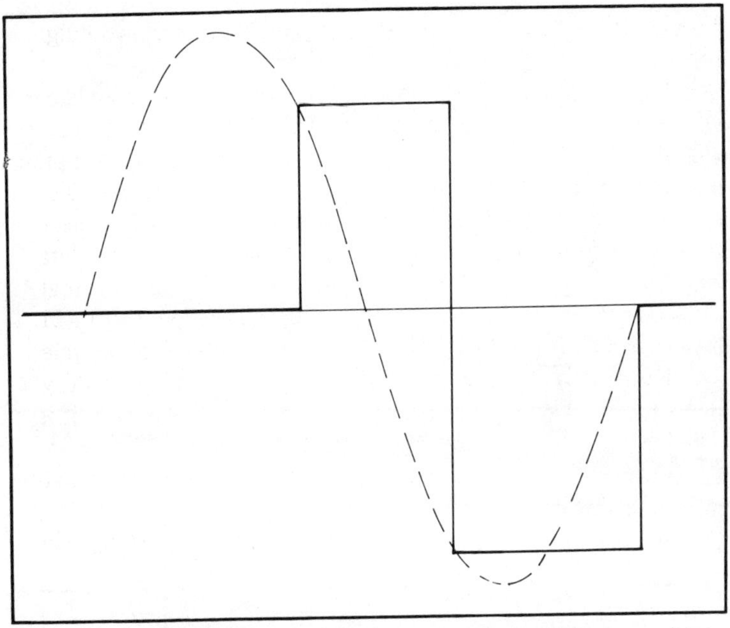

Fig. 14-21. If a signal is sampled just a few times per cycle, it will be impossible to determine the original input waveform.

times without leaking. However, it will take it longer to charge up to the input voltage value. If a large holding capacitor is used, it may not respond fully to sudden large changes in the input signal.

Practical S/H circuits use some forms of electronic switch for S1. A diode bridge is used for switching in Fig. 14-22. A floating (not referenced to ground) clock pulse signal is required.

The circuit in Fig. 14-23 uses a JFET as the electronic switching element. A MOSFET switch is used in the circuit of Fig. 14-24.

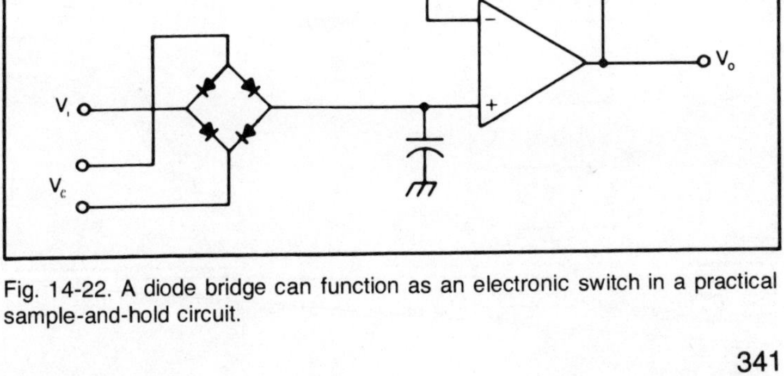

Fig. 14-22. A diode bridge can function as an electronic switch in a practical sample-and-hold circuit.

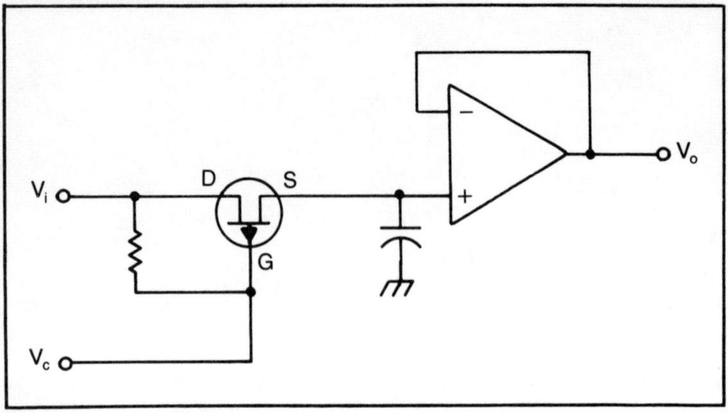

Fig. 14-23. A JFET can be used as an electronic switch in a sample-and-hold circuit.

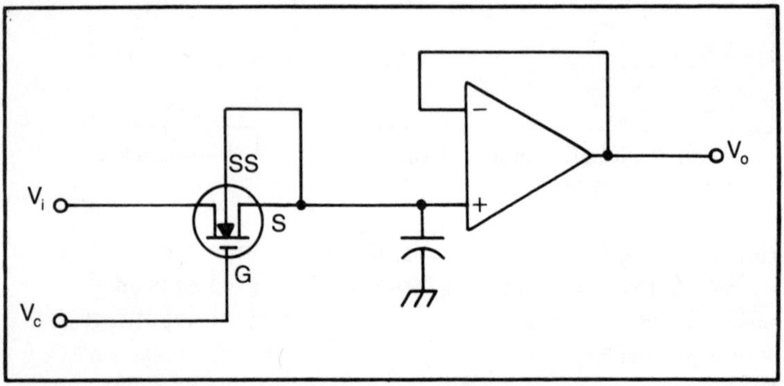

Fig. 14-24. Some sample-and-hold circuits use a MOSFET as the electronic switching device.

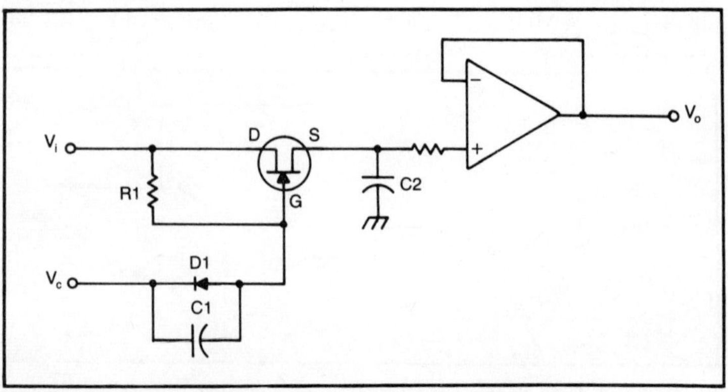

Fig. 14-25. Here is a practical sample-and-hold circuit.

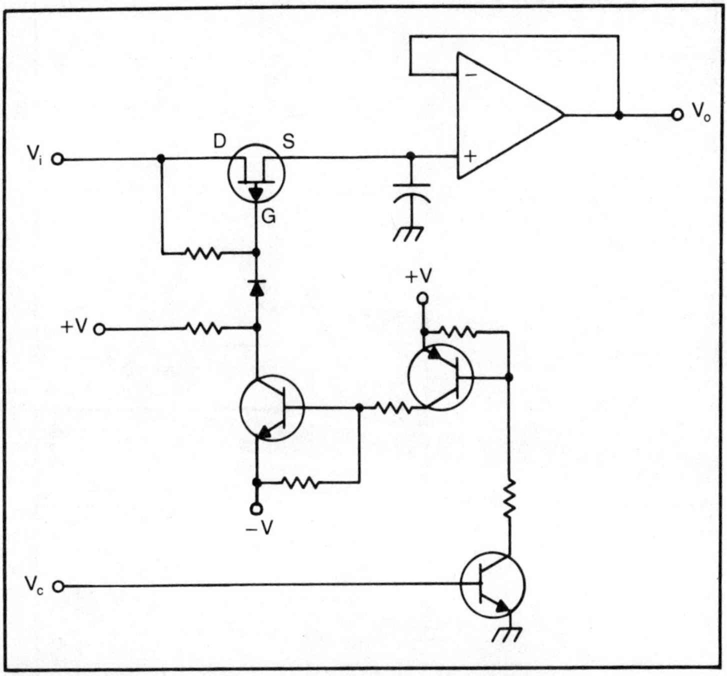

Fig. 14-26. This is another practical sample-and-hold circuit.

Three practical sample-and-hold circuits are shown in Fig. 14-25, Fig. 14-26, and Fig. 14-27. They include extra components to compensate for ac and dc errors.

AUDIO PREAMPLIFIER

The circuit shown in Fig. 14-28 is a preamplifier circuit that can

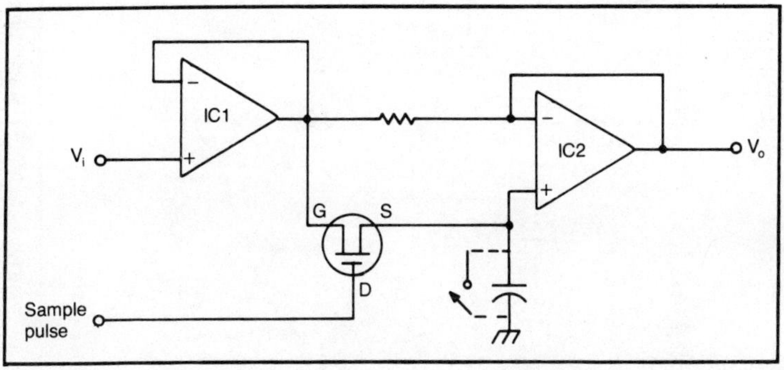

Fig. 14-27. One more practical sample-and-hold circuit is illustrated here.

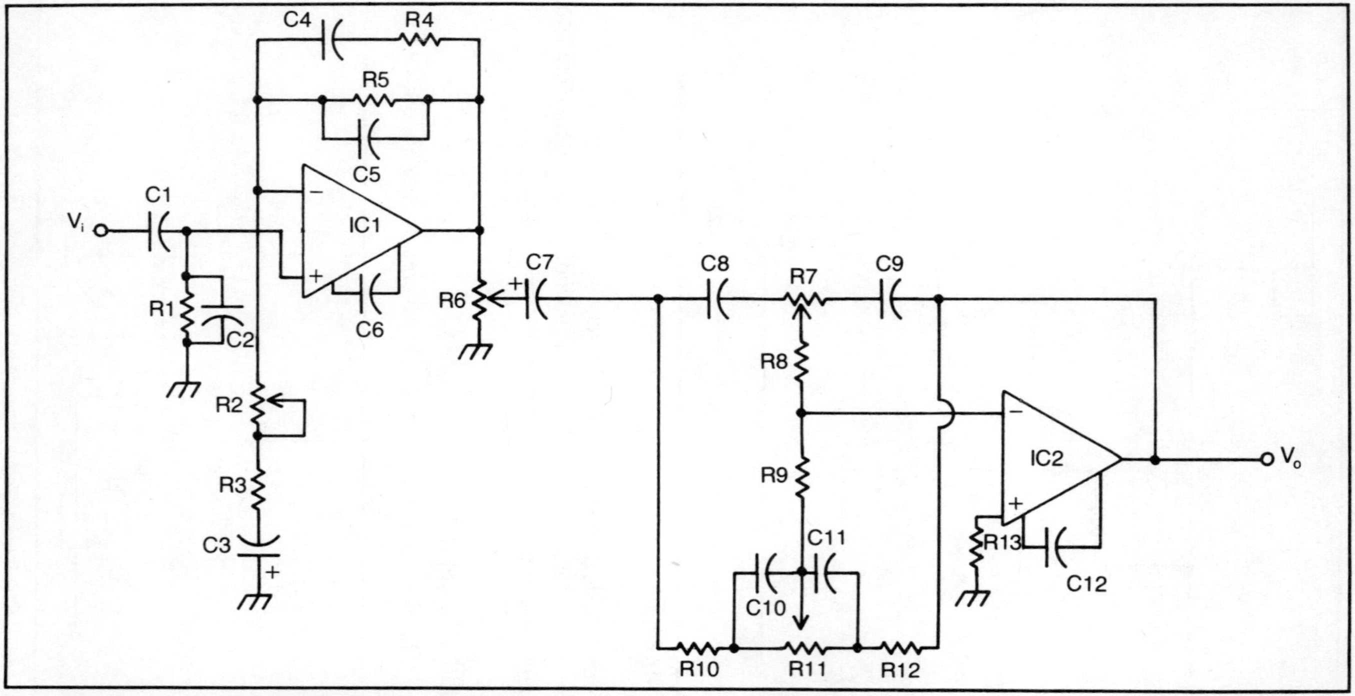

Fig. 14-28. This circuit is an audio preamp with bass and treble controls.

Table 14-1. Component Values for Fig. 14-28.

R1	47 K resistor
R2	100 ohm trimpot
R3	100 ohm resistor
R4	27 K resistor
R5	220 K resistor
R6	5 K potentiometer
R7, R11	100 K potentiometer
R8	3.3 K resistor
R9, R10, R12	10 K resistor
R13	68 K resistor
C1	1 μF capacitor
C2	50 pF capacitor
C3	68 μF electrolytic capacitor
C4	0.1 μF capacitor
C5, C8, C9	0.003 μF capacitor
C6	10 μF electrolytic capacitor
C7	47 μF electrolytic capacitor
C10, C11	0.03 μF capacitor
C12	15 pF capacitor

be used in many audio frequency applications. It can be used in radios, musical instrument amplifier, or stereo systems (a complete preamplifier circuit will be needed for each channel). The brute-force amplification would be performed by a separate power amplifier stage.

Circuit gain is set by potentiometer R6. Between the two op amp stages is a passive filter network that allows the user some control over the frequency response of the circuit. The high-frequency (treble) range can be cut back somewhat by adjusting potentiometer R7. Similarly, potentiometer R11 controls the low-frequency (bass) response.

For best results, high-quality, low-noise op amps such as the TL080 should be used. A standard op amp IC like the 741 might introduce a bit too much hiss and noise into the output signal for any true high fidelity applications. Table 14-1 list the compound values for this circuit. Trimpot R2 is adjusted for minimum output distortion.

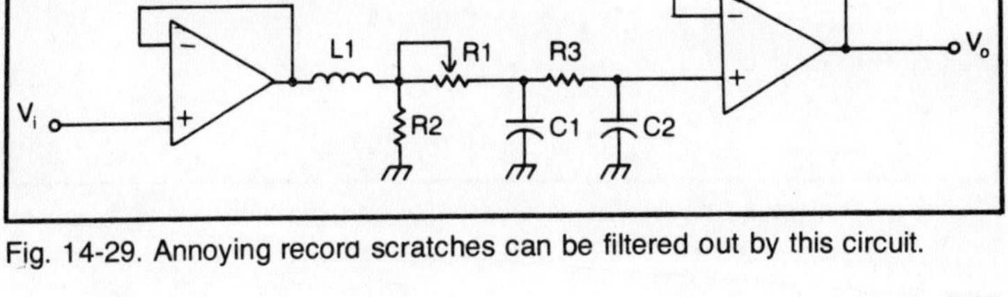

Fig. 14-29. Annoying record scratches can be filtered out by this circuit.

Table 14-2. Component Values for Fig. 14-29.

R1	25 K or 50 K potentiometer
R2, R3	47 K resistor
C1	0.002 μF capacitor
C2	200 pF capacitor
L1	50 mH coil

SCRATCH FILTER

It is very annoying to try to listen to old phonograph records that have so many scratches that much of the music is lost in a wall of noise. This can especially be a problem with old-fashioned shellac 78s. The circuit shown in Fig. 14-29 will help reduce much of the problem by filtering out high frequencies.

Suitable component values for this scratch filter circuit are listed in Table 14-2. The noise elimination can be adjusted to some degree by potentiometer R1.

For stereo systems you will need two complete scratch filter circuits, one for each channel. Install a bypass switch to cut the filter out of the system when it is not needed. On a fairly new, good condition recording, you don't want to filter out the high frequencies.

Standard quality op amps like the 741 may be used in the scratch filter. Any internal noise introduced by the op amp will be negligible compared to the noise in the original source.

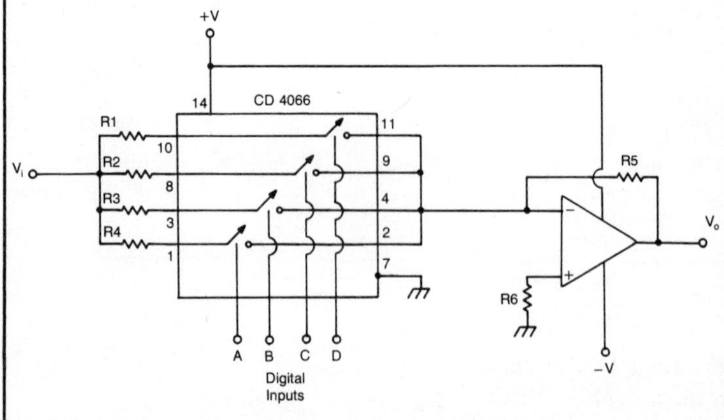

Fig. 14-30. A CD4066 quad bilateral switch can be used to create an amplifier circuit with digitally programmable gain.

PROGRAMMABLE GAIN AMPLIFIER

The unique circuit shown in Fig. 14-30 allows the gain of an inverting amplifier to be controlled by a 4-bit digital word.

The secret of this circuit is a special CMOS digital IC called the CD4066 quad bilateral switch. It is sometimes called a quad analog switch because analog signals may be fed through the internal switches.

A digital input for each internal switch controls its position. A logic 0 spans the appropriate switch, and a logic 1 closes it.

Using the circuitry illustrated in Fig. 14-30, the switches select the input resistor(s) that the input signal (V_i) passes through. If more than one switch is closed at one time, the resistors are in parallel. The effective value of the input resistance combines with the value of feedback resistor R5 to determine the gain of the inverting amplifier.

Observe how this circuit functions with the following component values:

R1 = 2.2 K
R2 = 4.7 K
R3 = 10 K
R4 = 18 K
R5 = 33 K
R6 = 15 K

Notice the staggered values of resistances R1 through R4. Each one is approximately twice the value of the preceding one.

If the digital inputs are all 0s (0000), all four switches are open, preventing any input signal from reaching the op amp. The circuit is effectively off.

Digital input D closes switch S1 when it is at logic 1 (0001). The input signal (V_i) is applied to the inverting amplifier through resistor R1. The gain of the circuit is defined by the R5/R1 ratio.

$$G = \frac{-R5}{R1} = \frac{-33{,}000}{2200} = -15$$

Switch S2 is controlled by digital input C (0010). The gain is defined by the R5/R2 ratio.

$$G = \frac{-R5}{R2} = \frac{-33{,}000}{4700} = -7$$

Switch S3 is controlled by digital input B (0100). In this case, R3 serves as the input resistance and sets the gain.

$$G = \frac{-R5}{R3} = \frac{-33{,}000}{10{,}000} = -3.3$$

Finally, digital input A controls switch S4, bringing input resistor R4 into play (1000).

$$G = \frac{-R5}{R4} = \frac{-33{,}000}{18{,}000} = -1.8$$

Now let's see what happens when more than one of the switches is closed. Let's say digital inputs A and C are at logic 1 while B and D are at logic 0 (1010). The circuit is equivalent to the one shown in Fig. 14-31. The input signal (V_i) passes through input resistors R2 and R4 in parallel. The parallel combination of these two components is calculated as follows.

$$R_p = \frac{R2 \times R4}{R2 + R4} = \frac{4700 \times 18{,}000}{4700 + 18{,}000} = \frac{84{,}600{,}000}{22{,}700}$$

$$\cong 3727 \text{ ohms}$$

The gain in this case is defined by $R5/R_p$.

$$G = \frac{-R5}{R_p} = \frac{-33{,}000}{3727} \cong -8.9$$

In a similar manner, I can determine the gain for each possible combination of digital inputs.

0000 $G = 0$, $V_0 = 0$ (circuit off)

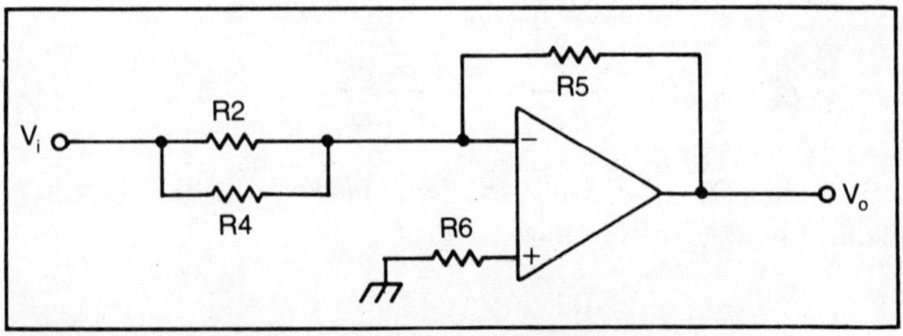

Fig. 14-31. This circuit simulates an input of digital 1010.

0001 $R_x = R1$

$$G = \frac{-R5}{R1} = -15$$

0010 $R_x = R2$

$$G = \frac{-\ R5}{R2} = -7$$

0011 $$R_x = \frac{R1 \times R2}{R1 + R2} = \frac{2200 \times 4700}{2200 + 4700}$$

$$= \frac{10,340,000}{6900} = 1499$$

$$G = \frac{R5}{R_x} = \frac{-33,000}{1499} \cong -22$$

0100 $R_x = R3$

$$G = \frac{-R5}{R3} = -3.3$$

0101 $$R_x = \frac{R1 \times R3}{R1 + R3} = \frac{2200 \times 10,000}{2200 + 10,000}$$

$$= \frac{22,000,000}{12,200} = 1803$$

$$G = \frac{-R5}{R_x} = \frac{-\ 33,000}{1803} = 18.3$$

0110 $$R_x = \frac{R2 \times R3}{R2 + R3} = \frac{4700 \times 10,000}{4700 + 10,000}$$

$$= \frac{47,000,000}{14,700} = 3197$$

$$G = \frac{-R5}{R_x} = \frac{-33,000}{3197} \cong -10.3$$

$$0111 \quad R_x = \frac{1}{\frac{1}{R1} + \frac{1}{R2} + \frac{1}{R3}}$$

$$= \frac{1}{\frac{1}{2200} + \frac{1}{4700} + \frac{1}{10,000}}$$

$$= \frac{1}{0.0004545 + 0.0002128 + 0.0001}$$

$$= \frac{1}{0.0007673} = 1303$$

$$G = \frac{-R5}{R_x} = \frac{-33,000}{1303} \cong -25.3$$

$$1000 \quad R_x = R4$$

$$G = \frac{-R5}{R4} \cong -1.8$$

$$1001 \quad R_x = \frac{R1 \times R4}{R1 + R4} = \frac{2200 \times 18,000}{2200 + 18,000}$$

$$= \frac{39,600,000}{20,200} = 1960$$

$$G = \frac{-R5}{R_x} = \frac{-33,000}{1960} \cong -16.8$$

$$1010 \quad R_x = \frac{R2 \times R4}{R2 + R4} = \frac{4700 \times 18,000}{4700 + 18,000}$$

$$= \frac{84{,}600{,}000}{22{,}700} \cong 3727$$

$$G = \frac{-R5}{R_x} = \frac{-33{,}000}{3727} \cong -8.9$$

1011 $R_x = \dfrac{1}{\dfrac{1}{R1} + \dfrac{1}{R2} + \dfrac{1}{R4}}$

$$= \frac{1}{\dfrac{1}{2200} + \dfrac{1}{4700} + \dfrac{1}{18{,}000}}$$

$$= \frac{1}{0.0004545 + 0.0002128 + 0.0000556}$$

$$= \frac{1}{0.0007229} \cong 1383$$

$$G = \frac{-R5}{R_x} = \frac{-33{,}000}{1383} \cong -23.9$$

1100 $R_x = \dfrac{R3 \times R4}{R3 + R4} = \dfrac{10{,}000 \times 18{,}000}{10{,}000 + 18{,}000}$

$$= \frac{180{,}000{,}000}{28{,}000} \cong 6429$$

$$G = \frac{-R5}{R_x} = \frac{-33{,}000}{6429} \cong -5.1$$

1101 $R_x = \dfrac{1}{\dfrac{1}{R1} + \dfrac{1}{R3} + \dfrac{1}{R4}}$

$$= \frac{1}{\frac{1}{2200} + \frac{1}{10,000} + \frac{1}{18,000}}$$

$$= \frac{1}{0.0004545 + 0.0001 + 0.0000556}$$

$$= \frac{1}{0.0006101} \cong 1639$$

$$G = \frac{-R5}{R_x} = \frac{-33,000}{1639} \cong -20.1$$

$$1110 \quad R_x = \frac{1}{\frac{1}{R2} + \frac{1}{R3} + \frac{1}{R4}}$$

$$= \frac{1}{\frac{1}{4700} + \frac{1}{10,000} + \frac{1}{18,000}}$$

$$= \frac{1}{0.0002128 + 0.0001 + 0.0000556}$$

$$= \frac{1}{0.0003683} = 2715$$

$$G = \frac{-R5}{R_x} = \frac{-33,000}{2715} \cong -12.2$$

$$1111 \quad R_x = \frac{1}{\frac{1}{R1} + \frac{1}{R2} + \frac{1}{R3} + \frac{1}{R4}}$$

$$= \frac{1}{\frac{1}{2200} + \frac{1}{4700} + \frac{1}{10,000} + \frac{1}{18,000}}$$

$$\frac{1}{0.0004545 + 0.0002128 + 0.0001 + 0.0000556}$$

$$= \frac{1}{0.0008229} \cong 1215$$

$$G = \frac{-R5}{R_x} = \frac{-33{,}000}{1215} \cong -27.2$$

A wide variety of gains are possible. One disadvantage of this circuit is that the gain does not increase with the control input digital word. If I rearrange the combinations according to gain, they would appear in the following order:

0000 G = 0	0001 G = −15
1000 G = −1.8	1001 G = −16.8
0100 G = −3.3	0101 G = −18.3
1100 G = −5.1	1101 G = −20.1
0010 G = −7	0011 G = −22
1010 G = −8.9	1011 G= −23.9
0110 G = −10.3	0111 G = −25.3
1110 G = −12.2	1111 G = −27.2

Still, this circuit can prove to be extremely useful for controlling amplifier gain under computer control. In a practical circuit, the internal resistances of the CD4066 switches will have an effect on the gain, as will the tolerances of the resistor values.

FUZZ BOXES

Rock musicians use a number of electronic circuits to create new sounds from their instruments, especially electric guitars. A perennial favorite among special effects is fuzz. This is actually a form of controlled clipping. It produces a brash, buzzy sound, rich in harmonics. Op amps make fine fuzz box circuits. A typical fuzz circuit is shown in Fig. 14-32. Component values are summarized in Table 14-3.

The op amp is operated by much lower than normal power supply voltages to ensure clipping on low-level signals. In this circuit, the positive and negative supply voltages do not necessarily have to be symmetrical about zero. In fact, some interesting effects can be achieved if they are unequal. Typical supply voltages for this circuit should be in the 1.5- to 4-volt range.

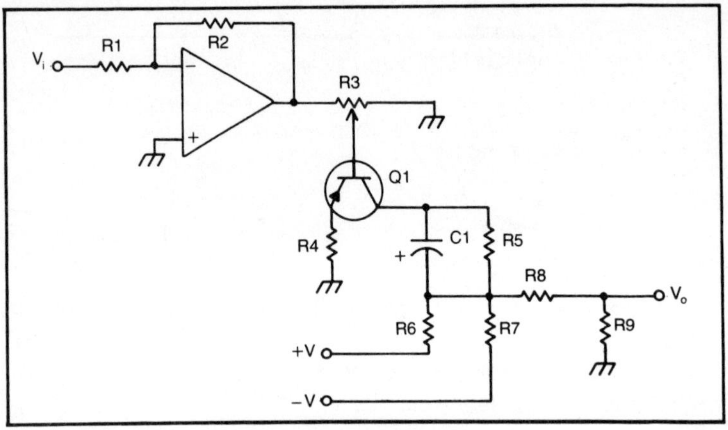

Fig. 14-32. This circuit can be used to create the popular fuzz distortion effect with an electric guitar, or other instrument.

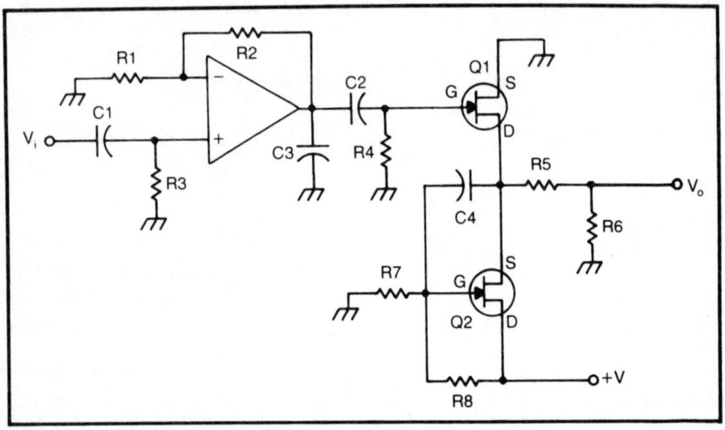

Fig. 14-33. Here is an alternative fuzz circuit.

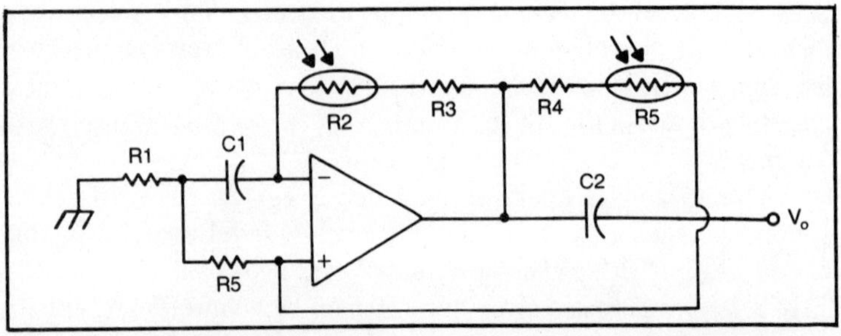

Fig. 14-34. The output frequency of this unique circuit is dependent on the amount of light striking the two photoresistors (R2 and R5).

R1	1 K resistor
R2	1 megohm resistor
R3	500 K potentiometer
R4, R5	10 K resistor
R6, R8	47 K resistor
R7	100 K resistor
R9	56 K resistor
C1	1.5 μF capacitor
Q1	pnp transistor

Table 14-3. Component Values for Fig. 14-32.

The values of resistors R8 and R9 determine the output level. An alternative fuzz circuit is illustrated in Fig. 14-33. The parts list is given in Table 14-4. The op amp in this circuit is powered by an ordinary dual polarity power supply.

LIGHT SENSITIVE SIGNAL GENERATOR

Figure 14-34 shows a very unique oscillator circuit. The output frequency is determined by the amount of light striking photoresistors R2 and R5. These components are designed to be light sensitive. They exhibit a fairly low resistance when exposed to a bright light source, and a much higher resistance when dark. For the most impressive results, these photocells should be mounted so that they may be independently lit or shaded.

Table 14-4. Parts List for Fig. 14-33.

R1	820 ohm resistor
R2, R4, R7, and R8	1 megohm resistor
R3	33 K resistor
R5	220 K resistor
R6	4.7 K resistor
C1, C2, C3, and C4	0.1 μF capacitor
Q1, Q2	JFET (2N5246, or similar)

If both photoresistors are completely darkened, the output frequency will be moderately low. Illuminating both of these devices will cause the output frequency to approximately double. Lighting one of the photoresistors while shading the other produces

R1	1 K resistor
R2, R5	photoresistor (see text)
R3, R4	33 K resistor
R6	100 K resistor

Table 14-5. Component Values for Fig. 14-34.

some fascinating results. If R2 is illuminated and R4 is kept dark, the output frequency will shoot way up to several kilohertz. If R2 is shaded and R4 is placed under a bright light, the output frequency will drop way down to just a few hertz, then rapidly climb back up to a moderate frequency.

The output of this circuit is basically a clipped sine wave, or a distorted square wave with excessive slew.

15

Experiments III

It's time once more to get a little practical, hands-on experience by experimenting with some of the circuits discussed in Chapters 10 through 14. The components and the equipment you will need to perform this series of experiments are listed in Table 15-1.

If you don't have access to an oscilloscope you will be unable to perform a few of the experiments. Many of the experiments in this chapter require only a VOM (volt-ohm-milliammeter), DMM (digital multi-meter), of VTVM (vacuum tube volt meter).

EXPERIMENT 1—COMPARATOR

The first circuit you will experiment with is the basic comparator. The parts and equipment required for this experiment are listed in Table 15-2. Carefully breadboard the circuit shown in Fig. 15-1. Remember to make the power supply connections to the IC, but only after double-checking all of the circuit connections. Use the following resistance values:

R1, R2	22 K
R3	100 K potentiometer
R4, R5	4.7 K

The inverting input of the op amp is fed through R4 a voltage set by voltage divider network R1-R2-R3. The noninverting input is referenced to ground through resistor R5. There is no feedback path.

Table 15-1. Components and Equipment List.

Breadboard system
- solderless socket
- dual polarity power supply (±15 volts)

Components
- 741 op amp IC (need 5)
- npn transistor (2N2222, or similar) (need 3)
- 1N914 diode (or similar) (need 4)
- 1N4001 diode (or similar) (need 3)
- 330 ohm resistor (need 4)
- 820 ohm resistor
- 1 K resistor (need 2)
- 2.2 K resistor (need 2)
- 2.7 K resistor
- 3.3 K resistor (need 2)
- 4.7 K resistor (need 2)
- 5.6 K resistor (need 4)
- 6.8 K resistor
- 10 K resistor (need 4)
- 12 K resistor (need 6)
- 15 K resistor (need 2)
- 18 K resistor
- 22 K resistor (need 4)
- 33 K resistor (need 2)
- 39 K resistor
- 47 K resistor (need 3)
- 82 K resistor
- 91 K resistor
- 100 K resistor (need 3)
- 120 K resistor
- 2.5 K potentiometer
- 10 K potentiometer (need 3)
- 50 K potentiometer (need 2)
- 100 K potentiometer
- 0.01 μF disc capacitor
- 0.1 μF capacitor (need 3)
- 0.47 μF capacitor (need 2)
- 1 μF 35 volt electrolytic capacitor
- LED (light emitting diode) (need 4)
- Normally open SPST push button switch
- SPDT switch (need 4)

Other
- hook-up wire
- voltmeter(s) (see text)
- oscilloscope

You should recognize this circuit as a ground referenced comparator (see Chapter 10). The varying input signal (fed to the inverting input) is compared to zero.

If you have two voltmeters handy, you will find this experiment easier if you connect one to the output, as shown in the diagram, and

Breadboard system
741 op amp
4.7 K resistor (need 2)
10 K resistor
22 K resistor (need 3)
33 K resistor
47 K resistor
91 K resistor
100 K potentiometer
hook-up wire
voltmeter(s) (see text)

Table 15-2. Parts List for Experiment 1.

connect the other to point A to monitor the input voltage. If you have only one voltmeter, you'll just have to move it back and forth between these points. You might want to add a high value resistor to load the output when the voltmeter is not connected to the output. This will give the circuit a finite rather than infinite load resistance to work with. This resistor is shown in dotted lines in Fig. 15-1. Its exact value is not critical. Use whatever's handy in the 47 K to 1 megohm range.

Adjust potentiometer R2 for a positive input voltage at point A, then measure the output voltage. It should be a fairly high negative value representing the negative saturation level of the op amp. Typically this will be a volt or two less than the −V power supply voltage.

For convenience in this discussion, I will assume that a ±15 volt power supply is being used and that the positive saturation level = 13.5 volts and the negative saturation values = −13.5 volts.

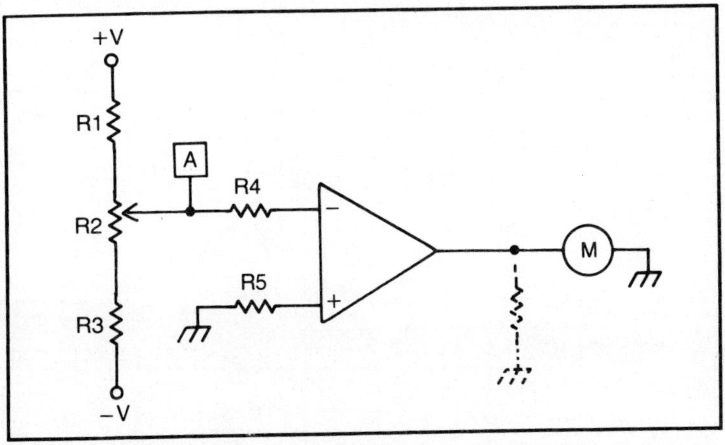

Fig. 15-1. In the first part of Experiment 1, I will be working with a comparator circuit that is referenced to ground.

The exact voltages in your circuit will probably differ somewhat, but the trends will be the same.

The op amp used as a comparator is in the open-loop mode, which means its gain is extremely high. This results in anything more than the smallest differential input voltage driving the op amp into saturation. Since in this case the voltage fed to the inverting input is greater than the reference voltage fed to the noninverting input (0 volts), the output will go negative.

Now, readjust potentiometer R2 for a negative input voltage at point A. What happens to the output voltage? It should go into positive saturation (+13.5 volts).

Next, carefully set potentiometer R2 for a reading of 0 volts at point A. The input signal now equals the reference voltage. The output of the circuit should be zero too (or very close to it, depending on various offset errors within the op amp itself).

Experiment with various settings of potentiometer R2. As long as the input voltage at point A is more than a few millivolts positive the output should be negative saturation. Whenever the input voltage (A) is more than a few millivolts negative, the op amp's output should swing into positive saturation.

If you graphed the input voltage against the output voltage, you should end up with something like the graph shown in Fig. 15-2.

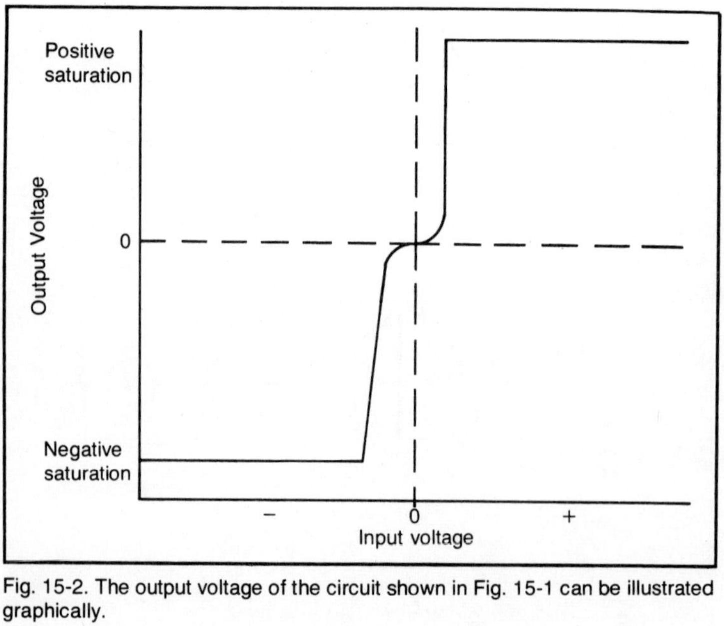

Fig. 15-2. The output voltage of the circuit shown in Fig. 15-1 can be illustrated graphically.

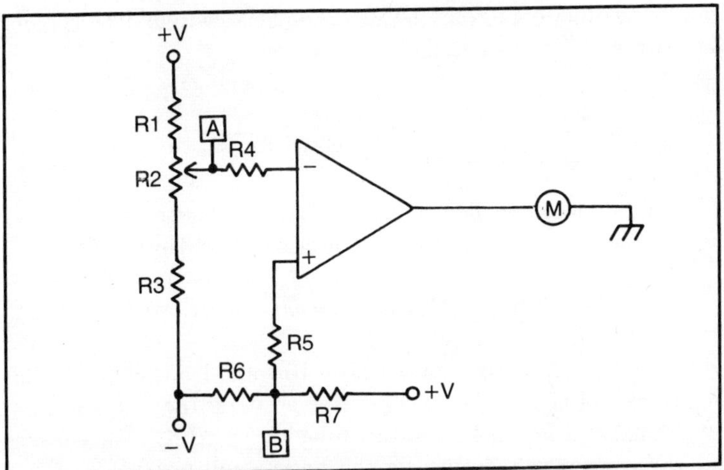

Fig. 15-3. A comparator with a nonzero reference point is used in the second part of Experiment 1.

Notice that if the input voltage is very close to zero, but slightly positive or negative, the output won't go completely into saturation. This crossover range is extremely small, and you'd probably find it difficult to locate using a standard potentiometer for R2.

Assuming that you are using a ±15 volt power supply and the component values specified above, the input voltage (A) can vary from approximately −10.4 to about +10.4 volts.

This comparator circuit uses zero (ground) as the reference voltage. Comparators with nonzero references are also possible and useful.

You can modify the circuit of Fig. 15-1 to set any desired reference voltage by adding a voltage divider. This is shown in Fig. 15-3. The voltage divider consists of resistors R6 and R7. The reference voltage can be measured at the point marked B on the diagram.

If resistors R6 and R7 have identical values the reference voltage will once again be zero. For nonzero reference voltages these two resistors must have unequal values.

For example, let's use a 10 K resistor for R6 and a 91 K resistor for R7. The total resistance equals the sum of R6 and R7.

$$R_t = R6 + R7 = 10{,}000 + 91{,}000 = 101{,}000 \text{ ohms}$$

The voltage drop across these resistors is a total of 30 volts since −15 volts is applied to one end of the divider, and +15 volts is fed to the other end. You can call on Ohm's Law to find the current flow through the voltage divider network.

$$I = \frac{E}{R_t} = \frac{30}{101{,}000} = 0.000297 \text{ amp} = 0.297 \text{ mA}$$

Since the current flow will be the same through each resistor in series, the voltage drop across R6 can also be found by Ohm's Law.

$$E = IR6 = 0.000297 \times 10000 = 2.97 \text{ volts.}$$

Point B is 2.97 volts away from −15 volts. In other words, the reference voltage is −12.03 volts. If you reverse the position of the two resistors (i.e., make R6 = 91 K and R7 = 10 K), you will still get the same absolute value, but the polarity will be reversed. That is, the reference voltage will become +12.03 volts.

Experiment with the following combinations. Adjust R2 to a number of different input voltages (A) for each reference voltage. If V_A is greater than V_{ref}, the output will be negative saturation. If V_A is less than V_{ref}, the output will be positive saturation. If V_A is equal to V_{ref}, the output should be zero (or near zero). Feel free to experiment with other resistor values.

R6 = 10 K	R7 = 47 K	V_{ref} = −9.74 volts
R6 = 22 K	R7 = 10 K	V_{ref} = +5.625 volts
R6 = 91 K	R7 = 22 K	V_{ref} = +9.16 volts
R6 = 22 K	R7 = 47 K	V_{ref} = −5.43 volts
R6 = 22 K	R7 = 33 K	V_{ref} = −3.0 volts
R6 = 47 K	R7 = 33 K	V_{ref} = +2.625 volts
R6 = 10 K	R7 = 33 K	V_{ref} = −8.02 volts
R6 = 91 K	R7 = 33 K	V_{ref} = +7.02 volts

EXPERIMENT 2—ZERO CROSSING DETECTOR

In this experiment, I will be using the noninverting zero-

Breadboard system
741 op amp
4.7 K resistor (need 2)
22 K resistor (need 2)
100 K potentiometer
hook-up wire
voltmeter(s)

Table 15-3. Parts List for Experiment 2.

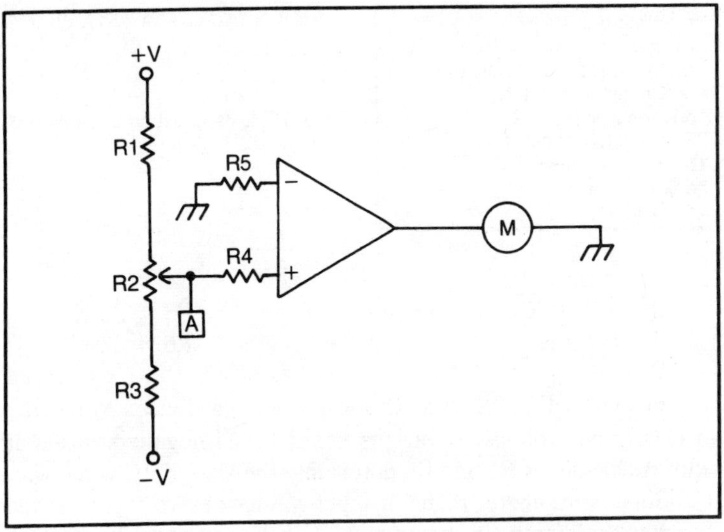

Fig. 15-4. This zero-crossing detector circuit is used for Experiment 2.

crossing detector circuit shown in Fig. 15-4. The components and equipment needed for this experiment are listed in Table 15-3.

Notice that this circuit isn't very different from the zero-referenced comparator circuit of Fig. 15-1. In fact, a zero-crossing detector is nothing but a zero-referenced comparator.

The big difference this time is the reference voltage (zero volts) is applied to the inverting input of the op amp, and the input voltage (A) is fed to the noninverting input.

Adjust potentiometer R2 for a variety of negative and positive voltages at point A. In each case, try to predict what will appear at the output. Then measure the voltage at the output, and see if you were right.

If the input voltage (A) is positive, the op amp should go into positive saturation. On the other hand, a negative input voltage should drive the op amp into negative saturation. Only if the input is zero (or very close to it) will the output be zero (or very close to it).

In other words, the zero crossing detector indicates the polarity of the input signal.

EXPERIMENT 3—WINDOW COMPARATOR

A window comparator indicates whether or not the input signal is within a specified range. The parts list for this experiment is given in Table 15-4. Carefully breadboard the circuit shown in Fig. 15-5 using the following resistance values:

Breadboard system
741 op amp
1N914 diodes (or similar) (need 4)
4.7 K resistor (need 2)
15 K resistor (need 2)
22 K resistor (need 4)
100 K potentiometer
hook-up wire
voltmeter(s)

Table 15-4. Parts List for Experiment 3.

R1 = 22 K
R2 = 100 K potentiometer
R3 = 22 K

R4 = 4.7 K
R5 = 22 K
R6 = 22 K

Resistors R1, R2, and R3 form a voltage divider to set the variable input voltage to various positive and negative values at point A. Resistors R5 and R6 determine the size of the window.

Set potentiometer R2 for an input voltage of zero at point A and measure the output. It should be zero volts (or very close to it).

Now raise the input voltage to 1 volt. Does the output voltage change? With the component values specified above, it shouldn't. It should remain at zero volts. The input voltage is still within the comparator window region.

Try various other positive input voltages. At what point does the output switch from zero to the positive saturation level. The

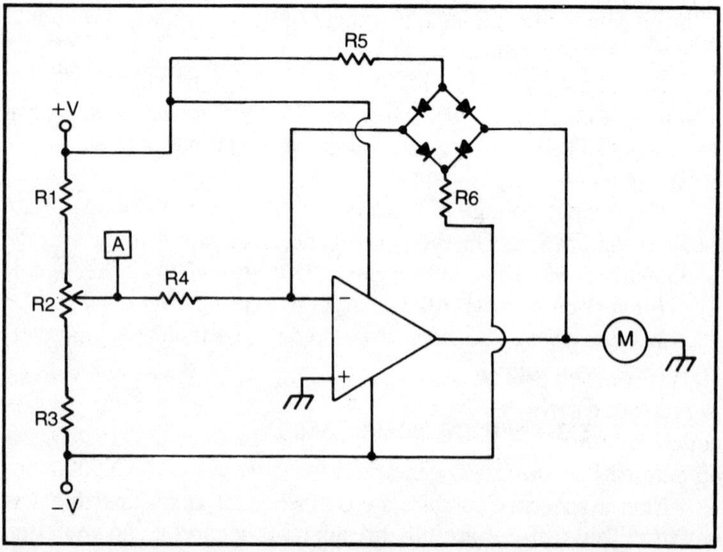

Fig. 15-5. In Experiment 3, I will be working with this window comparator circuit.

formula for the positive switchover point of a window comparator was given in Chapter 10.

$$V_i = \frac{R4(+V)}{R5}$$

Assuming a ±15 volt power supply and the component values listed earlier, you can expect the output switch from near zero to approximately +13.5 volts when the input voltage (A) exceeds 3.2 volts.

$$V_i = \frac{4,700 \times 15}{22,000} = \frac{70,500}{22,000} = +3.2 \text{ volts}$$

Since the power supply voltages are symmetrical (+V = −(−V)), and R5 = R6, the window should also be symmetrical around zero.

$$V_i = \frac{R4(-V)}{R6} = \frac{4700 \times -15}{22,000} = \frac{-70,500}{22,000} = -3.2 \text{ volts}$$

Experiment with a number of negative input voltages at point A.

Now turn off the power to the circuit and replace R6 with a 12 K resistor. Reconnect the power supply and run potentiometer R2 through its range for various input voltages (both positive and negative) at point A.

Since R5 and R6 are no longer of equal value, the comparison window will not be symmetrical around zero.

$$V_i = \frac{R4(+V)}{R5} = \frac{4,700 \times 15}{22,000} = 3.2 \text{ volts}$$

$$V_i = \frac{R4(-V)}{R6} = \frac{4,700 \times -15}{12,000} = \frac{-70,500}{12,000} = -5.875$$

As long as the input voltage(A) is between −5.875 and +3.2 volts, the output will be zero volts (or near to it). If the input voltage is more negative than −5.875 volts, the op amp should go into negative saturation. If the input voltage goes above +3.2 volts, the op amp will be switched into positive saturation.

Now, what would happen if you replaced R5 with another 18 K resistor? Try it and see if your prediction is correct.

You can reverse the asymmetry of the window that occurred in

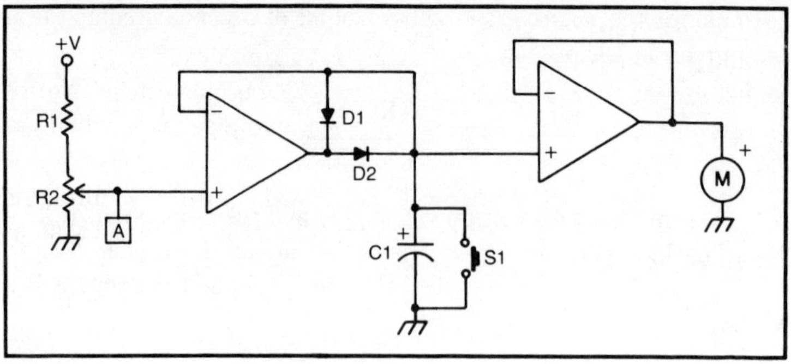

Fig. 15-6. A peak detector is examined in Experiment 4.

the second portion of this experiment by using an 18 K resistor for R5 and a 22 K resistor for R6.

EXPERIMENT 4—PEAK DETECTOR

For the fourth experiment in this series you will work with the peak detector circuit shown in Fig. 15-6. The parts list for this experiment is given in Table 15-5. Carefully breadboard the circuit using a 1 μF electrolytic capacitor for C1. Be sure to observe the polarity of this device. Do not install it backwards!

Resistor R1 and potentiometer R2 form a simple voltage divider to allow you to manually vary the input voltage at point A. Adjust R2 so that you get a zero volts reading at point A. Close push-button switch S1 for a moment to discharge capacitor C1. Turn the shaft of potentiometer R2 through its range. At its maximum setting, the input voltage (A) should be about +13.6 volts. Now turn the potentiometer shaft back through its range towards zero.

Look at the output and see what voltage is there. The peak value reached by the input voltage (+13.6 volts) should be held at the output, even though the input voltage (A) is again at zero.

Table 15-5. Parts List for Experiment 4.

Breadboard system	
741 op amp IC (need 2)	
D1, D2	1N914 diode (or similar) (need 2)
R1	10 K resistor
R2	100 K potentiometer
C1	1 μF 25 volt electrolytic capacitor
	0.01 μF disc capacitor
	Normally open SPST switch
hook-up wire	
voltmeter(s)	

Momentarily depress push-button switch S1. The output should go to zero, indicating that the capacitor is discharged. Slowly adjust potentiometer R2 while watching the output voltage. Notice how it directly follows the increasing input voltage. Stop when you get an output reading of about 5 volts.

Now slowly turn the potentiometer shaft in the other direction (towards zero) while watching the output voltage. Turn the shaft as far as it will go. The output voltage should change.

Next slowly turn R2 in the original (increasing voltage) direction. There should be no change in the output voltage until the input exceeds its former peak value (5 volts). The output will then track along with any further increases in output voltage.

Adjust R2 for an output voltage of about 7 volts, then turn the potentiometer shaft back to its minimum (zero input voltage) position. The output should hold steady at seven volts. Do not touch the circuit at this point. Just leave it alone and do something else for a few minutes. After several minutes, check the output voltage. It should start to drop off as leakage in the capacitor causes it to lose its charge. The output voltage should steadily decline towards zero at a slow rate.

Remove the power line from the circuit and replace capacitor C1 with a 0.01 μF disc capacitor. Now apply power and repeat the experiment. Notice that with the smaller capacitance, the output will react faster to short peaks in the input voltage. For the last part of the experiment, let the charge leak off the holding capacitor. This time it should happen much faster than with the larger (1 μF) capacitor you used earlier.

EXPERIMENT 5—LOGARITHMIC AMPLIFIER

For your next experiment, you will be working with the logarithmic amplifier circuit shown in Fig. 15-7. The parts and equipment you will need to perform this experiment are listed in Table 15-6. Carefully breadboard the circuit using the specified component values. Double-check everything before applying power.

Resistors R1, R7 and potentiometer R2 form a voltage divider to set the input voltage at point A. Assuming the power supply is putting out ±15 volts, the component values indicated in Table 15-6 will provide for an input voltage range of about 0.5 volt to 8 volts.

Adjust potentiometer R2 for an input voltage of 1 volt at point A. This will be our zero decibel reference point. The logarithm of 1 is 0.

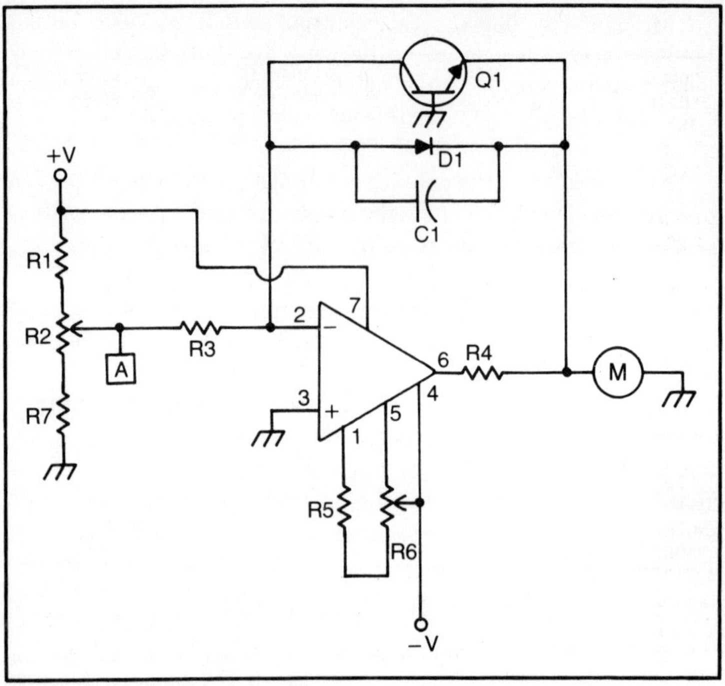

Fig. 15-7. This logarithmic amplifier circuit is used in Experiment 5.

Resistor R5 and potentiometer R6 are used to set the offset null of the op amp. With an input voltage of one volt at point A, adjust R6 for a zero volts output. Your logarithmic amplifier is now calibrated.

Readjust potentiometer R2 for an input voltage of 2 volts at point A. What is the output now? It should be the logarithm of 2, or

Breadboard system	
	741 op amp IC
D1	1N4001 diode (or similar)
Q1	npn transistor (2N2222, or similar)
C1	0.1 μF capacitor
R1	47 K resistor
R2	50 K potentiometer
R3	47 K resistor
R4	2.2 K resistor
R5	330 ohm resistor
R6	10 K potentiometer
R7	3.3 K resistor
hook-up wire	
voltmeter(s)	

Table 15-6. Parts List for Experiment 5.

Table 15-7. Worksheet for Experiment 5, Part 1.

Input Voltage (A)	Output Voltage	Input Voltage (A)	Output Voltage
0.50	______	4.75	______
0.75	______	5.00	______
1.00	0.00	5.25	______
1.25	______	5.50	______
1.50	______	5.75	______
1.75	______	6.00	______
2.00	______	6.25	______
2.25	______	6.50	______
2.50	______	6.75	______
2.75	______	7.00	______
3.00	______	7.25	______
3.25	______	7.50	______
3.50	______	7.75	______
3.75	______	8.00	______
4.00	______		
4.25	______		
4.50	______		

0.3 volts. Repeat this process for each of the input voltages listed in Table 15-7. Write down your results in each case.

Once you have tested each of the specified input voltages, compare your results with the calculated logarithms listed in Table 15-8.

Table 15-8. Calculated Values for Experiment 5, Part 1.

Input Voltage (A)	Output Voltage	Input Voltage (A)	Output Voltage
0.50	−0.30	4.75	0.68
0.75	−0.12	5.00	0.70
1.00	0.00	5.25	0.72
1.25	0.10	5.50	0.74
1.50	0.18	5.75	0.76
1.75	0.24	6.00	0.78
2.00	0.30	6.25	0.80
2.25	0.35	6.50	0.81
2.50	0.40	6.75	0.83
2.75	0.44	7.00	0.84
3.00	0.48	7.25	0.86
3.25	0.51	7.50	0.87
3.50	0.54	7.75	0.89
3.75	0.57	8.00	0.90
4.00	0.60		
4.25	0.63		
4.50	0.65		

Your measured values may deviate slightly from the calculated values, due to component tolerances, but they should be close. If you graphed the input and output voltages, you would end up with something that looks like Fig. 15-8. Notice how the curve becomes steeper as the input voltage is increased. Also notice how input voltages below the zero decibel reference level (1.00 volt) result in a negative output voltage.

The 1 volt input = 0 volt output calibration is entirely arbitrary. You can calibrate a logarithmic amplifier to any reference level you choose.

Set potentiometer R2 for an input voltage of 2.00 volts at point A. Recalibrate the circuit by adjusting potentiometer R6 for a zero

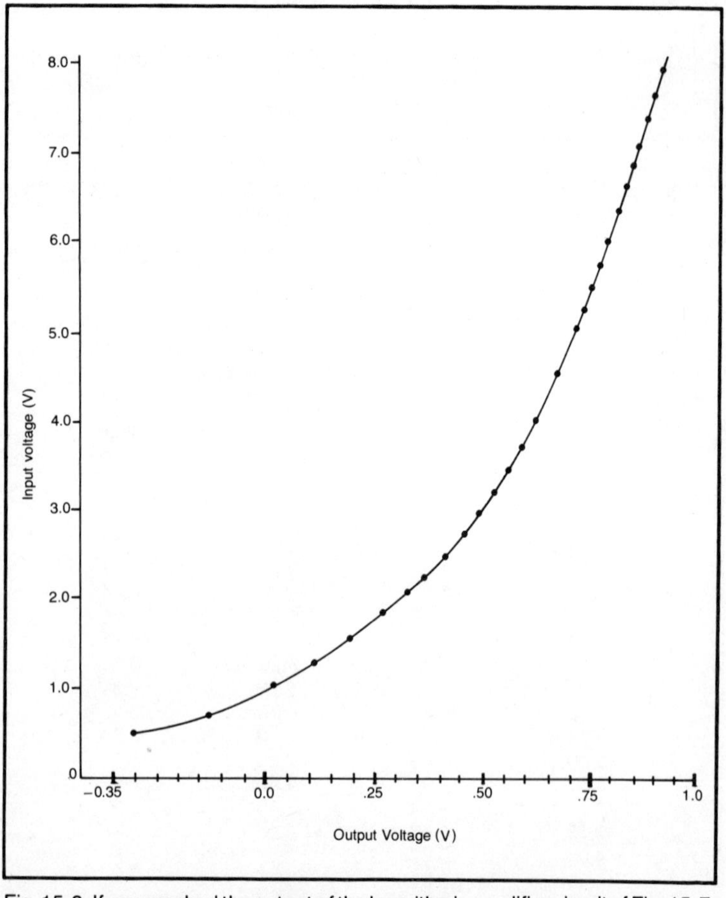

Fig. 15-8. If you graphed the output of the logarithmic amplifier circuit of Fig. 15-7, it should look something like this.

Table 15-9. Worksheet for Experiment 5, Part 2.

Input Voltage (A)	Units	Output Voltage
0.5	0.25	______
0.8	0.40	______
1.0	0.50	______
1.2	0.60	______
1.5	0.75	______
1.8	0.90	______
2.0	1.00	0.00
2.2	1.10	______
2.4	1.20	______
2.6	1.30	______
2.8	1.40	______
3.0	1.50	______
3.2	1.60	______
3.4	1.70	______
3.6	1.80	______
4.0	2.00	______
5.0	2.50	______
6.0	3.00	______
7.0	3.50	______
8.0	4.00	______

Table 15-10. Calculated Values for Experiment 5, Part 2.

Input Voltage (A)	Units	Output Voltage
0.5	0.25	−0.60
0.8	0.40	−0.40
1.0	0.50	−0.30
1.2	0.60	−0.22
1.5	0.75	−0.12
1.8	0.90	−0.05
2.0	1.00	0.00
2.2	1.10	0.04
2.4	1.20	0.08
2.6	1.30	0.11
2.8	1.40	0.15
3.0	1.50	0.18
3.2	1.60	0.20
3.4	1.70	0.23
3.6	1.80	0.25
4.0	2.00	0.30
5.0	2.50	0.40
6.0	3.00	0.48
7.0	3.50	0.54
8.0	4.00	0.60

volt output. You are now calibrated to a zero decibel reference level of 2.00 volts = 1 unit. The logarithm of one unit is still zero.

See what output voltages you get for each of the input voltages listed in Table 15-9. Compare your results with the calculated values listed in Table 15-10.

EXPERIMENT 6—ANTILOGARITHMIC AMPLIFIER

An antilogarithmic amplifier reverses the action of a logarithmic amplifier. The parts and equipment you will need for this experiment are summarized in Table 15-11.

For the first part of the experiment, carefully breadboard the circuit illustrated in Fig. 15-9. Notice that this is the same as the logarithmic amplifier of Fig. 15-8 except that the input and feedback components have reversed their positions.

The voltage divider made up of R1, R2, and R3 can be set for input voltages ranging from −1.67 volts to +3.89 volts at point A (assuming the power supply equals ±15 volts).

Adjust potentiometer R2 for an input voltage of zero volts at point A. Since the logarithm of one is zero, it follows that the antilogarithm of zero should be one. With a zero input, adjust the offset null control (potentiometer R6) for an output of +1 volt. Your antilogarithmic amplifier is now calibrated.

Now, set R2 so that the input voltage (A) of 0.30 volts. Can you guess what the output voltage will be before you look at the output voltmeter?

Refer back to Table 15-8. Notice that the logarithmic amplifier had an output of 0.30 volt when the input voltage was 2 volts. The

Table 15-11. Parts List for Experiment 6, Part 1.

Breadboard system	
	741 op amp IC (need 2)
D1	1N4001 diode (or similar) (need 2)
Q1	npn transistor (2N2222, or similar) (need 2)
C1	0.1 μF capacitor (need 2)
R1	100 K resistor
R2	50 K potentiometer (need 2)
R3	120 K resistor
R4	47 K resistor (need 3)
R5	330 ohm resistor (need 2)
R6	10 K potentiometer (need 2)
	2.2 K resistor
	3.3 K resistor
hook-up wire	
voltmeter(s) (one, two, or three)	

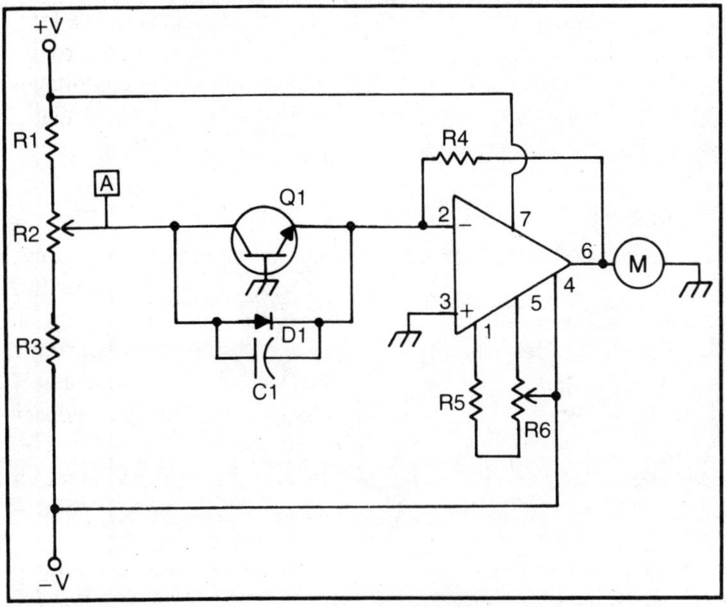

Fig. 15-9. The antilogarithmic amplifier of Experiment 6 is essentially the mirror image of the logarithmic amplifier used in Experiment 5.

logarithm of 2 is 0.30. The antilogarithm is the mirror image of a logarithm. Therefore, if the logarithm of 2 is 0.30, the antilogarithm of 0.30 should be 2. If the input voltage of the circuit of Fig. 15-9 is set to 0.30 volt (point A), the output should be 2.00 volts or very close to it.

Readjust R2 for each of the input voltages listed in the worksheet of Table 15-12. Compare your results to the calculated values shown in Table 15-13. Your results should be close to these calculated values, allowing for component tolerances, and minor errors in measurement.

Now construct the circuit shown in Fig. 15-10. IC1 and its associated components form a logarithmic amplifier. IC2 and its associated components form an antilogarithmic amplifier. Use the component values listed in Table 15-14.

Set potentiometer R2 to a point where the input voltage at point A is exactly one volt. Now adjust potentiometer R6 so that the output of IC1 (point B) is 0.00 volts. This calibrates the logarithmic amplifier portion of the circuit.

Now, a 0.00 volt signal is the input to the antilogarithmic amplifier stage (IC2). To calibrate this stage adjust R10 for an output reading of +1.00 volt.

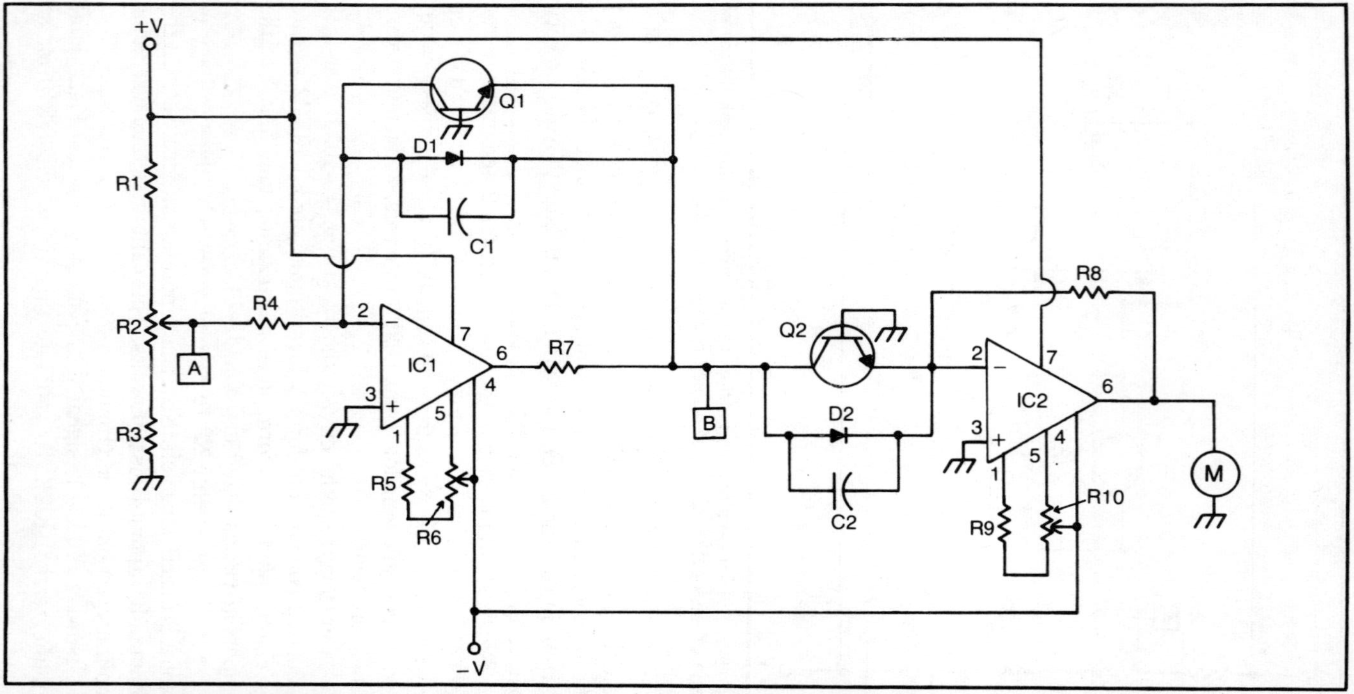

Fig. 15-10. If a logarithmic amplifier is followed by an antilogarithmic amplifier, the output should be equal to the input.

Table 15-12. Worksheet for Experiment 6, Part 1.

Input Voltage (A)	Output Voltage	Input Voltage (A)	Output Voltage
−1.50	________	0.60	________
−1.00	________	0.70	________
−0.75	________	0.80	________
−0.50	________	0.90	________
−0.25	________	1.00	________
0.00	1.00	1.10	________
0.10	________	1.25	________
0.20	________	1.50	________
0.30	________	2.00	________
0.40	________	3.00	________
0.50	________		

Table 15-13. Calculated Values for Experiment 6, Part 1.

Input Voltage (A)	Output Voltage	Input Voltage (A)	Output Voltage
−1.50	0.03	0.60	3.98
−1.00	0.10	0.70	5.01
−0.75	0.18	0.80	6.31
−0.50	0.32	0.90	7.94
−0.25	0.56	1.00	10.00
0.00	1.00	1.10	12.59
0.10	1.26	1.25	(13.5) *
0.20	1.58	1.50	(13.5) *
0.30	2.00	2.00	(13.5) *
0.40	2.51	3.00	(13.5) *
0.50	3.16		

* — positive saturation voltage

Table 15-14. Parts List for Experiment 6, Part 2.

IC1, IC2	741 op amp
D1, D2	1N4001 diode (or similar)
Q1, Q2	npn transistor (2N2222, or similar)
C1, C2	0.1 μF capacitor
R1, R4, R8	47 K resistor
R2	50 K potentiometer
R3	3.3 K resistor
R5, R9	330 ohm resistor
R6, R10	10 K potentiometer
R7	2.2 K resistor

Set R2 for each of the input voltages listed in Table 15-15 (at point A). Measure the voltage at point B (logarithm of A), and at the circuit output (antilogarithm of B). Fill in your results.

Your completed worksheet should look like Table 15-16. Since IC1 derives the logarithm, and IC2 immediately calculates the antilogarithm, the output should be equal to the original input signal (A) in all cases. Minor errors may show up here or there, but your results should be very close to the ones indicated here.

EXPERIMENT 7—MULTIPLICATION CIRCUIT

In Chapter 11, you learned how to multiply two voltages by adding their logarithms and deriving the antilogarithm of the resulting sum. You will prove that this method works in this experiment. The components and equipment you will need are listed in Table 15-17. The circuit is shown in Fig. 15-11. This is by far the most complex circuit you have dealt with in this book. It consists of five op amp stages.

Table 15-15. Worksheet for Experiment 6, Part 2.

Input Voltage (A)	LOG (B)	Antilog Output Voltage
0.50		
0.60		
0.70		
0.80		
0.90		
1.00		
1.10		
1.20		
1.30		
1.40		
1.50		
1.60		
1.70		
1.80		
1.90		
2.00		
2.25		
2.50		
2.75		
3.00		
3.25		
3.50		
3.75		
4.00		

Table 15-16. Calculated Values for Experiment 6, Part 2.

Input Voltage (A)	LOG (B)	Antilog Output Voltage
0.50	−0.30	0.50
0.60	−0.22	0.60
0.70	−0.15	0.70
0.80	−0.10	0.80
0.90	−0.05	0.90
1.00	0.00	1.00
1.10	0.04	1.10
1.20	0.08	1.20
1.30	0.11	1.30
1.40	0.15	1.40
1.50	0.18	1.50
1.60	0.20	1.60
1.70	0.23	1.70
1.80	0.26	1.80
1.90	0.28	1.90
2.00	0.30	2.00
2.25	0.35	2.25
2.50	0.40	2.50
2.75	0.44	2.75
3.00	0.48	3.00
3.25	0.51	3.25
3.50	0.54	3.50
3.75	0.57	3.75
4.00	0.60	4.00

Don't let the apparent complexity scare you off. Each of the five stages is simply one of the simple op amp circuits from earlier experiments. They are simply being combined to perform a more

Table 15-17. Parts List for Experiment 7.

Breadboard system	
	741 op amp IC (need 5)
D1, D2, D3	1N4001 diode (or similar)
C1, C2, C3	0.1 μF capacitor
Q1, Q2, Q3	npn transistor (2N2222, or similar)
R1, R8	100 K resistor
R2, R9	50 K potentiometer
R3, R10	3.3 K resistor
R4, R11, R19	47 K resistor
R5, R12, R20	330 ohm resistor
R6, R13, R21	10 K potentiometer
R7, R14	2.2 K resistor
R15, R16, R17, R18	10 K resistor
hook-up wire	
voltmeter(s)	

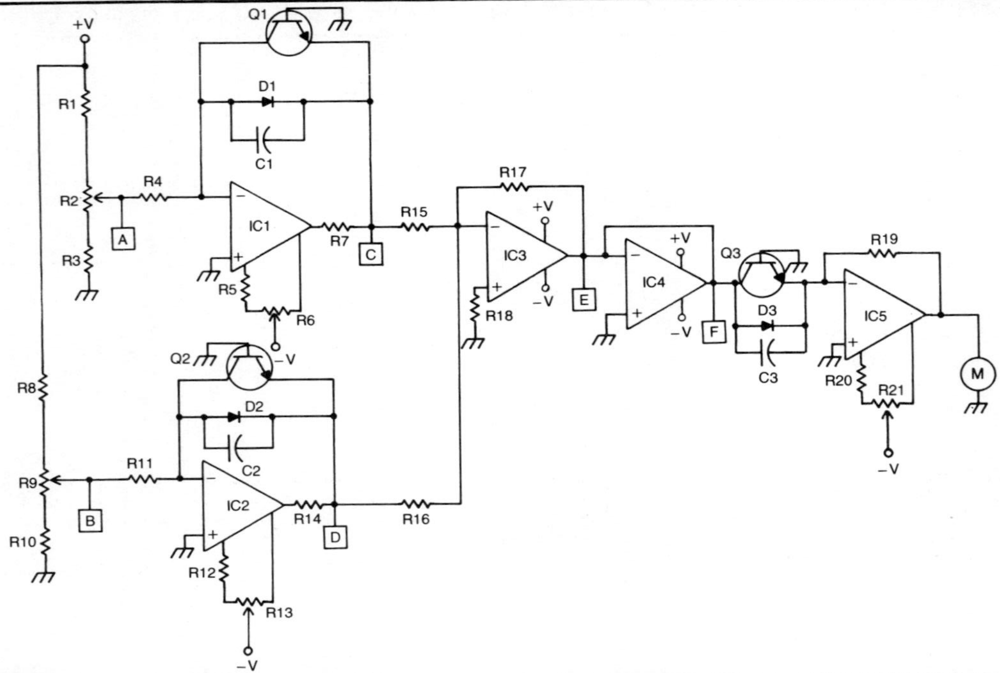

Fig. 15-11. A number of simpler op amp circuits are combined to form the multiplication circuit for Experiment 7.

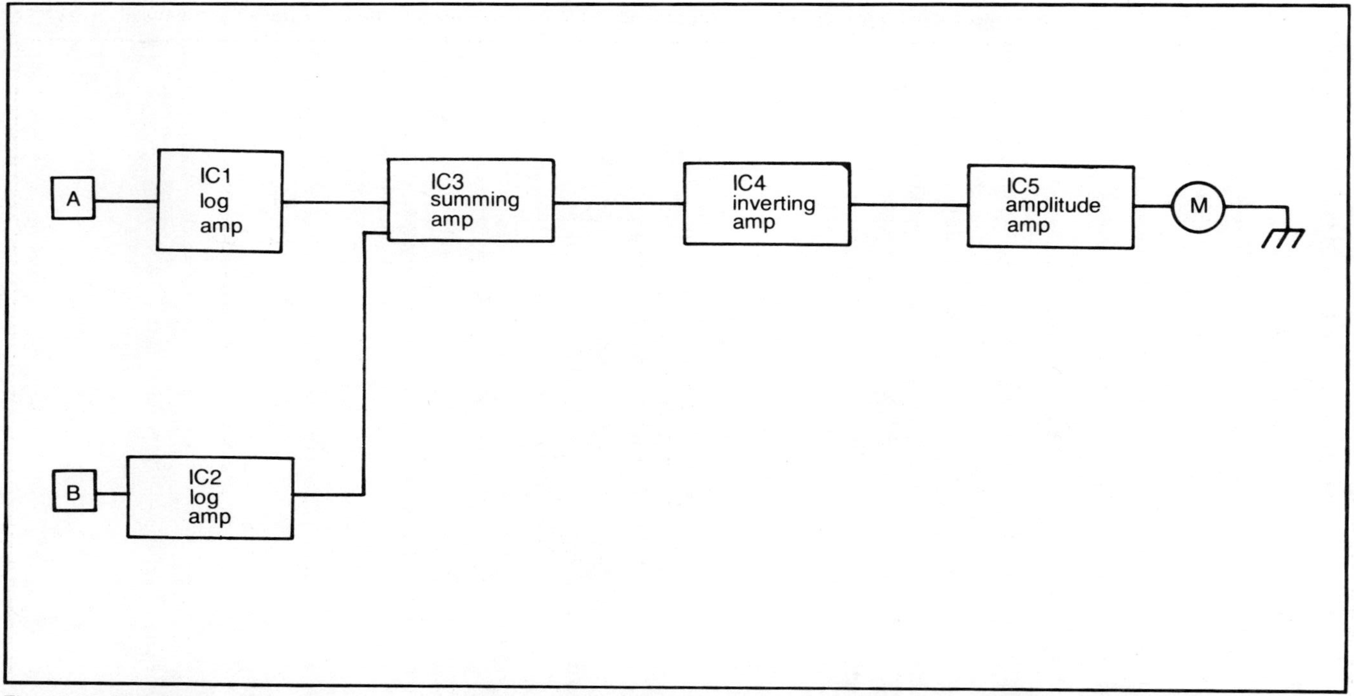

Fig. 15-12. The circuit shown in Fig. 15-11 can be more easily understood by breaking it down into a functional block diagram.

complex task. The function of each stage is broken down in the block diagram of Fig. 15-12.

Breadboard the circuit of Fig. 15-11, but instead of connecting the output of IC4 to point F, temporarily strap point A to ground. This applies an input of zero to antilogarithmic amplifier (E). Adjust potentiometer R21 to calibrate the antilogarithmic amplifier for an output of 1.00 volt for an input of 0.00 volts.

Now remove the jumper from point F to ground and connect the rest of the circuit as illustrated in the schematic diagram.

IC1 is a logarithmic amplifier. Its input (point A) is set by the voltage divider made up of R1, R2, and R3. Set potentiometer R2 for an input voltage of 1.00 at point A. Measure the op amp's output at point C and adjust potentiometer R6 for an output of 0.00 volts.

IC2 is an identical logarithmic amplifier stage. Adjust potentiometer R9 for an input voltage of 1.00 at point B. Monitor the output of this stage at point D and set calibration pot R13 for an output of 0.00. The circuit is now calibrated and ready to experiment with.

Leave potentiometers R2 and R9 at their present settings so that 1.00 volt appears at both point A and point B. The logarithmic outputs (points C and D) should both be 0.00, of course. The next stage (IC3) is a summing amplifier with unity gain. The inputs are 0 + 0, so the output (point E) is also zero. The unity gain inverting amplifier (IC4) has no effect on a zero volt signal. The input to the antilogarithmic amplifier stage (op amp E) is 0 volts at point F. The output will be 1.00 volt. This is the product of the two input signals (A and B). $1 \times 1 = 1$.

Now adjust R2 for an input voltage of 2.00 at point A. Set R9 for an input voltage of 1.50 at point B. The output of the entire circuit should be equal to $2.00 \times 1.50 = 3.00$ volts.

Trace your way through the circuit and see if it performs the way you would expect. IC1 converts its 2.00 volt input (point A) to its logarithm at point C. The output of this stage is 0.30. Meanwhile IC2 is taking the logarithm of 1.50 volt (point B). The voltage at point D is about 0.18 volt.

These two voltages (C and D) are combined in the summing amplifier (IC3). Since the inverting input of this op amp is being used, the output polarity is reversed;

$$E = -(C + D) = -(0.30 + 0.18) = -0.48 \text{ volt}$$

The next stage (IC4) is simply a unity gain inverting amplifier to correct for the polarity inversion of IC3. The voltage at point F is +0.48 volt. This voltage is fed to the antilogarithmic amplifier (IC5). This stage converts the signal into an output equal to the antilogarithm of 0.48, or 3.00 volts, which is the product of the original inputs.

Your exact measured values may vary slightly from these nominal values due to component tolerances, stage loading, and minor measurement errors. Still, the result should be close to the desired value.

Now, let's say input A is 0.75 volt, and input B is 3.25 volt. The logarithm of 0.75 volt is approximately −0.12 volt (point C). The logarithm of 3.25 volts is about +0.51 volt (point D). These logarithms are added by the summing amplifier.

$$E = -(C + D) = -(-0.12 + 0.51) = -0.39$$

This is inverted by IC4 to a positive value.

$$F = -E = -(-0.39) = 0.39 \text{ volt}$$

Finally, IC5 derives the antilogarithm of this value.

$$V_o = \text{antilog}(E) = \text{antilog}\ (0.39) = 2.4375 \text{ volt}$$

This should be equal to the product of the original inputs.

$$\text{Product} = A \times B = 0.75 \times 3.25 = 2.4375$$

Table 15-18. Worksheet for Experiment 7.

Inputs						
A	B	C	D	E	F	Output
0.75	0.75	______	______	______	______	______
1.10	1.60	______	______	______	______	______
2.00	2.00	______	______	______	______	______
2.00	1.00	______	______	______	______	______
3.00	2.00	______	______	______	______	______
3.00	1.75	______	______	______	______	______
3.00	1.50	______	______	______	______	______
3.00	1.25	______	______	______	______	______
2.00	1.25	______	______	______	______	______
2.00	0.75	______	______	______	______	______
2.00	0.50	______	______	______	______	______
0.80	1.20	______	______	______	______	______

Table 15-19. Calculated Values for Experiment 7.

Inputs						
A	B	C	D	E	F	Output
0.75	0.75	−0.12	−0.12	0.24	−0.24	0.5625
1.10	1.60	0.04	0.20	−0.24	0.24	1.76
2.00	2.00	0.30	0.30	−0.60	0.60	4.00
2.00	1.00	0.30	0.00	−0.30	0.30	2.00
3.00	2.00	0.48	0.30	−0.78	0.78	6.00
3.00	1.75	0.48	0.24	−0.72	0.72	5.25
3.00	1.50	0.48	0.18	−0.65	0.65	4.50
3.00	1.25	0.48	0.10	−0.58	0.58	3.75
2.00	1.25	0.30	0.10	−0.40	0.40	2.50
2.00	0.75	0.30	−0.12	−0.18	0.18	1.50
2.00	0.50	0.30	−0.30	0.00	0.00	1.00
0.80	1.20	−0.10	0.08	0.02	−0.02	0.96

Fill in the blank spaces for each combination of input values in Table 15-18. Compare your results with the calculated values listed in Table 15-19.

EXPERIMENT 8—ABSOLUTE VALUE CIRCUIT

The circuit shown in Fig. 15-13 will derive the absolute value (unsigned value) of the input voltage. The parts and equipment you will need for this experiment are listed in Table 15-20.

Carefully breadboard the circuit using the specified component values. The input voltage is set by the voltage divider network made up of R1, R2, R3. Using the resistance values listed in Table

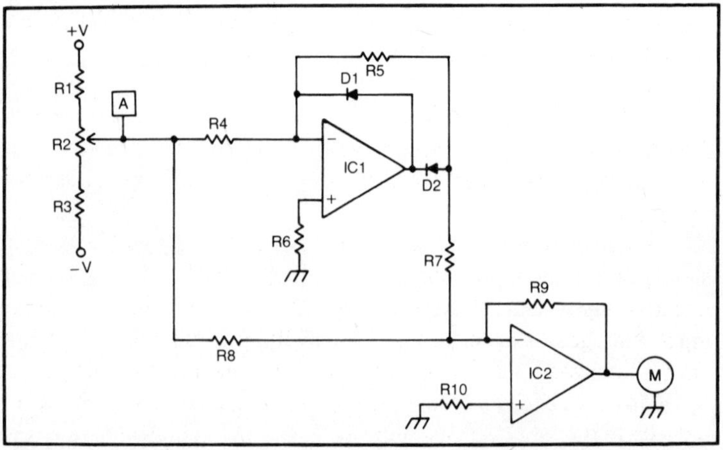

Fig. 15-13. This absolute value circuit is used in the first part of Experiment 8.

Table 15-20. Parts List for Experiment 8, Part 1.

Breadboard system	
	741 op amp IC (need 2)
D1, D2	1N914 diode (or similar)
R1, R3	1 K resistor
R2	10 K potentiometer
R4, R5, R8, R9	10 K resistor
R6, R7	4.7 K resistor
R10	2.7 K resistor
hook-up wire	
voltmeter(s)	

15-20, the input voltage may range from approximately −12.5 volts to about +12.5 volts.

Adjust potentiometer R2 so that the input voltage at point A is zero volts. The output voltage should also be zero. There may be some small deviation because of component tolerances and internal offsets within the ICs.

Now readjust R2 for a positive input voltage of about 1 volt. The output should follow the input voltage. Turn the shaft of the potentiometer slowly towards the maximum positive input voltage position while watching the output voltage. The output voltage should be equal to the input voltage throughout the positive range. There may be some clipping at the upper end of the range.

For positive input voltages the absolute value circuit behaves as a noninverting voltage follower with unity gain. The output voltage equals the input voltage.

Now slowly turn the shaft of the potentiometer (R2) in the opposite direction while monitoring the output voltage. The output should directly follow the input down to zero.

As the input voltage falls below zero (goes negative), the output should reverse direction, and start increasing in the positive direction again. When the input voltage is −1 volt, the output will be +1 volt. When the input is −2 volts, the output voltage will be +2 volts. And so forth.

Slowly turn R2 towards its most negative setting. The output should follow the input, but with reversed polarity, throughout the negative input range. There may be clipping at the higher negative input voltages. For negative input voltages the absolute value circuit behaves as an inverting voltage follower with unity gain.

Move R2 back and forth through its range. No matter what the polarity of the input voltage may be, the output signal will always be positive.

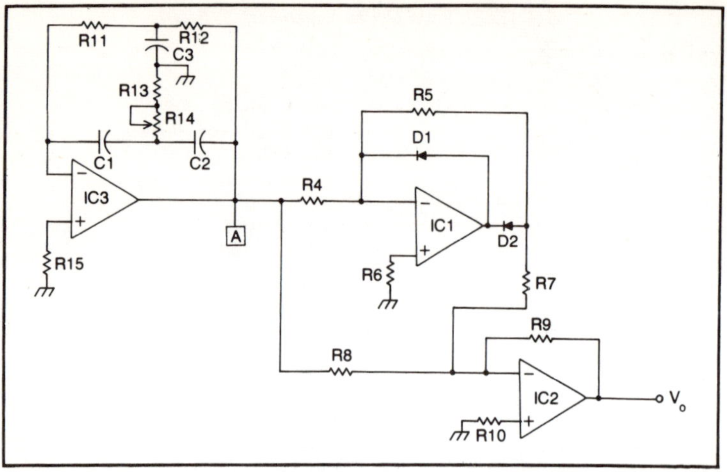

Fig. 15-14. The second part of Experiment 8 uses this modified absolute value circuit.

For the second part of this experiment, you will use the circuit shown in Fig. 15-14. This is the same absolute value circuit you just used, but instead of a dc voltage divider for an input, the input signal is taken from a sine wave oscillator. The components and equipment you will need for this part of the experiment are summarized in Table 15-21. Notice that an oscilloscope is required for this part of the experiment.

If you have a dual trace oscilloscope you can simultaneously monitor the input (A) and output signals. If you are using a single trace scope, you will have to move the input lead from point A to V_o.

Table 15-21. Parts List for Experiment 8, Part 2.

Breadboard system	
	741 op amp IC (need 3)
D1, D2	1N914 diode (or similar)
R1, R2, R3	not used
R4, R5, R8, R9	10 K resistor
R6, R7	4.7 K resistor
R10	2.7 K resistor
R11, R12	3.3 K resistor
R13	820 ohm resistor
R14	2.5 K potentiometer
R15	6.8 K resistor
C1, C2	0.47 μF capacitor
C3	1 μF capacitor
hook-up wire	
oscilloscope	

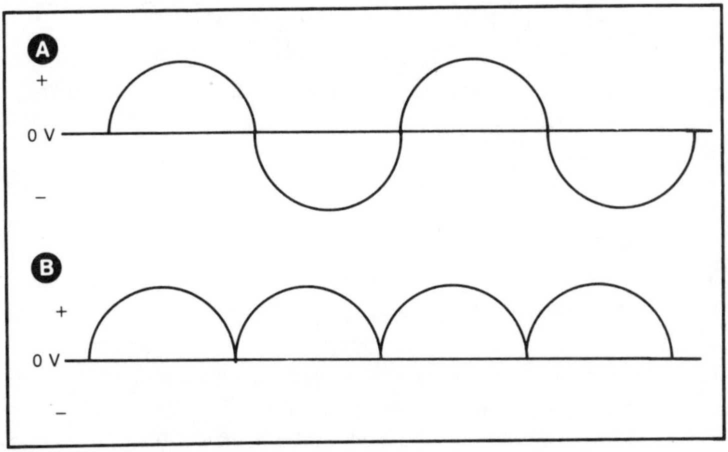

Fig. 15-15. An absolute value circuit can be used for precision full-wave rectification.

The input stage for this circuit (IC3 and its associated components) is a twin-T sine wave oscillator. See Chapter 6. This oscillator will have a frequency of approximately 100 Hz for the component values listed in Table 15-21.

The input signal at point A should be a sine wave, as illustrated in Fig. 15-15A. At the output the negative portion of the waveform will be flipped over and made positive, while the positive portion of the original waveform will be unchanged. The result will look something like Fig. 15-15B.

The absolute value circuit is sometimes used for precision full-wave rectification.

Parts
Breadboard system
741 op amp IC
220 to 470 ohm resistor (see text) (need 4)
5.6 K resistor (need 4)
10 K resistor (need 2)
12 K resistor (need 6)
15 K resistor
18 K resistor
22 K resistor
33 K resistor
39 K resistor
82 K resistor
LED (need 4)
SPDT switch (need 4)
hook-up wire
voltmeter

Table 15-22. Parts List for Experiment 9.

EXPERIMENT 9—DIGITAL-TO-ANALOG CONVERTER

For Experiment 9, you will be working with digital-to-analog converters. The parts and equipment that you will need to perform this series of experiments are listed in Table 15-22. The oscilloscope is optional this time.

The first thing you need to do is to devise some source of a digital signal to convert into an analog voltage. The simple circuit shown in Fig. 15-16 will create a 4-bit work dependent on the position of the switches. If a given bit's switch is in the upper position, it will tap off approximately five volts from the voltage divider. This will represent a logic 1.

Moving the switch to the lower position grounds out that bit. The output will be zero volts, or logic 0. You can set the four switches to create any 4-bit binary word from 0000 to 1111.

The LEDs and their associated resistors are optional. If a given LED is lit, that bit is at logic 1. A dark LED implies a logic 0 output for that bit. The resistors are used to limit the current through each LED.

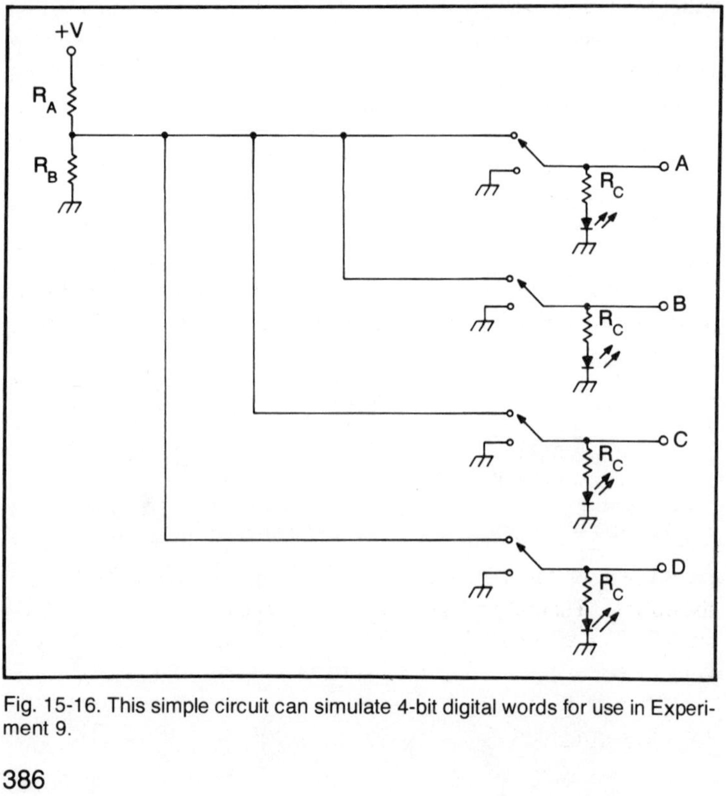

Fig. 15-16. This simple circuit can simulate 4-bit digital words for use in Experiment 9.

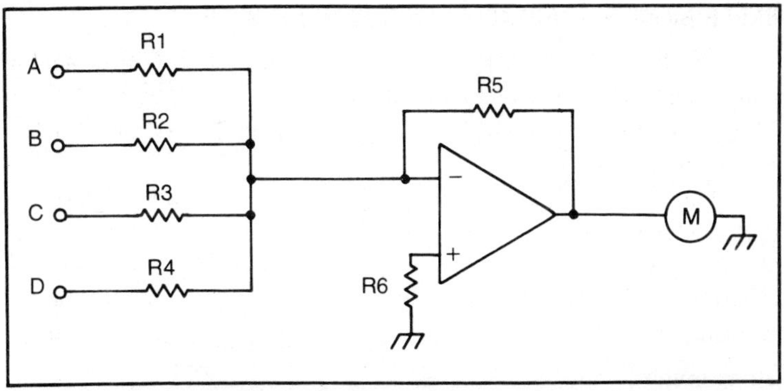

Fig. 15-17. This D/A converter circuit is explored in the first part of Experiment 9.

Three basic resistor values are used in this circuit. The resistor marked R_A should have a value of 33 K. The resistor marked R_B should have a value of 18 K. The resistors marked R_C may have values ranging from 220 to 470 ohms. A smaller value will cause the LED to glow more brightly. All R_C resistors should have the same value for a uniform display.

Breadboard the digital output simulator of Fig. 15-16 and the D/A converter circuit shown in Fig. 15-17. Use the following component values:

R1 = 82 K resistor	R4 = 10 K resistor
R2 = 39 K resistor	R5 = 10 K resistor
R3 = 22 K resistor	R6 = 15 K resistor

A is the least significant bit (2^0, or 1). D is the most significant bit (2^3, or 8). The 4-bit binary words will be written in the following order:

$$
\begin{aligned}
&\quad\;\; \text{D C B A} \\
&1\,0\,1\,1 = (1 \times 2^3) + (0 \times 2^2) + (1 \times 2^1) + (1 \times 2^0) \\
&\qquad\;\; = (1 \times 8) + (0 \times 4) + (1 \times 2) + (1 \times 1) \\
&\qquad\;\; = 8 + 0 + 2 + 1 = 11
\end{aligned}
$$

Any input bit at logic 0 will be feeding 0 volts into the input of the op amp. That bit will have no effect on the output. Set the binary switches for an input of 0000. The output should be zero volts, or very close to it.

Now raise the switch for input bit A, leaving the other three switches grounded. This makes the input word 0001. The D/A converter functions as an inverting amplifier with its gain set by the ratio of R5 and R1.

$$G = \frac{-R5}{R1} = \frac{-10{,}000}{82{,}000} = -0.12$$

The negative sign simply indicates that the output polarity is inverted. Since you will be using only positive input voltages in this application, the output will always be negative.

The exact input voltage for a logic 1 input bit is set by the power supply voltage and the values of R_A and R_B in Fig. 15-16. Assuming the supply voltage is ±15 volts, and R_A and R_B have the values specified earlier (33 K and 18 K respectively), this works out to approximately +5.3 volts. Component tolerances and variations in the supply voltage may cause slightly different values.

For an input of 0001 the output voltage should be equal to the input voltage (5.3 volts) times the gain (−0.12).

$$V_o = 5.3 \times -0.12 = -0.636$$

Measure the output. Do you get a value close to this? Return bit A to logic 0, and change bit B to logic 1 (0010). The input resistor for this bit is R2.

$$G = \frac{-R5}{R2} = \frac{-10{,}000}{39{,}000} = -0.26$$

Ideally this should be twice the gain of bit A (2 × −0.12 = −0.24), but since you are probably using standard resistor values, you can be satisfied with coming close. Component tolerances will cause deviation from the nominal values in any event. Once again, a logic 1 input bit equals 5.3 volts, so the output voltage for an input of 0010 is equal to V_iG.

$$V_o = 5.3 \times -0.26 = -1.36 \text{ volt}$$

Does the voltmeter read an output voltage close to this calculated value?

Ground bit B and raise bit C to logic 1 (5.3 volts). You should now have 0100 as your input binary work. The gain is set by R5 and R3.

$$G = \frac{-R5}{R3} = \frac{-10{,}000}{22{,}000} = -0.45$$

$$V_o = 5.3 \times -0.45 = -2.41 \text{ volts}$$

Next, set the input word to 1000. All switches are grounded (logic 0) except for D. What output voltage do you get now? It should work out according to the following equations:

$$G = \frac{-R5}{R4} = \frac{-10{,}000}{10{,}000} = -1$$

$$V_o = G \times D = -1 \times 5.3 = -5.3 \text{ volts}$$

What happens if more than one input bit is at logic 1? In this case, the circuit acts like a summing amplifier. The inputs are weighted by the different value input resistors. You learned way back in Chapter 3, the output of this circuit is equal to the following relationship:

$$V_o = \left(V_{ia} \times \frac{-R5}{R1}\right) + \left(V_{ib} \times \frac{-R5}{R2}\right) + \left(V_{ic} \times \frac{-R5}{R3}\right) + \left(V_{id} \times \frac{-R5}{R4}\right)$$

For the component values that you have been working with, the equation simplifies as shown here.

$$V_o = (V_{ia} \times -0.12) + (V_{ib} \times -0.26) + (V_{ic} \times -0.45) + (V_{id} \times -1)$$

Set the input binary word to 0011. That is, A and B are each at logic 1 (+5.3 volts) and C and D are at logic 0 (0 volts). The output in this case should be simply the sum of A and B calculated separately. Plug the appropriate voltage values into the formula, and find the nominal output voltage.

$$V_o = (5.3 \times -0.12) + (5.3 \times -0.26) + (0 \times -0.45) + (0 \times -1)$$

$$= -0.636 + -1.36 + 0 + 0 = -1.996$$

For a 4-bit binary word there are 16 possible combinations. Try the input to each combination and jot down your results in Table 15-23. Compare your results with the calculated values listed in Table 15-24.

For the next part of this experiment, you will use the R-2R ladder D/A converter circuit illustrated in Fig. 15-18. Again the input will be obtained from the digital simulator circuit of Fig. 15-16. Use the following component values:

Table 15-23. Worksheet for Experiment 9.

Input Word D	C	B	A	Output Voltage	Input Word D	C	B	A	Output Voltage
0	0	0	0	________	1	0	0	0	________
0	0	0	1	________	1	0	0	1	________
0	0	1	0	________	1	0	1	0	________
0	0	1	1	________	1	0	1	1	________
0	1	0	0	________	1	1	0	0	________
0	1	0	1	________	1	1	0	1	________
0	1	1	0	________	1	1	1	0	________
0	1	1	1	________	1	1	1	1	________

R1 12 K	R7 12 K
R2 12 K	R8 5.6 K
R3 22 K	R9 12 K
R4 5.6 K	R10 12 K
R5 12 K	R11 5.6 K
R6 5.6 K	

The most significant bit ($D = 2^3 = 8$) passes through R1 and R2 before being fed into the inverting input of the op amp. The gain for this bit depends on the ratio of R3 and R2 and the input voltage which passes through a voltage divider made up of R1, R4, R6, R8, and R10. The input voltage is tapped off at point E. The input signal passes through R1 and then is split off between R2 and ground through R4, R6, R8, and R10.

For a logic 0 input the voltage at point E will simply be 0. For a

Table 15-24. Calculated Values for Experiment 9.

Input Word D	C	B	A	Output Voltage	Input Word D	C	B	A	Output Voltage
0	0	0	0	0.00	1	0	0	0	−5.29
0	0	0	1	−0.64	1	0	0	1	−5.94
0	0	1	0	−1.36	1	0	1	0	−6.65
0	0	1	1	−2.00	1	0	1	1	−7.30
0	1	0	0	−2.41	1	1	0	0	−7.70
0	1	0	1	−3.05	1	1	0	1	−8.35
0	1	1	0	−3.76	1	1	1	0	−9.06
0	1	1	1	−4.41	1	1	1	1	−9.70

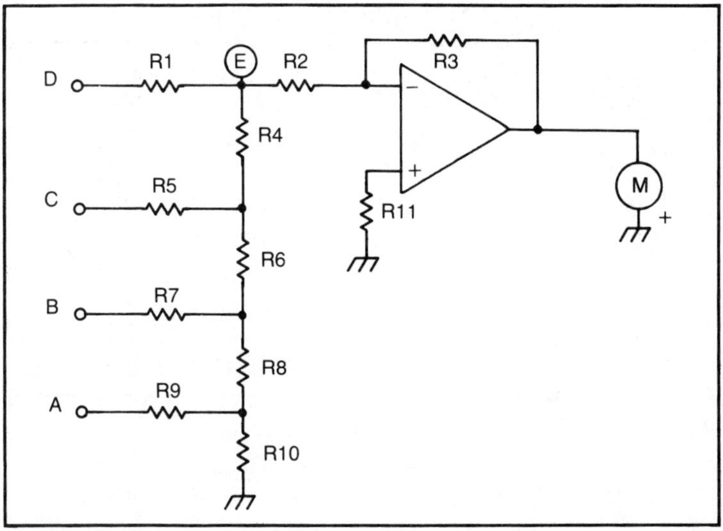

Fig. 15-18. The second half of Experiment 9 examines the operation of a R-2R ladder type D/A converter.

logic 1 input, a 5.3 volts signal is fed through the voltage divider. The total resistance to ground is a sum of the resistances.

$$\begin{aligned} R_t &= R1 + R4 + R6 + R8 + R10 \\ &= 12{,}000 + 5600 + 5600 + 5600 + 12{,}000 \\ &= 40{,}800 \text{ ohms} \end{aligned}$$

According to Ohm's Law, the current flow should be equal to .13 mA.

$$I = \frac{V_i}{R_t} = \frac{5.3}{40{,}800} = 0.00013 \text{ amp} = 0.13 \text{ mA}$$

The voltage drop across R1 equals 1.56 volts.

$$V_1 = IR1 = 0.00013 \times 12{,}000 = 1.56 \text{ volts}$$

Subtract this value from the original input voltage (5.3 voltage).

$$V_e = V_i - V_1 \times 5.3 - 1.56 = 3.74$$

This is fed into the inverting amplifier. The gain for the D input is calculated as follows:

$$G_D = \frac{-\ R3}{R2} = \frac{-\ 22{,}000}{12{,}000} = -1.8333 \cong -1.8$$

The negative sign simply indicates that the output polarity is reversed because the op amp's inverting input is being used.

The output voltage when input D is at logic 1 and the other three inputs are at logic 0 (1000) should be close to −6.86 volts.

$$V_o = V_E \times G_D = 3.74 \times -1.8 = -6.86 \text{ volts}$$

The situation is slightly more complex for the other input bits, but you can simplify matters by looking at each one individually. Figure 15-19 simplifies the schematic for input C. Inputs A, B, and D and their related components are ignored.

The input signal is fed through a voltage divider made up of R5, R6, R8, and R10. The input voltage is tapped off at point F between R5 and R6. The input resistance to the inverting amplifier is the series combination of R4 and R2.

Solve for the input voltage by first finding the current with Ohm's Law.

$$I = \frac{V_i}{R_t} = \frac{V_i}{(R5 + R6 + R8 + R10)}$$

$$= \frac{5.3}{(12{,}000 + 5600 + 5600 + 12{,}000)}$$

$$= \frac{5.3}{35{,}200} = 0.00015 \text{ amp} = 0.15 \text{ mA}$$

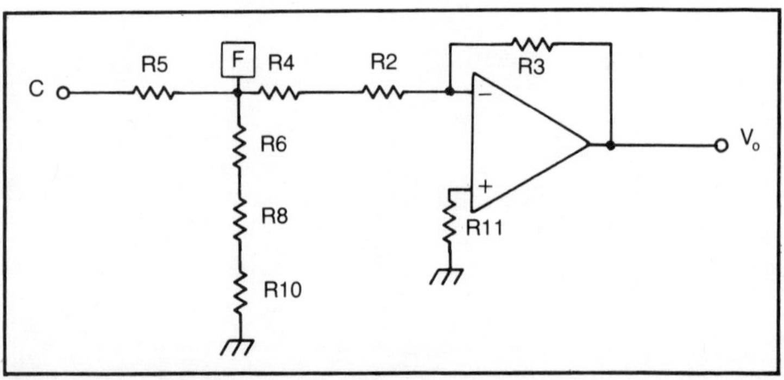

Fig. 15-19. Input C sees this equivalent circuit.

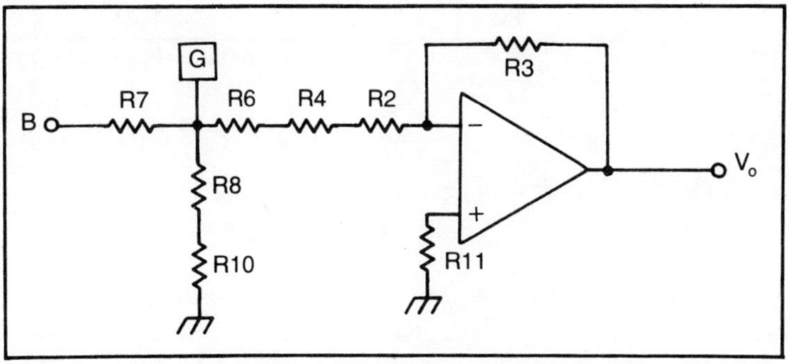

Fig. 15-20. This equivalent circuit is seen by input bit B.

$$V_F = V_i - V_C = V_i - IR5 = 5.3 - (0.00015 \times 12{,}000)$$
$$= 5.3 - 1.8 = 3.5 \text{ volts}$$

Next, find the gain for input C.

$$G_C = \frac{R3}{(R2 + R4)} = -\frac{22{,}000}{(12{,}000 + 5600)}$$
$$= \frac{-22{,}000}{17{,}600} = -1.25$$

So the output voltage for C = 1 (other bits = 0) (0100) works out to about −4.375 volts.

$$V_o = V_F \times G_C = 3.5 \times -1.25 = -4.375 \text{ volts}$$

Figure 15-20 shows the equivalent circuit for input B. Running through the equations for this input bit you find the following:

$$I = \frac{V_i}{R_t} = \frac{V_i}{(R7 + R8 + R10)} = \frac{5.3}{(12{,}000 + 5600 + 12{,}000)}$$

$$= \frac{5.3}{29{,}600} = 0.00018 \text{ amp} = 0.18 \text{ mA}$$

$$V_g = V_i - V_B = V_i - IR7 = 5.3 - (0.00018 \times 12{,}000)$$

$$= 5.3 - 2.15 = 3.15$$

$$G_B = -\frac{R3}{(R2 + R4 + R6)} = -\frac{22{,}000}{(12{,}000 + 5600 + 5600)}$$

$$= -\frac{22000}{23200} = -0.95$$

$$V_o = V_G \times G_B = 3.15 \times -0.95 = 2.99 \text{ volts}$$

Finally, solve the equations for bit A. See Fig. 15-21.

$$I = \frac{V_i}{R_t} = \frac{V_i}{(R9 + R10)} = \frac{5.3}{(12{,}000 + 12{,}000)}$$

$$= \frac{5.3}{24{,}000} = 0.00022 \text{ amp} = 0.22 \text{ mA}$$

$$V_h = V_i - V_A = V_i - IR9 = 5.3 - (0.00022 \times 12{,}000)$$
$$= 5.3 - 2.65 = 2.65 \text{ volts}$$

$$G_A = -\frac{R3}{(R2 + R4 + R6 + R8)}$$

$$= -\frac{22{,}000}{(12{,}000 + 5600 + 5600 + 5600)}$$

$$= -\frac{22{,}000}{28{,}800} = -0.76$$

$$V_o = V_h \times G_A = 2.65 \times -0.76 = -2.02 \text{ volts}$$

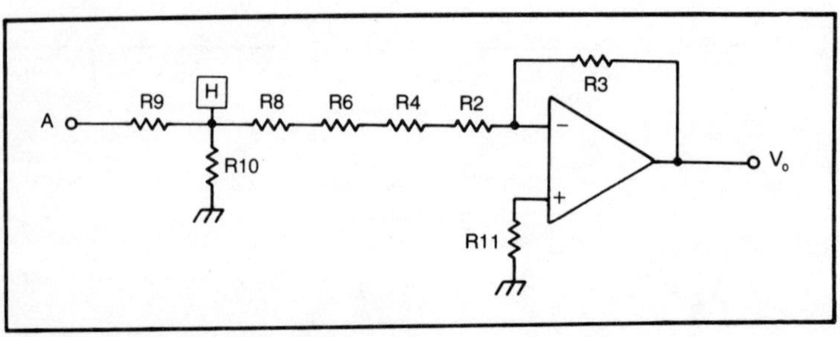

Fig. 15-21. Here is the equivalent circuit for input bit A.

Breadboard system
741 op amp IC
1 K resistor (need 2)
2.2 K resistor
3.3 K resistor
4.7 K resistor
10 K resistor (need 4)
22 K resistor (need 2)
33 K resistor (need 2)
100 K resistor (need 3)
voltmeter (15 volt range)
hook-up wire

Table 15-25. Parts List for Experiment 10.

Measure the output voltage for several combinations of digital inputs. There will be some imbalance of the steps because 12 K is not exactly twice 5.6 K, as required by the R-2R resistor network. For practical applications low tolerance precision resistors would be used.

EXPERIMENT 10—DC VOLTMETER

In this experiment, you will use an op amp in a dc voltmeter. The parts and equipment you will need are listed in Table 15-25.

You will start by breadboarding the circuit shown in Fig. 15-22. Resistors R1 through R7 form a voltage divider network to provide us with a number of voltages to measure during the experiment. Use the following resistor values in the voltage divider:

R1 33 K	R5 22 K
R2 2.2 K	R6 10 K
R3 10 K	R7 33 K
R4 1 K	

The op amp is connected as a simple unity-gain noninverting voltage follower with a voltmeter connected to the output. Connect

Table 15-26. Worksheet for Experiment 10, Part 1.

Input Point	Measured Output Voltage
A	________
B	________
C	________
D	________
E	________
F	________

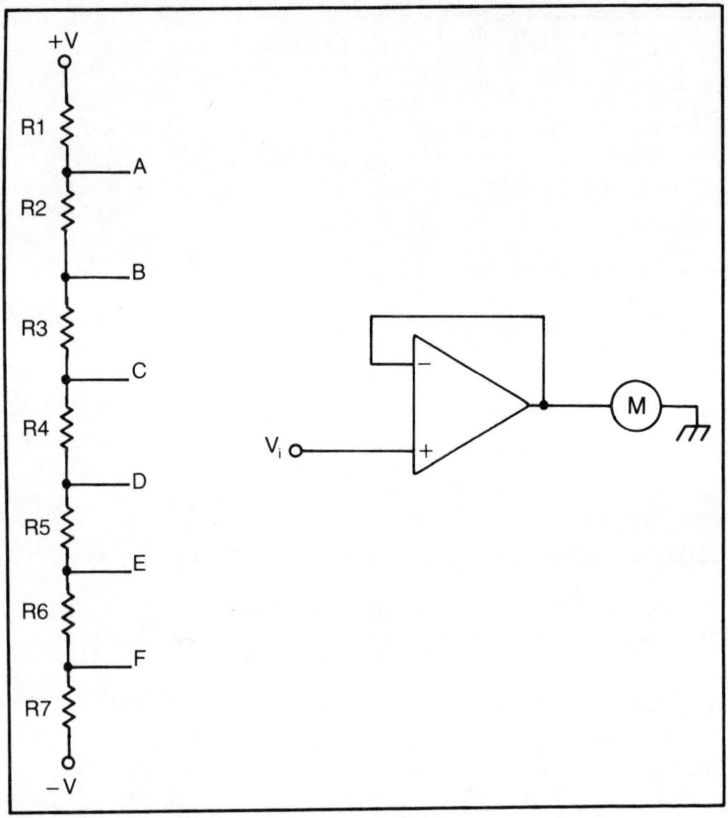

Fig. 15-22. This voltage divider network and unity gain dc voltmeter are used for the first part of Experiment 10.

the op amp's noninverting input (V_i) to each of the points in the voltage divider network (labeled A through F). Write each measured voltage in the appropriate space in Table 15-26.

Compare your results to the calculated values listed in Table 15-27. Your measured voltages should be close to the ones listed here. There may be some deviation from the nominal values due to component tolerances, op amp offsets, and minor measurement errors.

It's a simple matter to add gain to an op amp dc voltmeter simply by adding an input and a feedback resistor, as illustrated in Fig. 15-23. Remember to turn off the power supply while making changes to the circuit.

Use a 10 K resistor for R8 and a 4.7 K resistor for R9. Since this is a noninverting amplifier circuit, the gain is as follows:

Input Point	Calculated Output Voltage
A	+6.10 volts
B	+5.50 volts
C	+2.81 volts
D	+2.54 volts
E	−3.40 volts
F	−6.10 volts

Table 15-27. Calculated Values for Experiment 10, Part 1.

$$G = 1 + \frac{Rf}{R_i} = 1 + \frac{R9}{R8} = 1 + \frac{4700}{10,000}$$

$$= 1 + 0.47 = 1.47$$

Repeat the first part of the experiment, recording your values in the appropriate column in Table 15-28. Now change R9 to a 10 K resistor. Leave R8 alone. This makes the gain equal to Z.

$$G = 1 + \frac{10,000}{10,000} = 1 + 1 = 2$$

Repeat the measurements and jot down your results in Table 15-28. Next, use a 22 K resistor for R9. Again, the 10 K resistor is used for R8. This time the gain works out to 3.2.

$$G = 1 + \frac{22,000}{10,000} = 1 + 2.2 = 3.2$$

Repeat the measurements and record your results in the Table. You may have found some of the measurements forced the op

Table 15-28. Worksheet for Experiment 10, Part 2.

Input Point	Output Voltages		
	(R9 = 4.7 K)	(R9 = 10 K)	(R9 = 22 K)
A	________	________	________
B	________	________	________
C	________	________	________
D	________	________	________
E	________	________	________
F	________	________	________

Table 15-29. Calculated Values for Experiment 10, Part 2.

Input Point	Output Voltages		
	(R9 = 4.7 K)	(R9 = 10 K)	(R9 = 22 K)
A	+8.97 volts	+12.20 volts (*)	+19.52 (*)
B	+8.09 volts	+11.00 volts	+17.60 (*)
C	+4.13 volts	+5.62 volts	+8.99 volts
D	+3.73 volts	+5.08 volts	+8.13 volts
E	−5.00 volts	−6.80 volts	−10.88 volts
F	−8.97 volts	−12.20 volts *	−19.52 *

* The op amp may be driven into saturation, causing clipping of the output voltage.

amp into saturation. This is to be expected. You must be careful not to exceed the op amp's maximum output voltage in using a practical dc voltmeter. Compare your results with the calculated values listed in Table 15-29.

A dc amplifier with gain is obviously most useful when the voltage to be measured is quite small. Small voltages may be difficult to measure directly.

Return to the circuit shown in Fig. 15-22. This time use the following resistor values in the voltage divider:

R1 100 K
R2 1 K
R3 2.2 K
R4 3.3 K
R5 1 K
R6 4.7 K
R7 100 K

Try measuring the tap-off points on the voltage divider using

Table 15-30. Worksheet for Experiment 10, Part 3.

Input Point	Measured Output Voltages: Unity Gain	Measured Output Voltages: Gain = 11
A	________	________
B	________	________
C	________	________
D	________	________
E	________	________
F	________	________

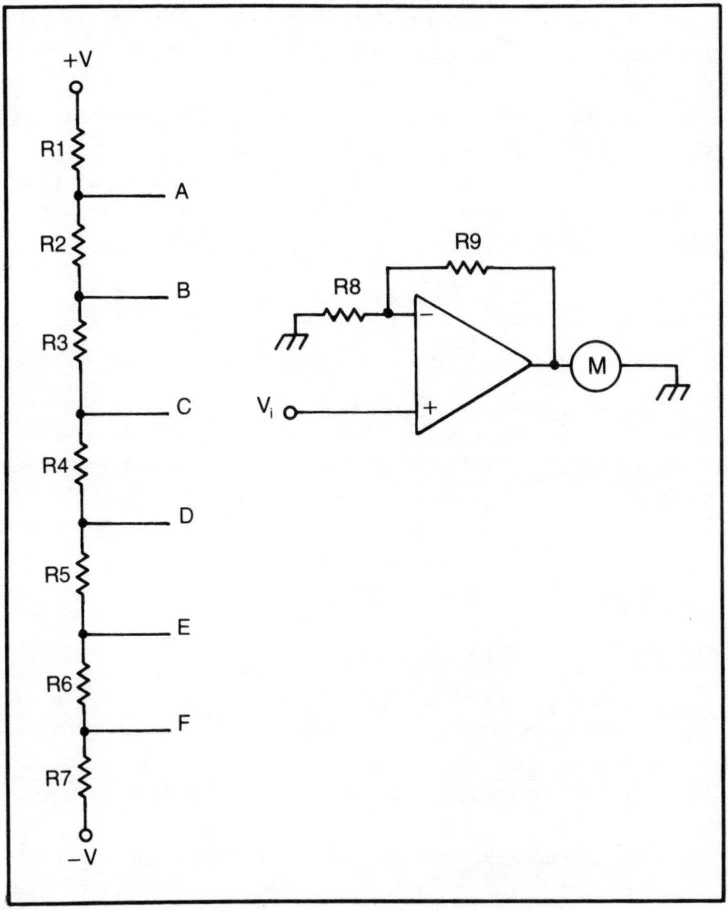

Fig. 15-23. Experiment 10 also demonstrates that a dc amplifier may have gain.

Table 15-31. Calculated Values for Experiment 10, Part 3.

	Measured Output Voltages	
Input Point	**Unity Gain**	**Gain = 11**
A	+0.86 volt	+9.46 volts
B	+0.72 volt	+7.92 volts
C	+0.41 volts	+4.51 volts
D	−0.06 volt	−0.62 volt
E	−0.20 volt	−2.18 volts
F	−0.86 volt	−9.46 volts

the unity gain dc voltmeter of Fig. 15-22. The analog voltmeter at the output should have a measurement range of about 15 volts.

Record your results in Table 15-30. You will probably have some difficulty reading the meter. The differences will be slight, and the meter's pointer won't move very far.

Now use the dc voltmeter with gain shown in Fig. 15-23. Use a 10 K resistor for R8 and a 100 K resistor for R9. This will give the following gain.

$$G = 1 + \frac{100,000}{10,000} = 1 + 10 = 11$$

Measure the voltage at each of the tap-off points on the voltage divider, and compare your results in Table 15-30. These voltages are much easier to read, aren't they? Compare your measured voltages with the calculated values shown in Table 15-31.

In a practical dc voltmeter, you would draw a new scale to place on the meter face so that you could read the input voltage directly, even though it is being multiplied by the gain of the op amp.

EXPERIMENT 11—NULL VOLTMETER

Another way to measure dc voltages is to use a null voltmeter. This is actually a variation on the comparator. The input voltage is compared with a reference voltage and the output device indicates if the input voltage is equal to, greater than, or less than the desired reference voltage.

The parts and equipment you will need to perform this experiment are summarized in Table 15-32. The circuit is illustrated in Fig. 15-24.

The reference voltage is set by the voltage divider made up of

Table 15-32. Parts List for Experiment 11.

Breadboard system	
741 op amp IC (need 2)	
LED (light emitting diode) (need 2)	
R1	100 K resistor
R2	47 K resistor
R3, R5	2.2 K resistor
R4	10 K potentiometer
R6, R7	1 K resistor
R8	820 ohm resistor
R9, R10	330 ohm resistor
hook-up wire	
voltmeter	

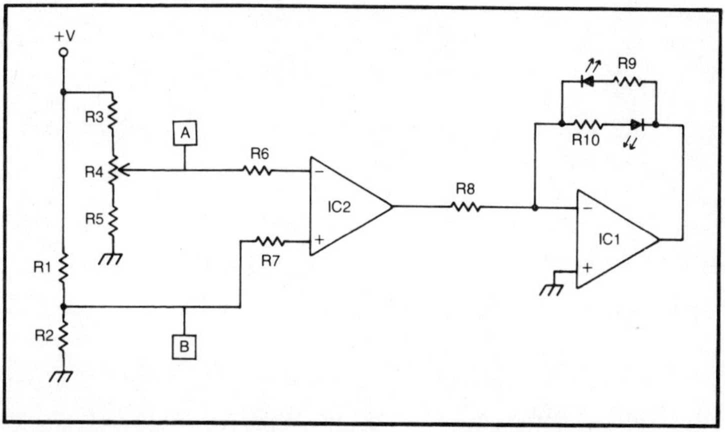

Fig. 15-24. This null voltmeter circuit is used for Experiment 11.

resistors R1 and R2. For the component values specified in the parts list with a ±15 volt power supply, the reference voltage should be about +4.8 volts. Use a voltmeter to measure the voltage at point B. It should be fairly close to this nominal value.

The input voltage to be measured is tapped off of the variable voltage divider made up of resistors R3 and R5 (2.2 K each) and potentiometer R4 (10 K). With these values, this input voltage can be varied from about 2.3 volts to around 12.7 volts.

Carefully breadboard the circuit of Fig. 15-24. Resistors R9 and R10 are included to limit the current flow through the LEDs to a safe level.

Rotate the potentiometer's shaft to one end of its range. One of the LEDs should be lit and the other should be dark.

Now rotate the potentiometer's shaft to its other extreme setting. The LEDs should reverse conditions. That is, the one that was originally lit should now be extinguished, and the one that was formerly dark should now be glowing brightly.

Slowly turn the potentiometer's shaft back and forth through its range. At some point both LEDs should be dark. They may not be completely extinguished, but the glow should be as dim as you can get it and should be equal for both of the LEDs. This indicates the null condition. The input voltage equals the reference voltage.

Use your voltmeter to measure the input voltage at point A. It should be equal to the reference voltage at point B.

16

Other Op Amp Devices

Throughout this book, I have concentrated on the 741 op amp IC. This device is good for most general op amp applications. It is inexpensive and widely available. But the 741 is by no means the only type of op amp IC. Literally hundreds of different devices are on the market today. In this chapter, I will briefly take a look at a few of these.

THE 747

The spec sheet and pinout diagram for the 747 dual op amp IC is shown in Fig. 16-1. This device contains two complete and independent 741 type op amps on a single IC chip.

The negative supply voltage (pin 4) is common to both of the internal op amp circuits, but independent positive supply voltages may be applied (pin 13 for op amp A, and pin 9 for op amp B). In most practical circuits, these two pins will be tied together, using a single positive voltage source.

Like the 741, the 747 features internal compensation. No external frequency compensation components are required in most applications.

Compare the spec sheet of the 747 with that of the 741 in Fig. 2-1. Notice that the electrical characteristics are identical. The two devices are directly interchangeable. In a circuit requiring two 741 op amps, a single 747 dual op amp may be used.

Whenever two or more circuits are etched onto a single silicon

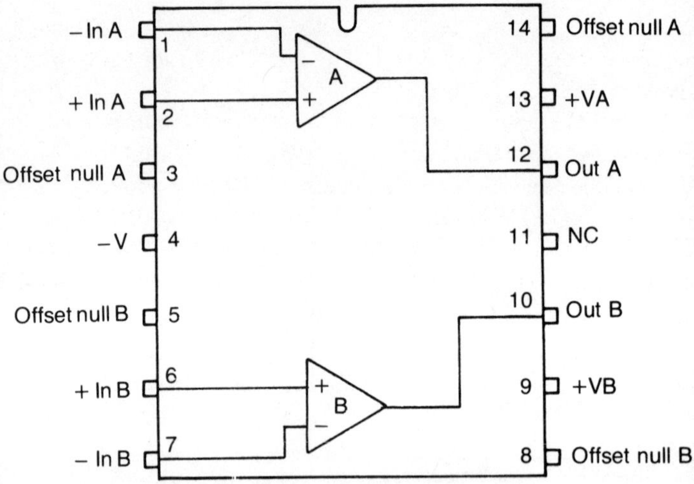

	Minimum	Typical	Maximum
Open-Loop Gain	50,000	200,000	--
Input Offset Voltage	--	1.0 mV	5.0 mV
Input Offset Current	--	20 nA	200 nA
Input Bias Current	--	80 nA	500 nA
CMRR	70 dB	90 dB	--
Slew Rate	--	0.5 V/μS	--
Input Resistance	1 megohm	6 megohm	--
Output Resistance	--	75 ohms	--
Differential Input Voltage	--	--	±30 volts
Channel Separation	--	120 dB	--

Fig. 16-1. The 747 contains two 741 type op amps in a single IC chip.

chip, there will be some leakage between them. The channel separation figure is a measurement of the isolation between the two op amps. If a signal is applied to op amp A, the same signal will pass through op amp B, but its amplitude will be reduced by 120 dB. Unless the input signal is very large, this leakage signal will be small enough to be ignored in the majority of applications.

THE 324

Another popular device is the 324. This device contains four independent op amp circuits. The pinout diagram and spec sheet are shown in Fig. 16-2. The op amps in a 324 IC can be used in place of 741s or 747s in most applications. The 324 features somewhat improved electrical characteristics over the 741 and the 747.

A single set of power supply connections are provided. These are common to all four of the op amps on the chip (pins 4 and 11). The 324 is rather unique among op amps in that it may be operated directly from a single-polarity power supply. Most op amps demand dual-polarity voltages.

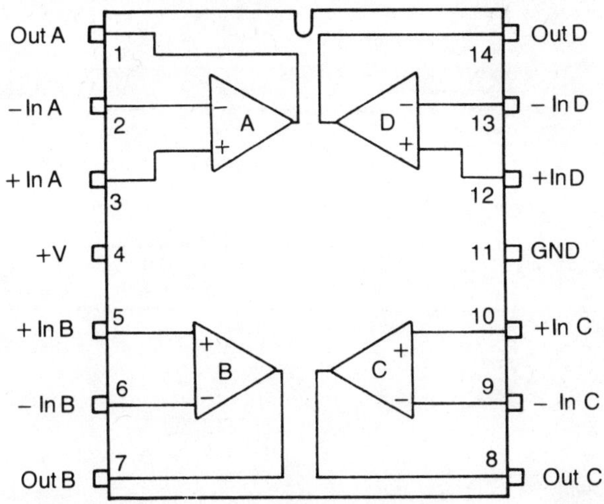

	Minimum	Typical	Maximum
Open-Loop Gain	--	100 dB	--
Input Offset Voltage	--	1.0 mV	2.0 mV
Input Offset Current	--	2 nA	10 nA
Input Bias Current	--	20 nA	50 nA
CMRR	70 dB	85 dB	--
Slew Rate	--	0.5 V/μS	--
Input Resistance	1 megohm	6 megohms	--
Output Resistance	--	75 ohms	--
Differential Input Voltage	--	--	±32 volts

Fig. 16-2. Four complete operational amplifier circuits are contained in the 324.

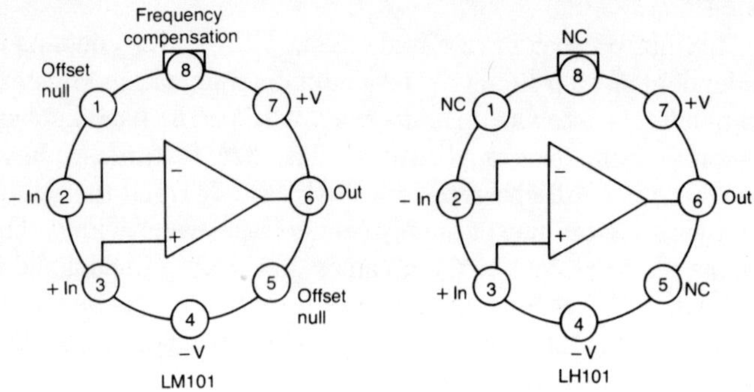

	Minimum	Typical	Maximum
Open-Loop Gain	50,000	160,000	---
Input Offset Voltage	--	1.0 mV	5.0 mV
Input Offset Current	--	40 nA	200 nA
Input Bias Current	--	120 nA	500 nA
CMRR	70 dB	90 dB	--
Slew Rate	--	0.5 V/μS	--
Input Resistance	300 K	800 K	--
Output Resistance	--	75 ohms	--
Differential Output Voltage	--	--	±30 volts

Fig. 16-3. Here is the pinout diagram and spec sheet for the LM101 op amp IC.

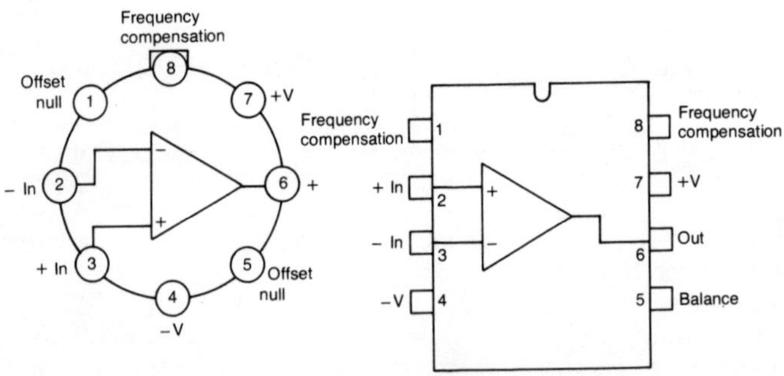

	Minimum	Typical	Maximum
Open-Loop Gain	50,000	200,000	--
Input Offset Voltage	--	1.0 mV	5.0 mV
Input Offset Current	--	20 nA	200 nA
Input Bias Current	--	80 nA	500 nA
CMRR	70 dB	90 dB	--
Slew Rate	--	0.5 V/μS	--
Input Resistance	--	N/A	--
Output Resistance	--	75 ohms	--
Differential Input Voltage	--	--	±30 volts

Fig. 16-4. The LH101 is an internally compensated equivalent for the LM101.

THE LM101/LH101

The pinout diagram and spec sheet for the LM101 op amp IC is shown in Fig. 16-3. This device exhibits somewhat higher input offset currents and input bias currents than the 741 family of operational amplifiers. The input resistance is also considerably lower.

Closely related to the LM101 is the LH101. The electrical characteristics of these two devices are identical. The difference between them lies in the fact that the LH101 is internally compensated, but the LM101 requires an external frequency compensation capacitor between pins 8 and 1. The pinout diagram for the LM101 op amp IC is also shown in Fig. 16-3.

THE 748

The 748 IC is a popular, moderately high-quality op amp device. It is closely related to the LM101 discussed above. The pinout is identical for these chips. The pinout diagram and spec sheet for the 748 op amp IC are shown in Fig. 16-4.

THE 709

The 709 op amp IC is one of the earlier devices to appear, but it

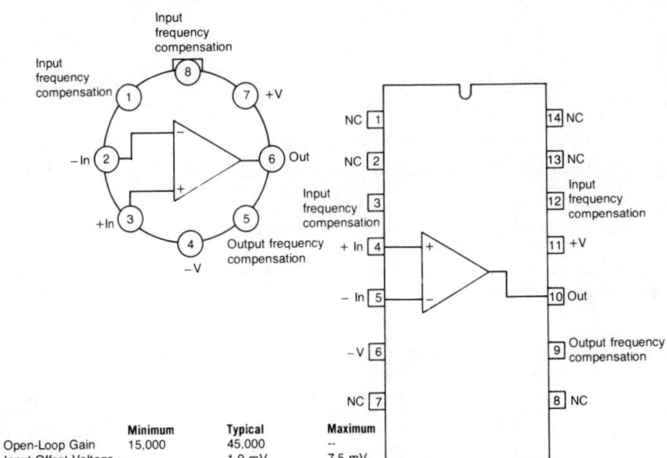

	Minimum	Typical	Maximum
Open-Loop Gain	15,000	45,000	--
Input Offset Voltage	--	1.0 mV	7.5 mV
Input Offset Current	--	50 nA	500 nA
Input Bias Current	--	200 nA	1500 nA
CMRR	65 dB	90 dB	--
Slew Rate	--	0.25 V/μS	--
Input Resistance	--	N/A	--
Output Resistance	--	150 ohms	--
Differential Input Voltage	--	--	±5 volts

Fig. 16-5. The 748 is basically an improved 101 op amp.

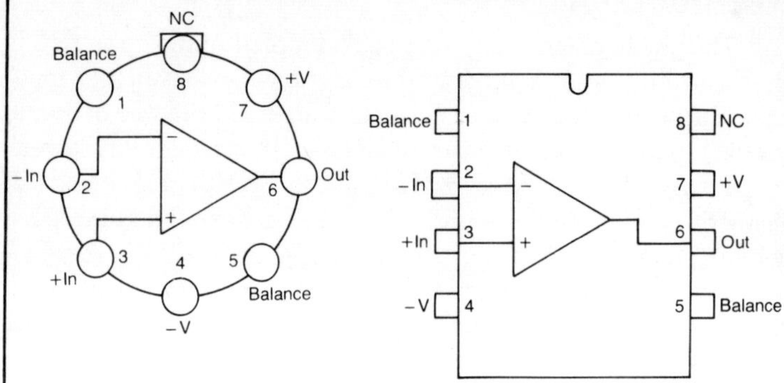

	Minimum	Typical	Maximum
Open-Loop Gain	—	106 dB	—
Input Offset Voltage	—	1.0 mV	2.0 mV
Input Offset Current	—	3 pA	10 pA
Input Bias Current	—	30 pA	50 pA
CMRR	85 dB	100 dB	—
Slew Rate	3 V/μS	5 V/μS	—
Input Resistance	—	10^{12}	—
Output Resistance	—	N/A	—
Differential Input Voltage	—	—	±40 volts

Fig. 16-6. The 709 has poorer specifications than most modern op amp ICs.

is still widely used. As the spec sheet, shown in Fig. 16-5, indicates, the electrical characteristics of this chip are quite inferior to the other op amp ICs presented in this chapter. For example, the 709 can be damaged or destroyed if the differential input voltage (the difference between the signal applied to the inverting input and the signal applied to the noninverting input) exceeds even five volts (positive or negative).

The limited abilities of this device preclude many applications. Still, the 709 is very inexpensive, and can be useful in noncritical, low-voltage applications.

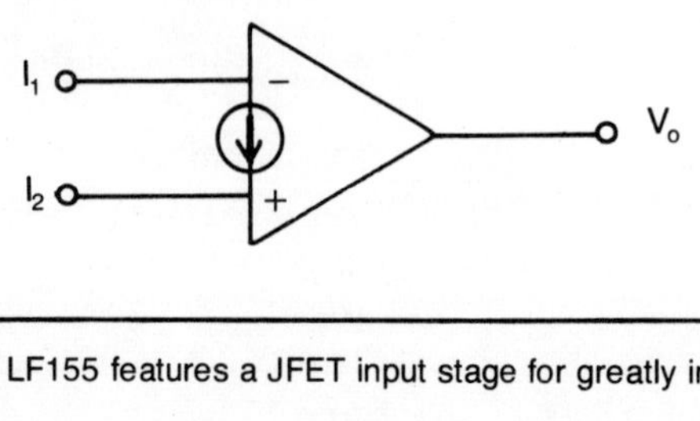

Fig. 16-7. The LF155 features a JFET input stage for greatly improved performance.

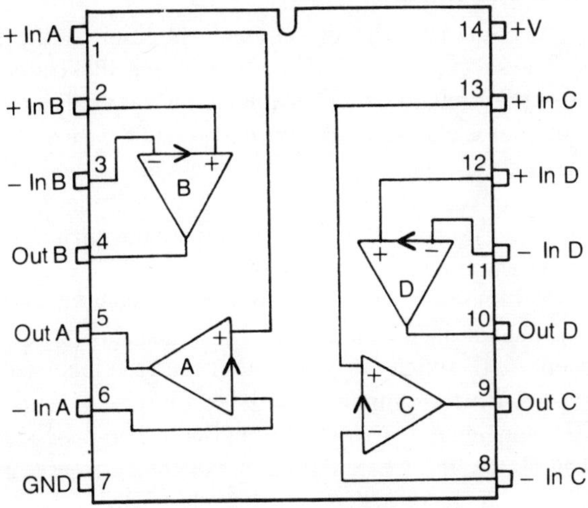

	Minimum	**Typical**	**Maximum**
Open-Loop Gain	--	70 dB	--
Input Offset Voltage	--	N/A	--
Input Offset Current	--	N/A	--
Input Bias Current	--	30 nA	200 nA
CMRR	--	N/A	--
Slew Rate	--	0.5 V/μS	--
Input Resistance	--	1 megohm	--
Output Resistance	--	8 K	--
Differential Input Voltage	--	--	N/A

Fig. 16-8. The Norton amplifier is similar to an op amp except the output responds to the differential input current.

THE LF155

In many precision applications, it is desirable to have an operational amplifier with a FET (field effect transistor) input stage. A typical monolithic FET input op amp is the LF155. The pinout diagram and spec sheet for this device appear in Fig. 16-6. Actually, the input stage for the LF155 is a JFET (junction field-effect transistor).

Notice the vastly improved electrical characteristics of this chip compared to the other op amp ICs described in this chapter. Especially notice the extremely high input resistance—10^{12}. This is 10,000,000,000,000 ohms, or 1 megohm squared! This exceptionally high input impedance ensures minimum loading of any input circuits.

THE NORTON AMPLIFIER

Closely related to the operational amplifier is the Norton amplifier. The schematic symbol for this device is illustrated in Fig. 16-7. Notice the similarity to the symbol for a standard op amp. The Norton amplifier is also known as the *current difference amplifier*, or CDA. A standard op amp, you'll recall responds to the differential input *voltage* between the inverting and noninverting inputs. The Norton amplifier responds to the differential input *current* between the inverting and noninverting inputs.

A Norton amplifier can be used in many applications similar to those using standard operational amplifiers, but the output voltage is dependent on the input currents rather than the input voltages.

A typical Norton amplifier IC is the LM3900. This device contains four independent Norton amplifiers. The spec sheet and pinout diagram for the Norton amplifier is shown in Fig. 16-8.

This chip may be powered from a single-polarity power supply from 4 to 36 volts, or from a dual-polarity supply of ±2 volts to ±18 volts. If a dual-polarity voltage supply is used, it should be symmetrical with respect to ground.

The LM3900 quad Norton amplifier IC features internal frequency compensation. External compensation circuits are not required.

Appendix

Hobbyist Power Supply Designs

One of the most pressing needs for most electronic hobbyists and experimenters is a stable, reliable source of dc power. Needed are bench power supplies, that can be used for many different projects, and dedicated power supplies that are incorporated into specific projects and thereafter remain with the project.

There are a number of sources of well-made commercial power supplies for both bench and service usage. In large quantities, these power supplies are quite low in cost and there would be a serious question of whether any given project or product should have its own custom-designed power supply or one of those so-called "O.E.M." power supplies. For the hobbyist who will build no more than one or two of a kind of any given project, the cost of those commercial power supplies might well be prohibitive.

Another aspect of homebrewing the power supply is the use of junkbox and surplus components. Almost all hobbyists and amateurs will eventually accumulate a respectable supply fo electronic components bought at bargain prices or salvaged from electronic equipment. When one uses these sources of components, it becomes possible to reduce the cost of a one-of-a-kind power supply to almost nil.

In the "old days," that is the time before integrated circuits (there *was* such a time, you know!), the design and construction of voltage regulators was a costly affair. Only a few high-quality power supplies would have voltage regulator circuits. Today, any number of manufacturers offer three and four-terminal integrated circuit voltage regulators that provide the best advantage of voltage regulation and none of the old headaches. In this appendix, the projects use IC voltage regulators.

It would be impossible to select a "universal" set of power supplies that will meet the needs of all electronic hobbyists (too wide is the range of overall interest)! But there are a few different sets of power supply ratings that are common to a large number of users. In this appendix, we will examine designs for the following power supplies:

5-volt dc @ 1-ampere.
5-volt dc @ 5-ampere.
12-volt dc @ 1-ampere.
± 12-volt dc @ 1-ampere (per voltage).
1 to 25-volt dc @ 5-ampere adjustable.

There are quite a number of other power supply designs we could examine, but they are really outside of the scope of this book. The 5-volt power supplies are intended for those users who experiment with digital electronic projects. The "universal" power supply for small digital projects is the 5-volt@1-ampere supply. The 5-ampere version (also at the standard "TTL" voltage of 5 volts) is used for small, single-board computers (i.e., KIM-1, Superboard II AIM-65 or Z80 Starter Kit). Higher-current, 5-volt power supplies are used in larger digital computers, but they are the topic of other books. The 12-volt @ 1-ampere power supply is used for a wide variety of transistor and linear IC applications, and for those CMOS digital projects in which V_{ss} is zero (ground potential).

For most readers who want to work with operational amplifiers, the most practical power supply will be the ±12-volt @ 1-ampere power supply. This circuit will provide up to 1 ampere of current at +12 volts and −12 volts (1 ampere from each).

5-VOLT DC, 1-AMP POWER SUPPLY

The 5-volt, 1-ampere power supply shown in Fig. 1 is usable for a large variety of TTL and CMOS digital projects, as well as those analog projects that will operate from this low voltage. The basis for this power supply is a three-terminal IC voltage regulator. The 5-volt type of regulator was one of the earliest available in three-terminal IC form. The LM-309K is now considered the venerable predecessor for all three terminal IC voltge regulator devices. In this application, you can use either the LM-309K, LM-340K-5 or the 7805, all three are approximately equivalent devices.

When selecting a voltage regulator, keep in mind the current

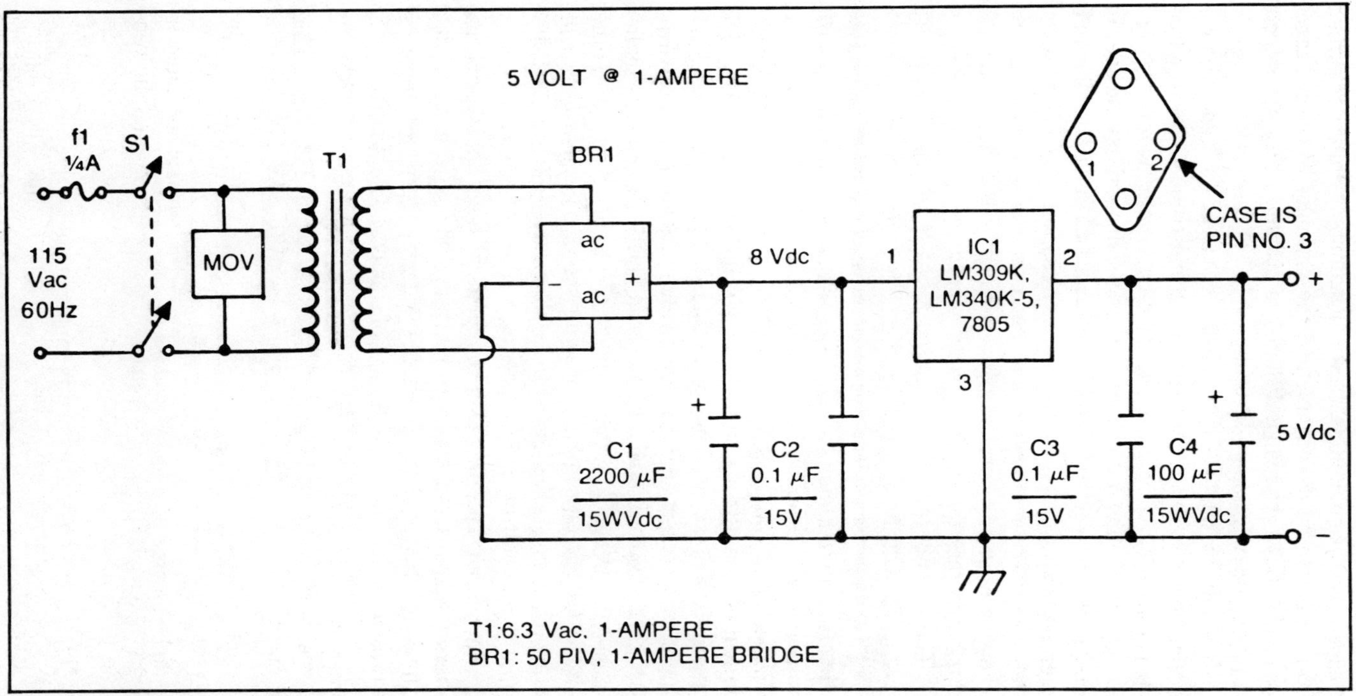

Fig. 1. A 5-volt, 1-ampere power supply.

rating that you are building for. Not all seemingly equal devices are usable in all cases. The problem is the package. There are three types. The "H" package (LM-309H) is usable in free-air without heatsinking to only 100 milliamperes. The "H" package is the same as the small, metal "T0-5" transistor can. The "K" package will support up to 1 ampere unheatsinked and some drive it to 1.5 amperes with suitable heatsinking (not recommended). The "K" package (LM-309K) is the same as the T0-3 diamond-shaped power transistor case. The popular "T" package (LM-340T-5) is a plastic power transistor case and it will safely support only 750 milliamperes without heavy heatsinking. For this power supply the "K" package devices are most heartily recommended. If the "T" devices are used, derate the output current accordingly.

The rectifier is a full-wave bridge rectifier rated at 50 volts PIV at 1 ampere of current. These rectifier stacks are available in black epoxy packages with lead wires or pins, depending upon your skeleton. Do not mount this rectifier in a manner that cuts off air circulation or overheating could result.

The filter capacitor at the output of the rectifier (C1) is selected according to the "rule-of-thumb" that requires 2000 μF per ampere of load current. Some authorities only require 1000 μF per ampere, but I recommend the higher value as good practice up to the point where the additional capacitance significantly adds cost or size (not a factor in small current supplies).

The voltage rating of the capacitor should be 15WVdc or more. The 15 volt rating is the lowest that will provide reliable operation. A 25-volt, 35-volt or more capacitor will not be out of line.

The output capacitor, C4, is optional and is used to improve the transient response of the power supply. There is a short time required for the regulator to react to changes in load current so it is possible to experience a "suck-out" of supply votlage in the face of a sudden heavy demand for current. The charge stored in capacitor C4 can be used to "dump" into the circuit in the event of a sudden need until the regulator can catch up with the demand (milliseconds).

The two 0.1 μF capacitors (C2 and C3) are used to improve the noise immunity of the voltage regulator. These small capacitors should be located physically as close as possible to the body of pins of the regulator. Otherwise, the effect of these capacitors will be minimal.

The transformer used in this supply is a 6-3-volt ac@1-ampere filament transformer (filaments are used in vacuum tubes, but the name persists despite the fact that filaments do not).

The "MOV device" is a *metal oxide varistor* that will provide better rejection of power line noise. The high voltage transients normal to power lines in many parts of the country will interrupt the operation of electronics devices, especially digital devices, and may cause damage to the regulator and rectifier. The MOV is a two-terminal device that is made by General Electric. It is readily available in electronic parts stores. It is strictly optional and the power supply will work nicely without it. Do not delete the MOV if you know the power line is noisy or if there is a local history of strange happenings on electronic equipment in your area. When computers just "bomb" for no reason, and do not show any damage to be repaired, is a good indication of power line transient noise.

Make sure that the switch used will support 110 volts ac across the pins. The purpose of the double-pole-single-throw switch is to break both sides of the power line (a safety feature).

5-VOLTS DC 5-AMP POWER SUPPLY

The circuit shown in Fig. 2 is a 5-ampere version of the first project. This supply is very similar to the 1-ampere version previously described. The principal difference between the supplies is in the ratings of the key components. The bridge rectifier, BR1, for example, must be rated at 5 amperes or more. I recommend that the builder select one of the 25-ampere types normally available to hobbyists. These rectifiers are often overrated, but they will run cool at 5 amperes despite the "real" rating. It is best if the rectifier is either mounted to the metal chassis or to a grounded heatsink. Use silicone heat transfer grease between the rectifier and the mounting surface.

The transformer is rated at 6.3 Vac at not less than 5 amperes with the 8-ampere Triad model preferred. The higher-rated transformer would allow the supply to run cooler (hence more reliably). The filter capacitor is selected according to the same rule as in the previous subject. Because the power supply is rated at 5 amperes, the capacitor should be 10,000 μF. If no single capacitor is available at that rating, then parallel two or more capacitors that add up to the 10,000 μF specified.

The voltage regulator (IC1) is the Lambda Electronics type LAS-1905. This regulator is rated at 5 volts, 5 amperes and is in a T0-3 transistor case. It is preferable that this regulator be mounted on a heatsink in order to reduce the temperature and increase reliability.

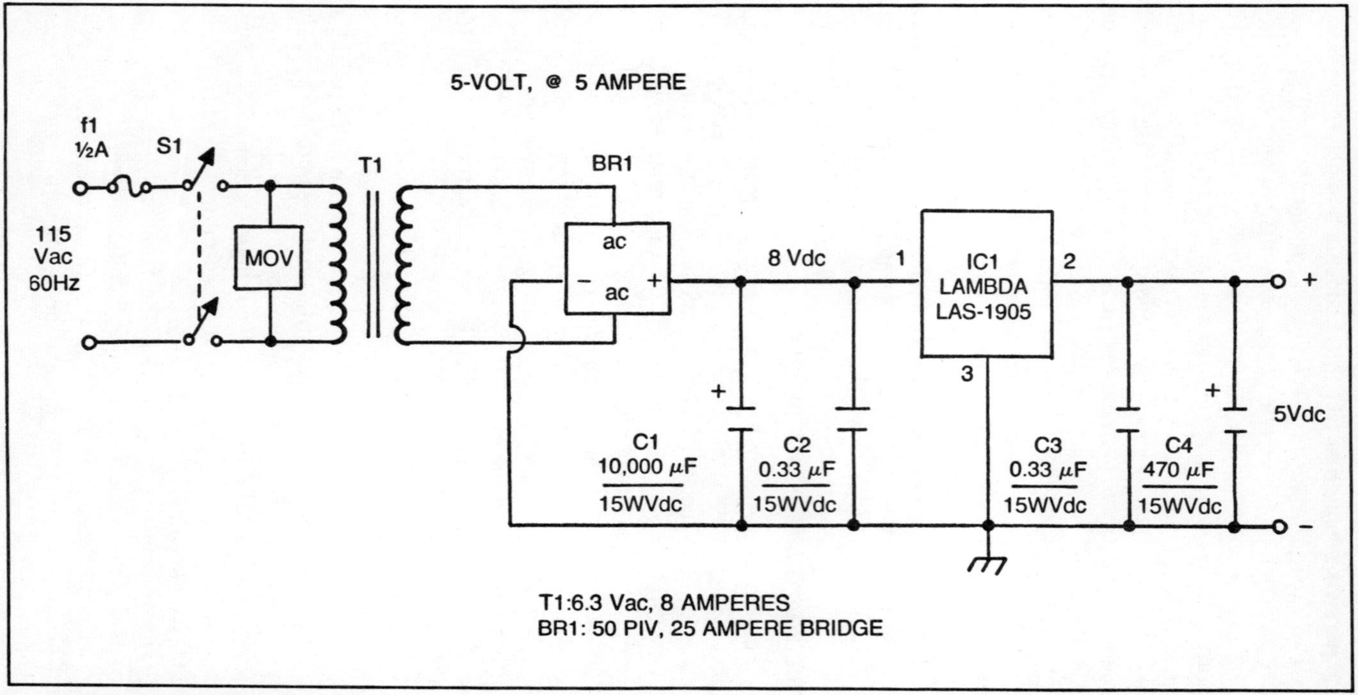

Fig. 2. A 5-volt, 5-ampere power supply.

+12-VOLTS DC, 1-AMP POWER SUPPLY

Another power supply like the first (refer to the information on the 5-volt, 1-ampere power supply) is the +12-volt, 1-ampere power supply shown in Fig. 3. This supply differs from the others in the ratings of its components. The transformer, for example, is a 12.6-volt ac, 1-ampere, "filament" transformer. Also, the rectifier has a high PIV voltage rating.

The integrated circuit voltage regulator (IC1) is a LM-340K-12 or a 7812 device. Again, the "K" package is preferred if a full 1 ampere of current is required. Otherwise, this power supply is the same as the other.

±12-VOLT DC, 1-AMP POWER SUPPLY

If we connect two single-polarity power supplies "back-to-back," we can make the form of dual-polarity power supply needed by operational amplifier circuits (and with other linear IC devices). But that method is wasteful of transformers and rectifiers and therefore is not well regarded.

The circuit shown in Fig. 4 is a better solution that will require only one transformer and one rectifier. The transformer is a 25.6 volt ac *center-tapped* type rated at not less than 2-amperes. The popular Triad 2.8 volt or the Radio Shack 3-ampere models are preferred. Note the center tap on the secondary of this transformer. No center tap is required on previous circuits so some readers might miss this essential connection!

The rectifier is a bridge stack that will pass at least 2 amperes without harm. *Note*: 3-amperes is a standard size, as is 1.5 amperes. The latter, however, will require de-rating of the overall output current rating of the power supply.

This power supply uses the transformer center tap as the zero potential reference for the power supply. Therefore, it is connected to the common line. The transformer is used such that one-half of the secondary supplies the positive side of the supply and the other half supplies the negative half. The rectifier is used in a manner that can be described as a pair of interconnected half-wave bridge rectifiers.

The filter capacitor is selected according to the 2000 μF/ampere rule. Note that purists will want to see more capacitance for C1 and C5 because the rectifiers are half wave (thereby requiring somewhat higher capacitance). But this isn't strictly necessary because the voltage regulator will smooth out much of the extra ripple.

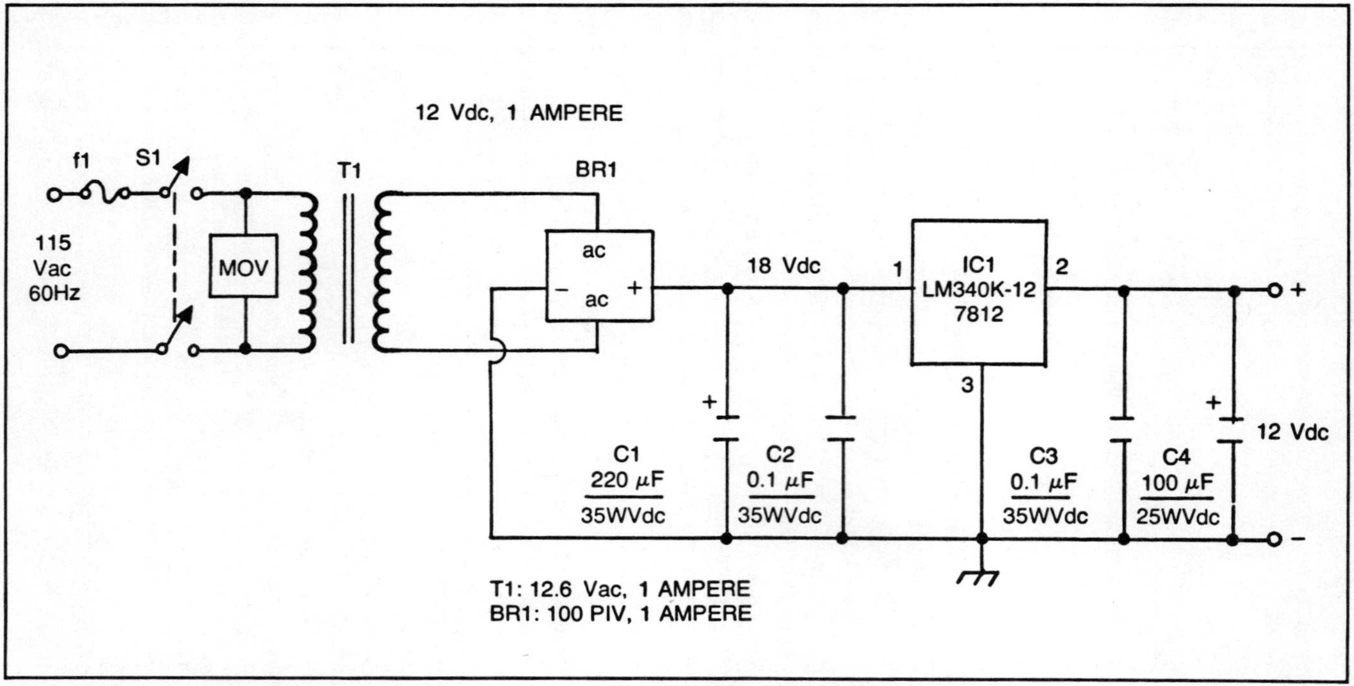

Fig. 3. A 12-volt, 1-ampere power supply.

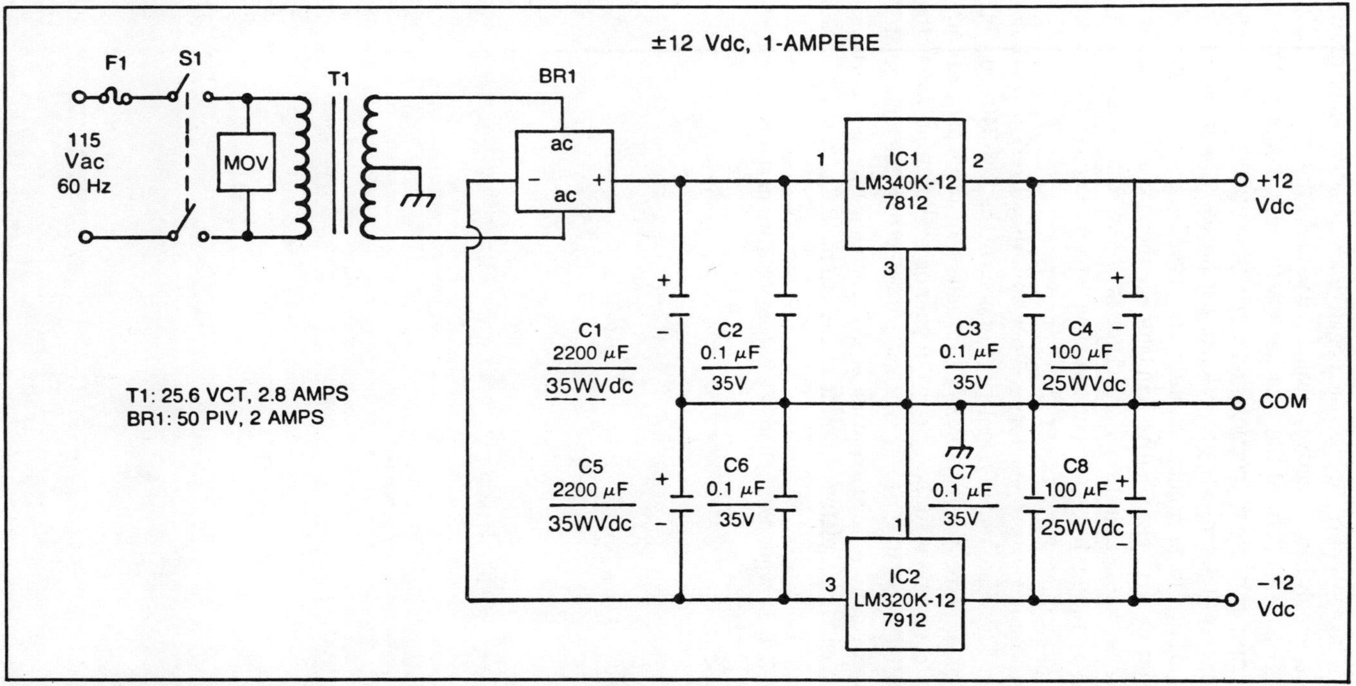

Fig. 4. A dual-polarity power supply.

The capacitors on both the negative and positive sides of the power supply have exactly the same use as in the first power supply in this appendix. Read the first section (5-volts, 1-ampere) for a description of their function. Like the first project, the output capacitors (C4 and C8) are optional, but highly recommended.

The voltage regulators are of the three-terminal, integrated-circuit variety. The positive regulator is either an LM-340K-12 or a 7812 device. Again, the "K" package is preferred for the full one-ampere rating. If the "T" package devices are selected, then count on 750 milliamperes rather than 1 ampere.

The negative regulator is either an LM-320K-12 or 7912 device. These are negative polarity versions of the LM-340 and 78xx devices, respectively. *Note*: The terminals for the input and common connections to the negative regulator are different than in the positive regulator. If you are used to inputting the unregulated power supply voltage to pin no. 1, and grounding the case (pin no. 3) on positive regulators, then it is easy to make the mistake when using a negative regulator. I did the first time around and the LM-320-12 went *pooofff* and became a silicon-to-carbon converter!

Index

Edited by Brint Rutherford